GRAPHITE AND PRECURSORS

World of Carbon

Series Editor: Dr Pierre Delhaès, *Centre de recherche Paul Pascal – CNRS, Pessac, France*

Volume 1

Graphite and Precursors
edited by Pierre Delhaès

This book is part of a series. The publisher will accept continuation orders which may be cancelled at any time and which provide for automatic billing and shipping of each title in the series upon publication. Please write for details.

GRAPHITE AND PRECURSORS

Edited by

Pierre Delhaès
Centre de recherche Paul Pascal – CNRS
Pessac, France

CRC Press
Taylor & Francis Group
Boca Raton London New York

CRC Press is an imprint of the
Taylor & Francis Group, an **informa** business
A SCIENCE PUBLISHERS BOOK

CRC Press
Taylor & Francis Group
6000 Broken Sound Parkway NW, Suite 300
Boca Raton, FL 33487-2742

First issued in paperback 2019

ISBN-13: 978-90-5699-228-6 (hbk)
ISBN-13: 978-0-367-39779-1 (pbk)

British Library Cataloguing in Publication Data

Graphite and precursors. – (World of carbon ; v. 1)
   1. Graphite   2. Graphite – History
  I. Delhaes, Pierre
  620.1'98

ISSN: 1560-8557

Cover illustration: Polyedric pores obtained after graphitization at 2800°C
(M. Villey, 1979).

Visit the Taylor & Francis Web site at
http://www.taylorandfrancis.com

and the CRC Press Web site at
http://www.crcpress.com

Contents

Introduction to the Series

The *World of Carbon* book series aims to propose different approaches to carbon materials which summarize the essential information regarding advances and results accumulated in basic and applied research during this century. Indeed, carbon associated with other atoms is a key element in nature and life. The focus of these books is, however, elemental carbons in a condensed phase, i.e. related to materials science.

Besides the natural forms of carbon, the artificial ones have led to manifold technical applications. They cover areas such as industrial chemistry and metallurgy, terrestrial transport as well as aircraft and aeronautics and environmental protection. These examples are related to the numerous old and new forms of technical carbons that we will present in greater detail below.

The field of research on carbon materials is a beautiful example of the strong interactions between science and technology, where back and forth activity has worked together for a long time. As with other scientific events, an historical approach shows that advances are step by step rather than linear with strong breakthroughs; different strata of knowledge are accumulated but sometimes with a loss of memory of the previous ones. It is crucial for scientific knowledge, as a part of human activity, that a basic synthesis is realized which summarizes the numerous publications appearing every year. The aim of this series is thus to provide short tutorial articles containing a comprehensive summary of the different subjects related to the science of carbon materials. They will be addressed to engineers, scientists and students who are seeking fundamental points without 'reinventing the wheel.'

World of Carbon will be devoted to specific topics which cover all forms of carbons: the old ones like graphite and diamonds, but also the new ones like fullerenes and nanotubes. Each volume will cover fundamental research in chemistry and physics, as well as current applications and future developments. Such is the case of the first volume, which is devoted to the different forms of graphite and their precursors.

Subsequent volumes will be related to the most important industrial uses of graphitic type carbons: fibers and composites, carbon blacks, foams and aerogels, and insertion and reactivity products. Other polymorphic forms will not be neglected either, such as diamonds and carbynes but also the molecular curved forms which for example open new avenues in nanotechnology.

We expect to present a collection of the articles at a level and in a style accessible to a large audience that will cover almost all aspects of carbon materials.

Pierre Delhaès
President, French Carbon Group

Foreword

Carbon reigns at the top of a column on the middle of a line of the periodic table, a privileged position, as observed by Primo Levi, who writes in *The Periodic Table* (Shocken Books, NY 1984):

> *. . . every element says something to someone (something different to each) like the mountain valleys or beaches visited in youth. One must perhaps make an exception for carbon, because it says everything to everyone, that is, it is not specific, in the same way that Adam is not specific as an ancestor. . . .*

Carbon is, first of all, immune to time. It remained on Earth as stone in the company of oxygen and calcium for millions of years. Plants expelled it along with oxygen, and much later Man used it to construct lime kilns.

Carbon is infinitely variable. It is rarely isolated and has a tendency to join with itself to form multiple constructions. In cubic form, it is called diamond which, when cut, embellishes women and is passed down from generation to generation. It is 'eternal' despite its instability. On the other hand, it is stable when it has a strongly bonded and weakly stacked hexagonal sheet. Much has been written about this very black graphite. These hexagonal planes can thus produce carbon fibers and nanotubes. Carbon, not to be content with these configurations, also likes to constitute long chains of conjugated bonds called carbynes.

Carbon recently exhibited its sensitivity to the mathematization of science by materializing as the Euler relation for polyhedrons in the form of fullerenes. Its solitude in these important structures does not exclude its conviviality with other atoms to create innumerable molecules in which is the majority, though. In addition, numerous hosts can easily penetrate it and therefore create insertion compounds. It is more discrete in the building blocks of living matter which cannot do without its power of hybridization. The carbon atom is thus exceptional. It is for this reason it has been discussed for such a long time (Le Chatelier's course at the Sorbonne in 1908 was called 'Lessons on Carbon') and continues to be so today.

The quest for scientific knowledge has become, since the middle of the twentieth century, a more collective effort rather than an individual one. Those who become interested in this subject by chance have united because of government-led and economic incentives.

This is how associations were created. The first one was very informal, created in the United States under the energetic impetus of Professor Mrozowski. In France, Professors Letort and Pacault founded the French Group for the Study of Carbon, which was succeeded by cooperative study under a program organized by the Centre National de Recherche Scientifique (CNRS, or National Center for Scientific Research). Among this group's activities, which for that matter continue today, the publication of a collective work should be noted: *Les carbones* (Masson, Paris, 1965). This treatise was followed by a series of monographs edited by Professors Walker and Throwers and published in the United States.

Recently, the discovery of new forms of carbon have opened up new avenues of research, as reported in *Le carbonl dans tous ses états* (Gordon and Breach Science Publishers, Amsterdam, 1997, to be published in English under the title *Carbon Molecules and Materials*). In this context of renewal, a collective and international work is necessary to encompass the knowledge that has been acquired and to make it encyclopedic. Such is the goal of this series of books, to which I wish much success.

Professor A. Pacault
Honorary Director of the Centre de Recherches Paul Pascal
Corresponding Member of the French Academy of Sciences

Preface

The natural forms of carbon such as solid, coal or graphite flakes have been known since antiquity. Human activity associated with this substance, what the Romans called *carbonis*, started long ago with chars from firewood used for ritual paintings and in primitive metallurgical processes.

The knowledge of the element carbon in chemistry and its technical applications on a larger scale are, however, a very recent development. The industrial revolution in Europe nearly two centuries ago began exploiting coal mines and resulted in numerous applications of these graphitic forms that are still actively used today.

To illustrate these points, *Graphite and Precursors*, the first volume of the *World of Carbon* book series, is devoted to these graphitic forms after recognizing the other allotropes such as diamond, a second natural form of the carbon element. For this purpose, three main parts have been developed in this book. The first one deals with the basic properties of crystalline graphite, a very peculiar solid that is black, shiny and conducts electricity; it exhibits an anisotropic behavior for all its remarkable physical properties, in particular the electronic ones associated with a two dimensional π electronic gas.

The second part demonstrates how the main applications of graphitic materials are related in connection to their refractory character; some outstanding examples are their use as electrical brushes, electrodes for electric furnaces or nuclear energy applications, for example in fission or fusion reactors.

Carbon precursors are presented in the third and final section with a twofold purpose. The first is to correlate the organic precursors from benzenic and small condensed aromatic derivatives, known as the basal structural units, for nanoscale graphitic organization. The main chemical steps are respectively pyrolysis, carbonization and graphitization processes. The second purpose is to control this chemical evolution to create more elaborate forms for these carbon precursors. Indeed, within the past several decades, new sophisticated production of carbon materials has lead to novel applications. In this volume, the basic concepts and the associated manufacturing processes for the most stable forms of graphitic materials will be described, and the exciting new applications will be presented in the forthcoming volumes of this series.

Contributors

H.P. Boehm
Universität München
Munich
Germany

S. Bonnamy
Centre de recherche sur le matiére divisée
Orléans
France

T.D. Burchell
Oak Ridge National Laboratory
Oakridge TN
USA

P. Delhaès
Centre de Recherche Paul Pascal
Pessac
France

G. Dresselhaus
MIT
Cambridge MA
USA

M.S. Dresselhaus
MIT
Cambridge MA
USA

S. Flandrois
Centre de Recherche Paul Pascal
Pessac
France

Y. Hishiyama
Musashi Institute of Technology
Tokyo
Japan

S-H. Hong
Kyushu University
Fukuoka
Japan

M. Inagaki
Hokkaido University
Sapporo
Japan

J-P. Issi
Université Catholique de Louvain
Louvain-la-Neuve
Belgium

Y. Korai
Kyushu University
Fukuoka
Japan

I. Mochida
Kyushu University
Fukuoka
Japan

A. Oberlin
Saint Martin de Londres
France

B. Rand
University of Leeds
Leeds
UK

J. Robertson
Cambridge University
Cambridge
UK

R. Saito
University of Electro-Communications
Tokyo
Japan

Y-G. Wang
Kyushu University
Fukuoka
Japan

1. Polymorphism of Carbon

P. DELHAÈS

Centre de recherche Paul Pascal, CNRS, Université Bordeaux I, avenue Albert Schweitzer, 33600 PESSAC, France

1.1 INTRODUCTION

Carbon, the lightest of the group IV elements of the periodic table, is certainly the most versatile and interesting with respect to materials properties because of its various forms and phases.

Different solid state varieties of carbon are existing naturally, either on earth or extraterrestrially but there are also artificial forms. For more than one century, starting with EDISON's experiment on hot incandescent carbonaceous filament many forms of carbon have been prepared or even discovered in laboratory. Currently, these old and new classes of carbon are respectively graphite (the most stable phase) and diamond on one hand, carbynes and fullerenes on the other. In this review, we will include recent breakthroughs due to these new molecular forms of carbons, such as C_{60} and its analogs as well as nanotubes, which are described as curved atomic surfaces (Dresselhaus *et al.*, 1995).

The review begins by summarizing our knowledge of the thermodynamically stable and metastable phases of elemental carbon over a wide range of pressure, temperature and, eventually, reaction conditions. In a second section, we will define the critical parameters that allow us to explain and classify carbon's rich polymorphism (Delhaès, 1997). Recent experimental and theoretical works have established and predicted new metastable crystalline phases. These points will be summarized in the third part of this general presentation. We will examine the influence of heteroatoms which are usually hydrogen, oxygen or nitrogen on non-crystalline precursor forms of graphitic or diamond like carbon. A general description and classification of these carbon materials will

TABLE 1. Schematic classification of different forms of carbons.

| Physical dimensionality and coordination number (z) | Typical chemical bonding | | | Solid state phases |
	hybridization	length (Å)	energy (eV·mole^{-1})	
3D (tetrahedral structures, $z = 4$)	sp^3–sp^3	1.54	15	diamonds
2D (lamellar structures, $z = 3$)	aro–aro	1.40	25	graphites (plane surfaces)
	sp^2–sp^2	1.33	26.5	fullerenes (curved surfaces)
1D (chains or rings, $z = 2$)	sp^1–sp^1	1.21	35	carbynes

be given. A complementary approach will be the influence of the neighbouring atoms in the Mandeleev table of elements, boron and nitrogen, which lead to doped carbon and solid solutions exhibiting a similar polymorphism with modified physical properties. Finally, without giving any detail on the interesting properties of these materials we sketch they are mainly divided in two broad classes, those with thermo-mechanical properties including hardness and those with functional properties associated with electronic attractive characteristics.

1.2 POLYMORPHISM OF CRYSTALLINE CARBONS

1.2.1 Basic Concepts

The nature of the chemical bonding between neighbouring carbon atoms is of prime interest. Its electronic hybridization ($1s^2$, $2s^2$, $2p^2$) leads to several types of covalent bonding.* As can be demonstrated with a quantum chemistry treatment on the hypothetical diatomic C_2 molecule, the linear combination of atomic orbitals (LCAO) leads to two-kinds of molecular orbitals (Atkins P.W., 1990): the σ-type orbital with a cylindrical symmetry about the internuclear axis, and the π-type orbital with nodal plane including the molecular axis. This orbital hybridization allows us to introduce two essential parameters for classifying the different forms of crystalline carbons (Table 1).

(i) The coordination number of a given carbon atom ($z = 4$, 3 or 2) in the solid state:
  – A carbon atom is bound to four equidistant nearest neighbours in a tetrahedral arrangement as in cubic diamond (Figure 1). This is simple σ-type bonding associated with sp^3 orbital hybridization.
  — A carbon atom is bound to three equidistant nearest neighbours 120° apart in a given plane as in hexagonal graphite (Figure 1). A double bond with both σ and π-types are present, inducing a shorter bond length than in the first situation (sp^2 hybridization in a planar symmetry).
  – A carbon atom is bound to two neighbours in a linear chain where it can form a single and a triple bond (sp^1 hybridization). This situation is encountered in polymeric forms of carbon called carbynes (see part 1.3.1).

---

*Three carbon isotopes are known ^{12}C, ^{13}C, ^{14}C that we do not detail in this general presentation.

a)

A

B

A

b)

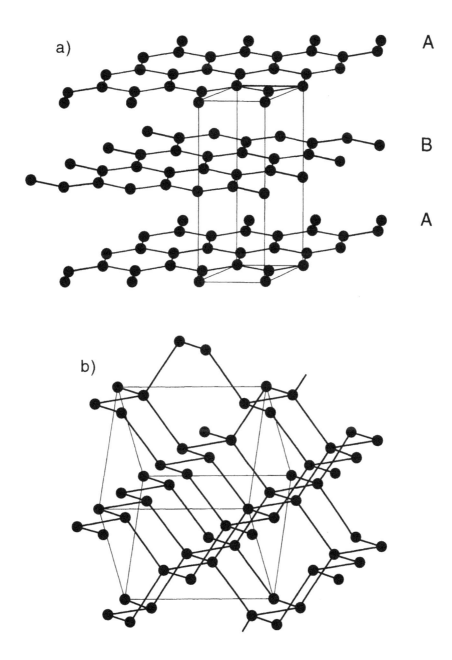

FIGURE 1. Crystallographic structures of the two most representative carbon allotrops: (a) hexagonal graphite with the layering sequence ABAB... (b) cubic face centered diamond.

(ii) The lattice dimensionality and the associated topological approach, in particular by
     defining the number of carbon atoms in polycyclic forms.

Indeed in classical euclidian geometry there is a direct relationship between the lattice
dimensionality (1D, 2D or 3D) and together the coordination number and the hybridization
orbitals. This is due to the invariance of translation and rotation symmetries associated with
usual atomic and molecular crystallographic structures.

In the case of topological changes such as the number of bonds in a given ring it has been
shown that the surface curvature as found in fullerenes opens the way to new allotropic
forms (Kroto *et al.*, 1985). One interesting point, because of the surface curvature is the
onset of a rehybridization process inducing a certain amount of σ character in a π type
orbital which modifies its aromatic character (Haddon, 1992).

To summarize this part (see Table 1) we observe that the type of chemical bonding is
a convenient way to classify the physical properties which are dependent on the type of
bonding and its energy. A simple σ-type bonding is sufficient to characterize the structural
thermal and mechanical properties whereas the presence of π orbitals will be crucial for
electronic and magnetic properties (Delhaès, 1997).

1.2.2 Thermodynamic Stability and Associated Phase Diagram

A stable thermodynamic state is associated with the absolute minimum of Gibbs free energy
expressed as a function of P and T, but the existence of local minima will induce the
possibility of metastable states. The probability of a phase transformation is determined by
the Gibbs free energy difference ΔG, between the two considered states and the possible
thermodynamic paths between them (Figure 2). Two main points are relevant:

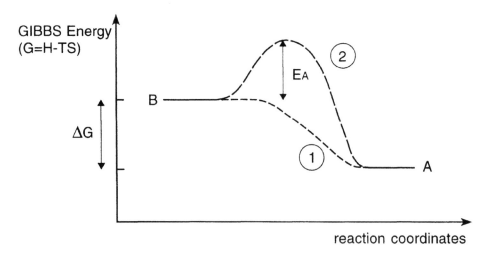

FIGURE 2. Schematic representation of the GIBBS energy change $|\Delta G|$ between a thermodynam-
ically stable state (A) and unstable or metastable one (B). If the path is of type 1 the state B will be
always unstable, but if the path has a large energy barrier, E_A, compared to the thermal energy, as
path 2, a metastable state B will be obtained and stabilized.

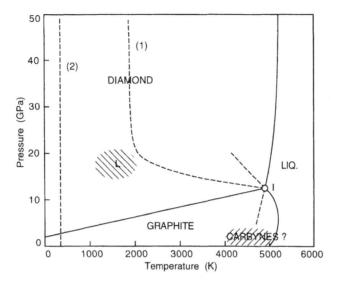

FIGURE 3. Phase diagram of carbon (Bundy *et al.*, 1996). Solid lines represent equilibrium phase boundaries and the associated triple point I; dotted lines 1 and 2 represent respectively the threshold of fast P/T cycles that convert graphite or hexagonal diamond (called L in the phase diagram) into cubic diamond, and the path along which an hexagonal graphite compressed at room temperature is transformed. Note that the question mark following carbynes indicate the possible existence region.

(i) The phase transformation between two thermodynamic states is governed by the existence or absence of an activation energy (E_A). If $E_A \sim 0$, then only an unstable state will exist. On the other hand if $E_A \gg kT$ (the thermal energy) the energy barrier will create a local minimum on the energy surface leading to the presence of a quenched kinetic state.

(ii) This second situation is favoured in such as the presence of large bonding energy (see Table 1) and high associated cohesion energy is found in carbon materials. When the carbon atoms are locked into a given phase configuration, a large amount of activation energy is required to produce a different stable phase. Therefore high temperatures and high pressures are necessary to initiate spontaneous phase transitions.

Following this general picture, a classical thermodynamic phase diagram has been established after several decades of experimental work (Bundy *et al.*, 1996). The (P, T) phase diagram for atomic carbon is presented in Figure 3. Its salient features are the following:

– The transition line between the stable graphite and diamond stable regions runs from 1.7 GPa/OK to the graphite-diamond-liquid triple point I (12 GPa/5000 K) (Berman and Simon, 1995). It should be mentionned that with thermal treatment under inert atmosphere diamond is completely transformed into graphite at about 1800°C.

- The melting line of graphite is extended from the triple point down to 0.011 GPa/5000 K (for the lower pressure region the vapor phase is present).
- The melting line of diamond runs to higher P and T above the triple point with a positive slope (Bundy *et al.*, 1996).

Beyond this general presentation several points which remain controversial are noteworthy:

- A hexagonal diamond structure called Lonsdaleite is known to be stable at high T and P (see dotted line in Figure 3).
- The carbyne phase could exist at high temperatures below the melting line of graphite.
- The liquid phase region could be divided in two parts, one would be an insulator and the other an electrical conductor (Van Theil and Ree, 1993).

To finish this phase diagram presentation, it is useful to point out the following points. These phase transformations are considered as theoretically reversible. For example, this means that we can observe a change between graphite and diamond and the transition enthalpy is under characteristic of a first order transition. Out of equilibrium conditions, such as shock waves, strong local pressures, or catalytic phase transformations are also realized but not represented on Figure 3 (Bundy *et al.*, 1996). Furthermore one step further can exists if some irreversible chemical reaction is occurring and we get a new defined compound associated with a reaction diagram. We assume that this situation is occurring in the new fullerene type phases which are molecular solids (C_n) as compared with the classical extended solids just described above (C_∞). These new phases will be described in part 1.3.2.

1.2.3 Theoretical Approaches and New Predicted Phases

The synthesis of cubic diamond thanks to different techniques which afford an excess of input energy (shock waves, high pressure or plasma chemistry) (Demazeau, 1997), has opened the way to unknown compounds. These experimental approaches are based upon theoretical calculations which predict non-existant forms of carbon and related compounds (Cohen, 1994). In the classical models a calculation of the excess cohesion energy at zero Kelvin compared to the stable thermodynamic phase is carried out, neglecting the entropic term of the Gibbs energy (see Figure 2). This approach has been developed by some authors (Cohen, 1994; Hoffmann *et al.*, 1983). Several models considered, based on quantum mechanical calculations for periodic solids having itinerant electrons. Concerning static properties, such as hardness, an equation of state (for instance Birch-Murnaghan's equation) is widely used where the cohesion energy is calculated as a function of the volume. The essential parameter is the bulk modulus B_0 at zero Kelvin and its pressure derivative for the equilibrium volume. A semi empirical expression for B_0 has been proposed as a useful starting point for calculations:

$$B_0 = -V_0 \left(\frac{dP}{dT} \right)_{T \Rightarrow 0} = \frac{\langle N_c \rangle}{4} (1972 - -220\lambda) d^{-3.5}$$

where $\langle N_c \rangle$ is the average coordination number for the compound considered and λ is an empirical ionicity factor which is zero for pure carbon solids. It is clear from this relation that short bond lengths with a large bond energy are best for getting a large B_0. Indeed the highest density of strong covalent bonds will lead to superhard compounds associated to a very low compressibility factor. The presence of single (σ-type) or multiple bonds ($\sigma + \pi$-types) in the carbon network will also be essential for the understanding of electronic properties (Dresselhaus *et al.*, 1998).

1.3 REAL AND VIRTUAL FORMS OF CARBONS

Considering the large number of carbon polymorphs these solid phases will be classified based on their coordination number but not from the type of hybridization that was previously introduced. For each class of crystalline structure with a fixed coordination number we will present the different real or virtual (i.e. predicted but not found experimentally) allotropic forms with different polytypes. The unavoidable presence of covalent chemical bonds in four, three or two directions induce a physical dimensionality for all the properties. This topological approach will be pursued by analyzing the new forms of carbon described on curved surfaces (fullerene type). In the last section we will examine the more exotic structures which have been predicted based on mixed coordination numbers.

1.3.1 Structures With a Fixed Coordination Number

1.3.1.1 Carbynes

The linear arrangement of elemental carbon as in other conjugated polymers (see polyacetylenes or polydiacetylenes for example) has been mentioned in the literature (Kudryautsev *et al.*, 1997). Basically, this polymeric form can be derived either from a chain with alternating single and triple bonds (polyyne) with alternate bond lengths of 1.20 Å and 1.58 Å) or alternatively from double bonds (polycumulene) with a uniform bond length: 1.28 Å (see Table 2).

The first polyyne configuration appears to be the most stable because of the stabilization caused by the opening of a gap at the Fermi level for the π electronic gas. This is referred to as the Peierls distortion effect (Kastner *et al.*, 1995).

However the existence of pure carbyne forms as the so called α and β forms is still controversial. A complete X-ray diffraction analysis of any linear species is not yet available due to the lack of a suitably sized carbyne single crystal, but several structures have been announced. There are contuining reports on its synthesis (Kudryavtsev *et al.*, 1997) and its natural occurrence. Firstly, natural carbynes have been reported in exotic environments such as interstellar space, meteorites and meteoritic craters (Heimann *et al.*, 1984). Secondly, several attempts have been carried to synthesize long chains. It has been shown that species up to 30–300 atoms could be prepared (Kavan, 1998) using different synthetic routes or physical methods. For example, thermally stable acetylenic carbon species capped with inert groups present a chain length of more than 300 atoms (Lagow *et al.*, 1995). Another example is the "carbolite" compound, prepared by quenching of the carbon vapours from

TABLE 2. Experimental and calculated symmetries and bond lengths, bulk moduli, and excess cohesion energy, at zero Kelvin and under atmospheric pressure, calculated from the Birch-Murnaghan equation of state (Yin and Cohen, 1983; Spear et al., 1990).

Phases	Crystal symmetry and C-C bond length $d(\text{Å})$	Specific mass (g/cm^3)	Bulk modulus (GPa)	Difference of cohesion energy ΔE_c (eV.atom^{-1})
(1) Tricoordinated graphite	hexagonal ($d = 1.42$)	2.26	280	0
	rhomboedral ($d = 1.42$)		–	(small)
bct-4	tetragonal ($d = 1.44$)	2.96	360	+1.1
H-6	hexagonal ($d = 1.46$)	3.16	370	+1.7
(2) Tetracoordinated diamond	cubic ($d = 1.54$)	3.51	440	+0.30
	hexagonal ($d = 1.54$)	3.51	440	+0.33
BC-8	cubic (centered)	4.0	410	+1.0

arc discharge, which consists of a parallel array of carbon chains (Palnichenko and Tanuma, 1996). It appears that the structural and physical properties of this high temperature carbon allotrop are progressively emerging thanks to the improvement of new synthesis techniques for stabilizing these so-called "white" forms of carbon.

1.3.1.2   Graphites

This basic form of carbon is constituted with sp^2 type aromatic bonds which form condensed hexagonal cycles. This planar, and supposed by infinite, atomic sheet is called graphene (see Table 2). The bulk forms of graphite type allotrops are formed by the stacking of the graphene layers. Except for the possibility of the hypothetical carbon six-rings stacked with the identity period of 1 (1H-polytype) (Heimann et al., 1984) the low temperature and low pressure forms are the well known hexagonal and rhombohedral unit cells. The lattice structure of a single crystal of graphite is hexagonal (space group P6$_3$/mmc and polytype 2H): the graphene layers are stacked in translational ...ABAB... sequence with an in-plane nearest neighbor distance of 1.421 Å and perpendicular interplane distance of 3.354 Å at room temperature (Figure 1) (Bacon, 1948). A second possibility is a stacking periodicity ...ABCABC... as discovered in a rhombohedral phase. The hexagonal phase is thermodynamically stable but in particular thanks to grinding processes a large percentage of rhombohedral form can be obtained (Boehm and Hoffmann, 1955).

More generally, weak disorder can result from stacking faults. These departures from the ideal translation give rise to a small increase of the graphite interlayer distance. The graphene layers become both uncorrelated with their neighbour and possess a mean finite size, in which case they are called turbostratic carbons or graphites (see in the following part 1.4).

The electronic structure of these two dimensional graphene structures is a zero gap semiconductor (contact point between the valence and conduction π bands) which are transformed for an ideal hexagonal graphite, to a semi-metal with a small π band overlap (Dresselhauss et al., 1998).

New metastable phases have been proposed which consist entirely of threefold coordinated carbon in a rigid three dimensional lattice, as in a diamond type phase (Hoffmann *et al.*, 1983). The proposed goal is to combine the electronic properties of graphite with the structural ones of diamond. Two series of theoretical analysis on hypothetical crystalline phases have been carried out consisting of layers of polyene chains joined by bonds parallel to the c axis. The orientation of the chains rotates about the c-axis either by 60° (H-6 structure) (Tamor and Hass, 1990) or by 90° (bct-4 structure) (Liu and Cohen, 1992) (see Figure 4).

From theoretical calculations it appears that the excess cohesive energy for this metastable phases, which could be low-compressive metals with an hardness comparable to diamond, is not prohibitive for expecting some metastable phase (see Table 2). It should be mentioned nevertheless that no experimental approach has been reported so far.

1.3.1.3 Diamonds

In the ideal diamond structure every carbon atom is surrounded by four other carbon atoms at the corner of a regular tetrahedron forming strong covalent sp^3 bonds. The crystal structure is cubic face centered (Fd3m space group) (Figure 1). An hexagonal form of diamond called Lonsdaleite exists (Bundy and Kasper, 1967) with the same crystal symmetry as hexagonal graphite but with different site locations and a smaller interplanar separation. It also should be mentioned that in artificial diamonds a series of polytypes analogous to those well known in silicon polytypes can be described through their vibrational spectra (Spear *et al.*, 1990). Two other metastable phases have been quoted by Russian workers but not yet confirmed. They used very energetic techniques thanks to ionic discharges. They report a new cubic centered form with a calculated density exceeding by 15% the density of cubic diamond (Matyushenko *et al.*, 1979) and on the other hand a new f.c.c. structure with a monoatomic unit cell of a lower density ($d = 1.59$ g.cm^{-3}). This so called γ-carbon could be metallic but quite instable because of longer bond lengths compared to those in diamond (Palatnik *et al.*, 1985).

Besides the esthetic appearance, the cubic diamond phase is characterized by its unique hardness and its stability under high pressure. Since experimentalists are attempting to reach very large pressures, it becomes important to investigate theoretically the possible phase transformations under very high pressure. Indeed it has been calculated that as a function of the unit cell volume five different phases of tetracoordinated carbons could be expected (Yin and Cohen, 1983). The only phases which could be obtained under high hydrostatic pressure are a simple cubic one and a body-centered cubic structure (BC8). It is interesting to note that a parent to BC8, a superdense carbon called supercubane (Figure 5) was proposed independently (Johnston and Hoffmann, 1989) to explain the polycrystalline material found experimentally with a plasma technique (Matyoshenko *et al.*, 1979).

The last point relative to some new predicted phases is the transformation into higher-coordinated structures at very large pressure. A theoretical simulation has shown that under terapascal pressures diamond could collapse into a metallic sixfold coordinated structure, named SC4, despite large activation barriers (Scandolo *et al.*, 1996). To conclude, it appears that the search for new hard phases is an exciting novel subject which needs sophisticated experimental approaches.

(a)

(b)

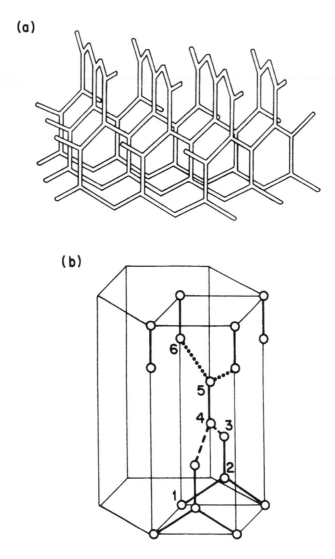

FIGURE 4.   Perspective view (a) and unit cell (b) of the H-6 structure showing only the bonds between nearest neighbour atoms (from Tamor and Hass, 1990).

1.3.2   Structures on Curved Surfaces

1.3.2.1   Historical Outline

A new chemistry of carbon materials started a few years ago with the possible synthesis of a molecular-cage arrangement consisting of sixty carbon atoms with the form of a soccer ball (Kroto *et al.*, 1985). This new perspective has been confirmed by the chemical isolation of

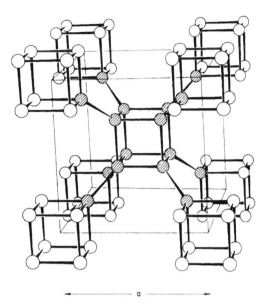

FIGURE 5. Proposed supercubane structure with a cubic unit cell of length a (shaded atoms) (from Johnston and Hoffmann, 1989).

C_{60} thanks to a plasma generated in a simple arc discharge and its structural characterization (Krätschmer *et al.*, 1990). Indeed the nearly spherical molecule C_{60}, as predicted by Euler's rule (see Figure 6), and the so called related fullerenes have attracted a great deal of interest in the recent years. Hollow cage structures with different numbers of atoms have been observed but also a new tubular form called nanotubes, are formed upon the addition of catalysts inside the carbon electrodes (Ijima, 1991). It appears immediatly that the cylindrical portions of the tubules consist of rolled graphene sheets which can be either capped or not depending on the diameter and the experimental conditions. The fundamental point associated with these new forms is that the graphene type sheets are very flexible, as has also been found for other lamellar compounds such as BN or some chalcogenides. It appears that the trigonal carbon network can form new molecular carbon species. This general mechanism may be driven by the energetic gain involved in the competition between the elimination of dangling bonds and the plane curvature, but the details of the growth processes are not fully elucidated.

During their formation not only hexagons, but also pentagons (or eventually heptagons or octogons) are formed which induce a curvature of the graphene sheets (Ebbesen, 1994). It appears, therefore, that for this peculiar class of surface allotrops, a general topological classification is useful.

1.3.2.2   Topological classification

It is necessary to apply non-euclidian geometry for curved surfaces (Schwarz, 1890). Two basic parameters are useful to define these fullerene type structures, the mean curvature H

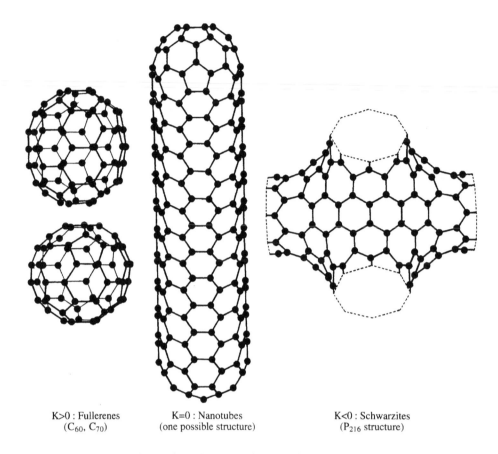

K>0 : Fullerenes   K=0 : Nanotubes   K<0 : Schwarzites
  (C$_{60}$, C$_{70}$)   (one possible structure)   (P$_{216}$ structure)

FIGURE 6. Examples of curved graphene varieties classified through their Gaussian curvature K (from Delhaès, 1997).

and the Gaussian curvature K:

$$H = \frac{K_1 + K_2}{2}, \quad K = K_1 . K_2$$

where K_1 and K_2 are two local curvatures at a given point on the surface. The gaussian curvature, homogeneous to a surface inverse, allows us to classify the following geometrical shapes (Terrones and Mackay, 1992):

– $K > 0$ (sphere): fullerenes C$_{60}$ and homologs
– $K = 0$ (plane or cylinder if $H \neq 0$): nanotubes
– $K < 0$ (saddle): Schwarzites (see in the following)

- The Fullerenes

The 60 carbon atoms are known to form a truncated icosahedron where all carbon sites are equivalent. This truncated icosahedron one of the perfect polyhedres described by Platon, has 20 hexagons and 12 isolated pentagons consistent with Euler's theorem. It crystallizes at room temperature in a cubic face centered system. Because of this high molecular symmetry C_{60} has attracted a lot of interest in physics and chemistry. However the synthesis of larger molecular weight fullerenes by arc discharge techniques is also noteworthy. The next fullerenes beyond C_{60} that satisfy the isolated pentagon rule are C_{70} and C_{76}. Currently significant or small quantities of C_{70}, C_{76}, C_{78}, C_{82}, C_{84} and C_{86} have been isolated (for C_{78} and higher compounds, several isomers are expected). The structure and properties of these higher mass fullerenes are now under study. It should be mentioned that larger carbon aggregates up to the range C_{600}–C_{700} have been observed in mass spectra along with small onion-like particles (Dresselhaus *et al.*, 1995). Finally, a C_{36} molecule has also been detected experimentally which shows that new molecular forms of carbon have yet to be discovered (Piskoti *et al.*, 1998).

- The Nanotubes

The synthesis of hollow tubes consisting of concentric cylindrical sheets proceed in different ways. The first one was by using d.c. electric discharge at high temperature (locally around $3600°C$) which furnish a mixture of different types of carbon. The second way has been the catalytic decomposition of hydrocarbons with transition metal catalyts (Fe, Co, Ni) at low temperatures (below $1000°C$) as already developed for microfibers (Baker and Harris, 1978). More recently a laser ablation technique has been proposed which leads to better defined nanotubes (Thess *et al.*, 1996). Indeed typical carbon nanotubes are 1–25 nm across with a length up to 10 μm. They consist of 1–50 graphene cylinders wrapped around each other along a common long axis.

The experimental evolution has been to synthesize at the beginning multiwalled and then single wall nanotubes, which are formed by the laser method for which a very narrow diameter distribution is detected. As shown by transmission electron microscopy (TEM) these nanotubes with different diameters can be either capped or open. For the thinner ones ($\varnothing = 7$ Å) a half-fullerene C_{60} capping can exist (Figure 6) or for larger sizes, and sometimes after heat treatment a conical shape is observed. Indeed there is currently a large variety of nanotubes and the main pending problem is to prepare a homogeneous batch. The purification of single wall nanotubes, which form in aligned bundles, is a necessary technical improvement. Nevertheless, the structural and physical properties (electronic and mechanical characteristics) are under study for these defect free single wall nanotubes which offer different folding symmetry including arm-chair, zig-zag, chiral nanotubes (Dresselhaus *et al.*, 1995). Indeed the understanding of gas phase growth mechanisms and the associated topological and atomic defects are of prime interest for defining different morphologies. Nanoparticles are also formed under these experimental conditions as nanocones (Sattler, 1995) helix shape tubules and microfibers (Ihara and Itoh, 1995) or even onions (Ugarte, 1995). Currently a large development is occurring relative to the potential nanotechnologies. There is currently much interest in using nanotubes to develop potential nanotechnologies.

• The Schwarzites

It has been also suggested that graphene type structures with a negative gaussian curvature may be possible by introducing seven-or eight-membered rings in addition to the usual six-membered rings (Terrones and Mackay, 1992; Lenosky *et al.*, 1992). In fact these negatively curved carbon networks belong to the class of periodic minimal surfaces (Schwarz, 1890) and it has been suggested that these novel tridimensional carbon structures be called "Schwarzites".

Several models of periodic graphite-like surfaces have been investigated theoretically; These models consist of different kinds of periodic surfaces (P,D,G types) with big unit cells containing from 24 to 216 atoms (see one example called P.216 on Figure 6). Following the general principles outlined in paragraph 2.3 calculations of the cohesive energy and bulk modulus have been carried out for these extended systems. It has been shown that the cohesive energy appears to be greater than in C_{60}. This means that these forms could exist as a metastable state. One interesting case is the polybenzene model with only 24 carbon atoms in a simple cubic unit cell (Huang *et al.*, 1993). Nevertheless in spite of different attempts no clear experimental evidence for their existence has been presented so far.

1.3.2.3 Physical properties and phase stability

A few remarks relative to the physical properties of these surface allotropes should be made. The surface curvature induces a bond distortion because of the presence of a local strain which produces a rehybridization of the molecular orbitals (Haddon, 1992). Indeed the hybridization is no longer a π-orbital of sp^2 type, but there is a mixing with the σ-orbital which is significant for C_{60}, C_{70} and small radius nanotubes. This effect will modify the electronic properties and the chemical reactivity of the smallest molecular compounds and will become less and less important for more extended systems.

The π-type electronic dimensionality of these forms complements their structures (Dresselhaus *et al.*, 1998). C_{60} and its homologues can be considered zero dimensional, the nanotubes are one dimensional, a graphene sheet is two dimensional, and the predicted Schwarzites would be rather three dimensional. The physical properties of fullerenes and the many nanotubes isomers can be associated with quantum size effects, which is one of the reason why there is such excitement for probing and manipulating these nanoscale materials (Odom *et al.*, 1998; Issi, 1998).

The temperature and pressure stabilities and the phase transformations of C_{60} are a key point of interest when considering the C_{60} molecule as an independent molecular species that could be transformed under T or P to another surface or volume allotropes. This reaction diagram (Figure 7) has been investigated by different authors (Nunez-Regueiro *et al.*, 1995; Blank *et al.*, 1997) and the main features are the following:

– At moderate temperature under hydrostatic pressure several oligomeric and polymeric phases of C_{60} have been isolated and characterized (see orthorhombic and rhomboedral phases).
– With increased temperature an irreversible chemical transformation to a graphitic phase is observed.

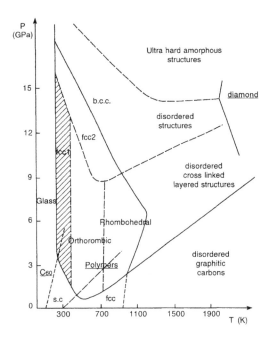

FIGURE 7. Sketch of the pressure-temperature reaction diagram showing the various phases of C_{60} created under different experimental constraints (from respectively Blank *et al.*, 1998 and Nunez-Regueiro *et al.*, 1996).

– At very high pressures (up to 20 GPa) a diamond phase is formed along with other ultrahard phases which are not yet completely characterized.

Indeed, a quite large number of more or less crystalline materials have been discovered which are not fully characterized (see Figure 7), but these experiments open the way to a new field of investigation.

1.3.3 Exotic Structures with Variable Coordination Numbers

An alternative is to predict new forms of carbon with mixed coordination numbers i.e. with $z = 2$ and 3 or $z = 3$ and 4. The pursued goal is to combine the physical properties of the classical allotropes such as hexagonal graphite and cubic diamond, with rational space filling structures. These topological approaches are based on Wells's seminal work on the structure of two and three dimensional nets and polyedra forms (Wells, 1977). It must be noticed, however, that almost none of these theoretical predictions have been experimentally confirmed which is why they are called "exotic".

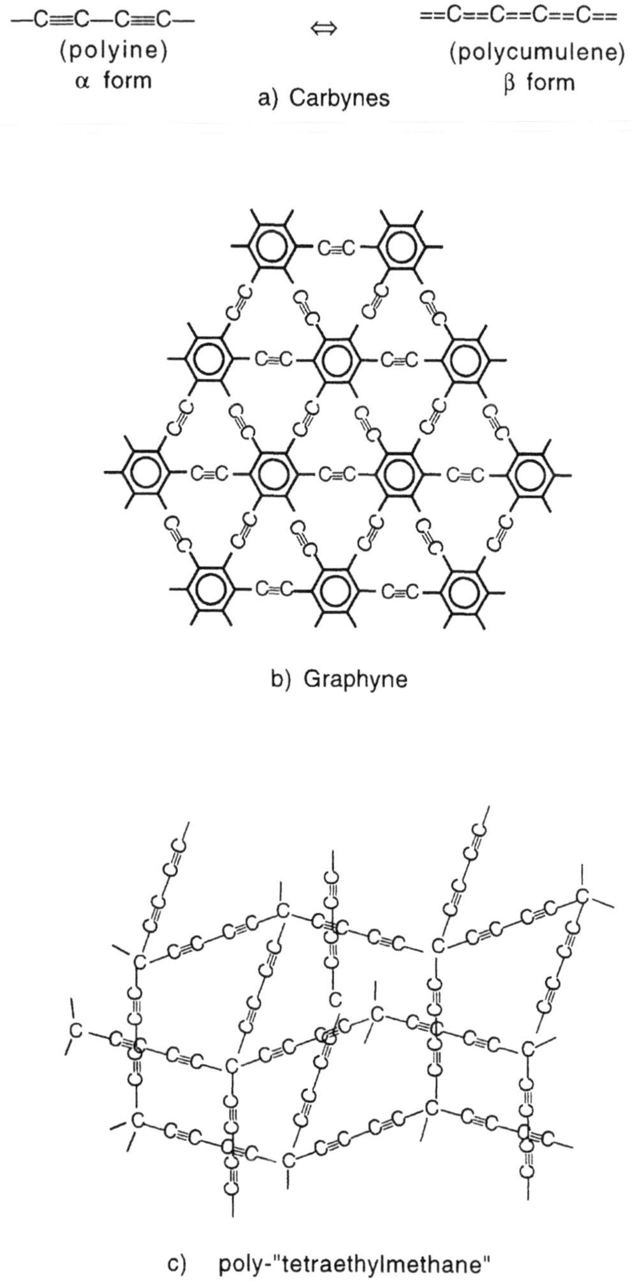

16   P. DELHAÈS

a) Carbynes

b) Graphyne

c)   poly-"tetraethylmethane"

FIGURE 8. Examples of proposed acetylenic polymeric carbons with different space fillings:
(a) 1D: linear polyynes "carbynes"; (b) 2D: layer of graphyne (from Baughman *et al.*, 1987);
(c) 3D: possible superdiamondoid network (from Diederich, 1994).

1.3.3.1 (2-3) Connected carbon nets

These systems would present an intermediary between carbynes and graphene as conjugated polymers with an average of more than one π electron per carbon atom. Different types of molecular organisations, linear or layered, can be envisaged with carbons in both sp^2 (or sp^3) and sp states.

The first of these interconnected chains, to that reinforce the thermostability and mechanical properties of poly-p-phenylene or poly(-p-phenylene) xylylidene, to form a crystalline cubic state (Baughman and Lui, 1993). Conjugated enediynes also have examined because a 3D organization could be obtained (Figure 8) there are serious synthetic problems to get ordered polymers (Gleiter and Kratz, 1993; Diederich, 1994).

Secondly, the prediction of structure and properties for planar structures appears more attractive with a reasonably low formation energy. This proposal is based on molecular sheets with the same planar symmetry as graphene, which can be formally viewed as resulting from the replacement of one third of the aromatic carbon-carbon bonds by $-C{\equiv}C-$ linkages (Baughman *et al.*, 1987). This material called graphyne should behave as a quite large bandgap semiconductor (see Figure 8).

1.3.3.2 (4-3) Connected carbon nets

The goal is to propose new compounds with an intermediate valency between graphite and diamond, i.e. less than one p electron per carbon atom. Planar systems with a variable number of carbon atoms inside a ring (four or ten-numbered rings for example) are selected. They form a 2D plane nets which can be interconnected by tetracoordinated carbons to obtain a 3D lattice (Merz *et al.*, 1987).

More recently a 3–4 connected net containing trigonal and tetrahedral atoms in a 2:1 ratio, built upon the 1-4 cyclohexadienne motif, has been proposed (Bucknum and Hoffmann, 1994). An hypothetical tetragonal form based on a spiroconjugation (Figure 9) could give rise to a new conducting and dense form of carbon ("Glitter model"). All these predicted forms will need the control of specific synthetic chemistry to be effectively realized in the future.

1.4 NON-CRYSTALLINE AND DOPED CARBONS

Up to now we have presented real or virtual crystals, but non-crystalline or doped forms of carbon exist when some positional disorder is present in homogeneous or inhomogeneous solids. The different types of disorder which have been recognized (Delhaes and Carmona, 1981) are the following:

– mixtures of chemical bondings in pure carbon (tetracoordinated and tricoordinated atoms),
– lattice defects associated with the loss of a long range order with two main types of local organization (glasses or microcristallites),
– presence of linked heteroatoms (H,H,N,O,...) or so-called impurities in insertional or in substitutional positions (as for example B or N).

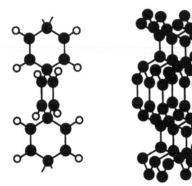

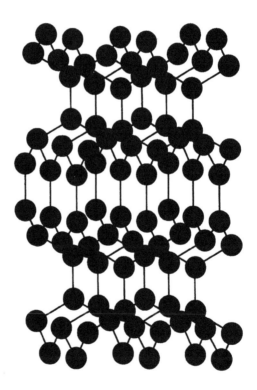

FIGURE 9. A hypothetical 3–4 connected carbon net built from the 1–4 cyclohexadiene molecule (from Bucknum and Hoffmann, 1994).

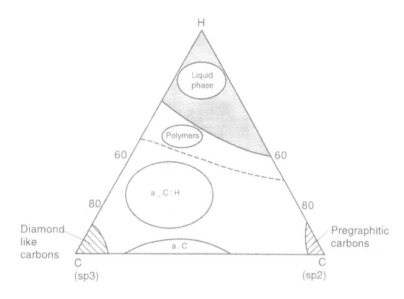

FIGURE 10. Ternary compositional diagram with a schematic classification of the different kinds of non-crystalline carbons.

In these non-crystalline solids the Gibbs free energy is in principle larger than for the already defined crystalline forms because of the presence of larger entropic term at a given finite temperature associated with some structural disorder. If a local Gibbs energy minimum, as presented on Figure 2, exists under particular experimental conditions, then different non-crystalline metastable phases will result. This pseudo-polymorphic character will depend crucially on the preparation conditions, which are on one hand the nature of the precursor (phase, composition) and on the other hand how the excess of energy is provided to the material, i.e. the proceeding technique (see paragraph 4.3). We will briefly review the different varieties of pseudo polymorphic, carbons (pure or impure).

1.4.1 Classification of Non-Crystalline Carbons

We concentrate on the case where the initial stable phase is an hydrocarbon which contains only hydrogen and carbon. When this organic molecule is submitted either to a pyrolysis at high temperature followed by a carbonization process, or to some ionic bombardment process, different phases are obtained. A schematic classification is presented on the ternary diagram (Csp^2, Csp^3, H) (Figure 10).

A continuum of compounds exists but a few main solid phases are usually recognized:

– Polymers which are essentially condensed planar polyaromatic compounds considered as precursors of a graphene sheet (Robertson, 1998).

– Hydrogenated amorphous carbons (a-C:H) prepared, for example, by plasma chemistry of a gaseous hydrocarbon; films of different compositions are deposited by plasma enhanced chemical vapor deposition. These a-C:H films contain from 40% to less than 10% hydrogen, with different ratios of sp^3 to sp^2 hybridized carbons exhibiting different physical properties (McKenzie *et al.*, 1983; Bubenzer *et al.*, 1983).

It is noteworthy that for special purposes other amorphous films with different heteroatoms can be prepared, for example, with nitrogen, a-C:N (Ricci *et al.*, 1994) or fluorine, a-C:F (Endo and Tatsumi, 1996).

– Diamond like carbons (DLC): this is a quasi homogeneous phase with a large majority of tetracoordinated atoms as in cristalline diamond (Aisenberg and Chabot, 1971). Depending on the experimental process, different films with microcrystalline or more glass type structures can be obtained (Angus and Hayman, 1988). These DLC are characterized by a very high hardness and a low coefficient of friction.
– Amorphous carbons (a-C): this contains tetracoordinated, tricoordinated (and eventually some bicoordinated atoms) either free of hydrogen or with a very small amount. This type of material was noticed a long time ago (Kakinoki *et al.*, 1960), but more recent works have confirmed its existence (Weissmantel *et al.*, 1980: these authors called it i-C). These unhydrogenated a-C are thought to be organized in small aromatic clusters embedded in an amorphous diamond like phase. This mixing controls their mechanical and electronic properties (Kelires, 1994).
– Pregraphitic carbons: they are formed with condensed polyaromatic compounds the size of which increases with the temperature of treatment. Simultaneously the hydrogen content decreases. Their structure is characterized by the presence of basal structural units (BSU) made with a few coronene type polyaromatic compounds and included in crystallites. Usually two main evolution steps toward the hexagonal graphite are recognized called respectively (primary, secondary) carbonisation and graphitation processes (Oberlin and Bonamy, 1998).

1.4.2 *Doped Carbons and Solid Solutions*

The deliberate introduction of foreign atoms into any covalent carbon networks is a wide subject of interest. The purpose is to change either the physical properties or the chemical ones. In this short presentation we look at two aspects of these new compounds.

– Doped carbons with essentially boron or nitrogen atoms in substitional position. In hexagonal graphite and pregraphitic materials a low content of doping can be realized (at thermodynamic equilibrium up to 2.35% of boron). They modify considerably the electronic properties associated with their π-donor or π-acceptor character (Marchand, 1971). More recently, doping of the other allotrops, such as cubic diamond, as well as fullerenes and nanotubes are in progress.
– Solid solutions also have been investigated. Binary or ternary compounds with boron and nitrogen exist as disordered solid solutions in quenched metastable states in a graphitic type lattice (Delhaès, 1997). They do not form a defined chemical compound as usually

predicted by theoretical calculations for graphene sheets or tubular forms. It appears, therefore, that in this area, there is a large gap between the experimental facts and these predictions.

Finally, it should be mentioned that parent compounds, such as boron-nitride (BN) or carbonitrides (for example C_3N_4), which are isomorph of the carbon allotrops, have been discovered and are currently under studies (Demazeau, 1997).

1.4.3 Morphologies of Non Crystalline-Carbons

One complementary point which is crucial for applications, is that different forms or morphologies of non-crystalline carbons are produced. These different growth forms are functions of the experimental conditions, in particular the mechanisms of nuclei formation in homogeneous or not phase, and in the latter situation the deposition onto a substrate. One short way to present these different forms of graphitic carbons is to classify them following a three dimensional geometrical arrangement of BSU and crystallites (Oberlin, 1989):

- planar symmetry: film deposits as for example, diamond-like or pyrocarbons;
- cylindrical symmetry: from multiwalled nanotubes to wiskers up to fibers;
- spherical symmetry: as mesophase microbeads, onions or carbon black particles.

These carbonaceous forms are either dense, as pyrocarbon or glassy carbons or porous as for example felts or aerogels.

These different morphologies, which occur at very different nanometric to micrometric scales, are observed by electronic and optical microscopies, as described in numerous reviews and books (Dresselhaus *et al.*, 1995).

1.5 CONCLUSION

In this general presentation we have shown that, starting from a quantum chemistry basis and the hybridization of carbon atomic orbitals, a carbon atom can exhibit several coordination numbers. This specific characteristic induces a strong polymorphism of these solids associated with different topological situations. Currently about ten different crystalline phases have been observed to which can be added the non-crystalline forms. A recent break-through has been the discovery of curved forms, such as fullerenes and nanotubes. One step further in this direction is the theoretical prediction of several other allotropes which are unknown in nature. We have given several examples of such metastable phases, but the scientific challenge is to find the appropriate way to prepare these new compounds.

These carbon based materials have already had a major impact in science and technology. In this review, however, we do not developp the complementary aspects of the synthetic routes and processes, nor the forecasted chemical and physical properties. Concerning synthesis, it is essential to recall that two points are relevant: the nature and composition of the precursor phase and the technique which is used to supply the excess of energy (Delhaès and Carmona, 1981). It is these variables which will induce the key intermediary species. For example with pure carbon submitted to some ionic or photonic beam different size clusters are detected which show that ionic chains, rings and fullerenes are present when the number of atoms increases (Handschuh *et al.*, 1995). On the other hand, classical chemical vapor deposition of hydrocarbons gives rise to polyaromatic units then to graphene like carbons. It turns out that the different reactive processes necessary to obtain a definite phase are not completely understood because the chemistry of carbon is complex and difficult to control.

The last point concerns the different properties which are presented by these polymorphic forms which range from ceramics to polymers science. Diamond is the hardest known material with a high thermal conductivity combined with its transparent and insulating characters. These properties are widely exploited in several technological applications. All the forms of graphite, depending on their morphologic forms and shapes give rise to many industrial applications, in particular the production of carbon blacks, fibers and composites. The main additional characteristics are due to the electrical properties associated with its lamellar character which are interesting, for example, in electrochemistry or some surface applications related to environmental problems.

One significant point concerning the theoretical approaches is the combination of the properties of both major phases. For example, a dream phase, as postulated for example in Hoffmann's papers, would be to synthesize a new carbon materials with the mechanical properties of diamond and the electronic ones of graphite. It turns out, however, that the latest relevant advances are those concerning the carbon nanoparticles with fullerenes and multiple or single wall nanotubes. A new branch of material science is appearing at the interface of molecular compounds and nanophases with new possible technological applications. Indeed new but also old form of carbons still are of great interest in several areas of science and technology.

References

Aisenberg, S. and Chabot, R. (1977) *J. Appl. Phys.*, **42**, 2953.

Angus, J.C. and Hayman, C.C. (1988) *Science*, **241**, 913.

Atkins, P.W. (1990) *Physical Chemistry* (fourth edition) Oxford University Press.

Bacon, G.E. (1948) *Acta Cryst.*, **1**, 337.

Baker, R.T.K. and Harris, P.S. (1978) *Chemistry and Physics of Carbon*, **14**, 8.

Baughman, R.H., Eckhardt, H. and Kertesz, M. (1987) *J. Chem. Phys.*, **87**, 6687.

Baughman, R.H. and Lui, C. (1993) *Synthetic Metals*, **55–57**, 315.

Berman, R. and Simon, F. (1955) *Z. Elektrochem.*, **59**, 333.

Blank, V.D., Buga, S.G., Dubitsky, G.A., Serebryanaya, Popov, M.Y. and Sundqvist (1998) *Carbon...*

Boehm, H.P. and Hofmann, V. (1995) *Anorg. Allgem. Chem.*, **278**, 58, 299.

Bubenzer, A., Dischler, B., Brandt, G. and Koidl, P. (1983) *J. Appl. Phys.*, **54**, 4590.

Bucknum, M.J. and Hoffmann, R. (1996) *J. Am. Chem.*, **116**, 11456.

Bundy, F.P., Basset, W.A., Weathers, M.S., Hemley, R.J., Mao, H.K. and Goncharov, A.F. (1996) *Carbon*, **34**, 141.

Bundy, F.P. and Kasper, J.S. (1967) *J. of Chem. Phys.*, **46**, 3437.

Cohen, M.L. (1994) *Sol. St. Comm.*, **92**, 45.

Delhaès, P. (1997) *in Le Carbone dans tous ses états*, editors P. Bernier and S. Lefrant (Gordon and Breach).

Delhaès, P. and Carmona, F. (1981) *Chemistry and Physics of Carbon*, **17**, 89, editors P.L. Walker and P.A. Thrower (Marcel Dekker).

Demazeau, G. (1997) *in Le carbone dans tous ses états*, editors P. Bernier and S. Lefrant (Gordon and Breach).

Diederich, F. (1994) *Nature*, **369**, 199.

Dresselhaus, M.S., Dresselhaus, G. and Ecklund, P.C. (1995) *Science of fullerenes and carbon nanotubes* (Academic Press).

Dresselhaus, M.S., Dresselhaus, G. and Saito, G. (1998) *World of Carbon I*.

Ebbesen, W. (1994) *Annu. Rev. Mat. Sci.*, **24**, 235.

Endo, K. and Tatsumi, T. (1996) *Appl. Phys. Lett.*, **68**, 2864.

Gleiter, R. and Kratz, D. (1993) *Angew. Chem. Int. Ed. Engl.*, **38**, 842.

Handschuh, H., Gantefor, G., Kessler, B., Bechthold, P.S. and Eberhardt (1995) *Phys. Rev. Lett.*, **74**, 1095.

Haddon, R.C. (1992) *Accounts of chemical research*, **25**, 127.

Heimann, R.B., Kleiman, J. and Salansky, N.M. (1964) *Carbon*, **22**, 147.

Hirsch, A. (1994) *The chemistry of the fullerenes* (Thiene Medical Publ. New York).

Hoffmann, R., Hughbanks, T. and Kertsesz, M. (1983) *J. Am. Chem. Soc.*, **105**, 4831.

Huang, M.Z., Ching, W.Y. and Lenosky, T. (1993) *Phys. Rev. B*, **47**, 1593.

Ihara, S. and Itoh, S. (1995) *Carbon*, **33**, 931.

IIjima, S. (1991) *Nature*, **354**, 56.

Issi, J.P. (1998) This volume chapter.

Johnston, R.L. and Hoffmann, R. (1989) *J. Am. Chem. Soc.*, **111**, 810.

Kakinoki, J., Katada, K., Hanawa, T. and Ino, T. (1960) *Acta Crystallogr.*, **13**, **171&13**, 448.

Kastner, J., Kuzmany, H., Kavan, L., Bousek, F.P. and Kurti, J. (1995) *Macromolecules*, **28**, 344.

Kavan, L. (1998) *Carbon* (to appear).

Kelires, P.C. (1994) *Phys. Rev. Lett.*, **73**, 2460.

Kroto, H.W., Heath, J.R., O'Brien, S.C., Curl, R.F. and Shallgy, R.E. (1985) *Nature*, **318**, 162.

Krätschmer, W., Lamb, L.D., Fostiropoulos, K. and Huffman, D.R. (1990) *Nature*, **347**, 354.

Kudryavtsev, Y.P., Evsyukov, S.E., Guseva, M.B., Babaev, V.G. and Khvostov (1997) *Chemistry and Physics of Carbon*, **25**, ed. P.A. Thrower (M. Dekker New-York) 25, 1–69.

Lagow, R.J., Kampa, J.J., Wei, H.C., Battle, S.L., Genge, J.W., Laude, D.A., Harper, C.J., Bau, R., Stevens, R.C., Haw, J.F. and Munson, E. (1995) *Science*, **267**, 362.

Lenosky, T., Gonze, X., Teter, M. and Elser, V. (1992) *Nature*, **355**, 333.

Liu, A.Y. and Cohen, M.L. (1992) *Phys. Rev. B*, **45**, 4579.

Marchand, A. (1971), *Chemistry and Physics of Carbons* (Ed. by P.L. Walker and P. Thrower) **7**, 155.

Matyushenko, N.N., Strel'nitskii, V.E. and Gusev, V.A. (1979) *JETP Letters*, **30**, 199.

McKenzie, D.R., Mcphedran, R.C., Bavides, N. and Botten, L.C. (1984) *Philos. Magazine B*, **48**, 341.

Merz, K.M., Hoffmann, R. and Balaban, A.T. (1997) *J. Am. Chem. Soc.*, **109**, 6742.

Nunez-Regueiro, M., Marques, L., Hodeau, J.L., Bethoux, O. and Perroux, M. (1995) *Phys. Rev. Lett.*, **74**, 278.

Oberlin, A. (1978) *Chemistry and Physics of Carbons*, vol. 17 (Ed by P.L. Walker and P. Thrower)

Oberlin, A. (1989) *Chemistry and Physics of Carbons*, vol. 22, 1 (Ed by P.L. Walker and P. Thrower)

Oberlin, A. and Bonamy, S. (1998) This volume chapter 9.

Odom, T.W., Huang, J.L., Kim, P. and Lieber, C.M. (1998) *Nature*, **391**, 62

Palatnik, L.S., Guseva, M.B., Babaev, V.G., Savchenko, N.F. and Fal'ko, I.I. (1984) *Sov. Phys. JETP*, **60**, 520.

Palnichenko, A.V. and Tanma, S. (1996) *J. Phys. Chem. Solids*, **57**, 1163.

Piskoti, C., Yarger, J. and Zettl, A. (1998) *Nature*, **393**, 771.

Ricci, M., Trinquecoste, M. Auguste, F., Canet, R., Delhaès, P., Guimon, C., Pfister-Guillouzo, G., Nysten, B. and Issi, J.P. (1993) *J. Mat. Res.*, **8**, 480.

Robertson, J. and O'Reilly, E.P. (1997) *Phys. Rev. B*, **35**, 2946.

Sattler, K. (1995) *Carbon*, **33**, 915.

Scandolo, S., Chiarotti, G.L. and Toslatti, E. (1996) *Phys. Rev. B*, **53**, 5051.

Schwarz, H.A. (1890) *Gesamunelte Mathematische Abhandlungen* (Bd1), Springer-Verlag Ed.

Spear, K.E., Phelps, A.W. and White, W.B. (1990) *J. Mater. Res.*, **5**, 2277.

Tamor, M.A. and Hass, K.C. (1990) *J. Mater. Res.*, **5**, 2273.

Terrones, H. and Mackay, A.L. (1992) *Carbon*, **30**, 1251.

Thess, A., Lee, R., Nikolaev, B., Dai, H., Petit, P., Robert, J., Xu, C., Lee, Y.H., Kim, S.G., Rinzler, A.G., Colbert, D.T., Scuseria, G.E., Tomanek, D., Fischer, I.E. and Smalley, R.E. (1992) *Nature*, **273**, 483.

Ugarte, D. (1995) *Carbon*, **33**, 989.

Van Theil, M. and Ree, F.M. (1993) *Phys. Rev. B*, **48**, 3591.

Weissmantel, C., Bewilogna, K., Dietrich, D., Erler, H.J., Klose, S., Nowick, W. and Reisse, G. (1980) *Thin Sol. Films*, **72**, 29.

Wells, A.F. (1977) Three dimensional nets and polyedra (Wiley Ed.).

Yin, M.T. and Cohen, M.L. (1983) *Phys. Rev. Letters*, **50**, 2006.

2. Electronic Band Structure of Graphites

M.S. DRESSELHAUS[1], G. DRESSELHAUS[2] and R. SAITO[3]

[1]*Department of Electrical Engineering and Computer Science and Department of Physics, Massachusetts Institute of Technology, Cambridge, Massachusetts 02139, USA*
[2]*Francis Bitter Magnet Laboratory, Massachusetts Institute of Technology, Cambridge, Massachusetts 02139, USA*
[3]*Department of Electronics Engineering, University of Electro-Communications, Chofugaoka, Chofu, 182 Tokyo, Japan*

2.1 INTRODUCTION

The wide range of structures that occur in common graphites implicitly implies a range of electronic structures. At one extreme is the ideal single crystal graphite and at the other extreme are the highly disordered graphites where the mean free path of charge carriers is only as long as a few unit cell dimensions. In between these two extremes, disorder of various kinds modifies the electronic structure in several ways.

Some of these changes in the electronic structure are unique to carbon, and result from the fact that graphite is a narrow band overlap semimetal. In the absence of interplanar interactions, graphite would be a zero-gap semiconductor, as a result of symmetry. Thus the small band overlap in graphite results from interactions between planes. These interplanar interactions can be modified, and effectively reduced to zero, by introducing stacking disorder, resulting in *turbostratic* carbon, where the carbon atom sites on adjacent layers are uncorrelated and where the individual graphene sheets are not explicitly bonded to each other. When long range order in the planes is disrupted, this degeneracy is lifted, and, in effect, a narrow band gap semiconductor results, in which there are localized states in the gap near the band edges. This disorder can be extended continuously to describe amorphous carbon as a limiting case. Experimentally, heat treatment or annealing can be used to increase the amount of order in a disordered graphite, and ion implantation or neutron irradiation can be used to increase the disorder. The heat treatment temperature (T_{HT}) at fixed residence time has been used as a rough characterization parameter for the amount of disorder (Oberlin, 1984).

To account for the electronic properties of carbon, Mrozowski (Mrozowski, 1971) developed a simple model to explain the major changes in the electronic structure which occur as T_{HT} is increased in graphitizable carbons. Several distinct T_{HT} regions were identified to delineate characteristic features in the electronic structure (Dresselhaus, 1988)

(1) In the low T_{HT} region ($T_{HT} < 1000°C$), where the impurity content and defect density is large, an energy gap exists between the occupied and unoccupied states. Hall effect measurements indicate that the predominant structural defects produce holes, so that the electrochemical potential lies near the valence band edge. Long ago, Coulson *et al.* (Coulson, 1957) considered the energy levels of aromatic molecules (planar molecules with carbon hexagons and edge hydrogens to passivate dangling bonds), and showed that the density of electronic states near the bonding-anti-bonding energy gap began to approximate that of a graphene sheet when the number of carbon atoms exceeded ∼50 where the band gap was estimated at ∼5 eV. As T_{HT} increases, the excitation energy gap ΔE decreases and the transition from an insulator to a metal occurs for $T_{HT} \sim 1000°C$ by the percolation of electrons (or holes) between regions having sufficiently small values of ΔE (Delhaès, 1981). Most carbons in this range of T_{HT} would have such small mean free paths, that they would not be classified as a graphite. From an electronic standpoint, a criterion for considering a carbon to be a graphite is that $kl \gg 1$, where k is the wave vector of the carriers and l is the mean free path.

(2) In the region $1000°C \leq T_{HT} \leq 2500°C$, where graphites exhibit negative magnetoresistance associated with weak localization phenomena, the electronic properties are strongly influenced by the finite size of the hexagonal carbon planar networks. In this range of T_{HT}, various authors have tried to explain the electronic transport properties through graphene ribbons of width W and of infinite length L (Dresselhaus, 1988; McClure, 1982; Nakada, 1996). Recent results for the electronic structure of graphene ribbons are reviewed in Section 2.4. In the upper part of this range of T_{HT}, the electronic structure of turbostratic carbon ($2200°C \leq T_{HT} \leq 2500°C$) is modeled as a collection of uncoupled two-dimensional graphene layers, as discussed in Section 2.2.

(3) In region $T_{HT} > 2500°C$, the 2D graphite develops from turbostratically stacked uncoupled interlayer sheets to three-dimensional (3D) graphite. For $T_{HT} \geq 3500°C$, the structure of graphitizable carbons can approach that of single crystal graphite, in which both the in-plane ordering and the interlayer site correlations become well-developed. The electronic structure of single crystal graphite is described in Section 2.3. Pregraphitic carbons intermediate between 2D and 3D graphite will have a lower carrier density, a smaller effective band overlap energy, and enhanced carrier scattering by defects relative to 3D graphite.

Because of the large anisotropy of the crystal structure (Wyckoff, 1964), most models for the electronic structure of graphite start from a two-dimensional approximation, and explicitly treat the intraplanar interaction between the $2s$, $2p_x$, $2p_y$ atomic orbitals (commonly denoted as sp^2 bonding) to form strongly coupled bonding and antibonding trigonal orbitals. These trigonal orbitals give rise to three bonding and three antibonding

σ-bands which, respectively, lie $\sim$10 eV below and above the Fermi level E_F in the two-dimensional (2D) graphite band structure. In these models, the weakly coupled p_z atomic wave functions correspond to two π-bands which are degenerate by symmetry at the six corners of the two-dimensional Brillouin zone (see Figure 1) through which the Fermi level passes. Models for 2D graphite are appropriate for describing the properties of turbostratic graphites (Spain, 1973, 1981).

Whereas the two-dimensional band models give the in-plane dispersion relations for the two π-bands for a graphene sheet, the three-dimensional Slonczewski–Weiss–McClure (SWMcC) model, which is reviewed in Section 2.3, gives three-dimensional dispersion relations for the four π-bands corresponding to the full graphite unit cell containing four crystallographically distinct atoms based on the ABAB stacking sequence of crystalline graphite. In the 3D crystalline phase, graphite behaves like a semimetal with a band overlap of $\sim$40 meV between the valence and conduction bands, so that the transport properties of graphites are strongly dependent on the stacking arrangement of the graphene sheets. In the ground state ABAB stacking arrangement, half of the hexagon edges on one layer lie over the centers of hexagons on the adjacent layers (Wyckoff, 1964), since the number of edge sites in the honeycomb lattice is twice the number of center sites. It is interesting to note that different stacking arrangements give rise to significant differences in the energy dispersion relations $E(\vec{k})$ of the π-bands near the K-point, thereby resulting in differences in the properties of the Fermi surface near the edges of the Brillouin zone (McClure, 1969; Samuelson, 1980). Since disordered graphites have many stacking faults, and since $E(\vec{k})$ is quite sensitive to the stacking of the graphene sheets, it is difficult to describe the electronic dispersion relations for commercial (disordered) graphites quantitatively. Therefore, it is common to use idealized models in limiting cases to describe the electronic properties of highly ordered 3D graphite quantitatively, and to describe the properties of disordered graphites with more approximate models.

2.2 TWO-DIMENSIONAL GRAPHITE

A single layer of crystalline graphite is commonly called a graphene sheet or simply two-dimensional (2D) graphite. Even in 3D graphite, the carbon-carbon (C-C) interaction between two adjacent layers is small compared with intra-layer interactions, since the layer-layer separation of 3.35 Å is much larger than the nearest-neighbor distance between two carbon atoms, $a_{C-C} = 1.42$ Å. Thus the electronic structure of 2D graphite provides a good first approximation to that for 3D graphite, and directly pertains to physical graphite materials where the interlayer coupling goes to zero (known as the turbostratic limit). For these reasons, there have been many calculations of the electronic structure of 2D graphite (Wallace, 1947; Corbato, 1957; Bassani, 1967; Doni, 1969; Painter, 1970; Zunger, 1978; Ohno, 1979; Tatar, 1982; Fretigny, 1989), with some of the calculations also treating the 3D case.

In Figure 1 we show (a) the unit cell and (b) the Brillouin zone of two-dimensional graphite as a dotted rhombus and shaded hexagon, respectively (Saito, 1998), where $\vec{a}_1$ and $\vec{a}_2$ are unit vectors in real space, and $\vec{b}_1$ and $\vec{b}_2$ are the corresponding reciprocal lattice vectors. In terms of the x, y coordinates shown in the Figure 1, the real space unit vectors $\vec{a}_1$ and $\vec{a}_2$ of the

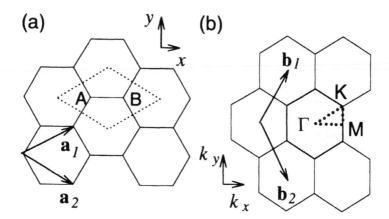

FIGURE 1. (a) The unit cell and (b) Brillouin zone of two-dimensional graphite are shown as a dotted rhombus and shaded hexagon, respectively. $\vec{a}_1$ and $\vec{a}_2$ are unit vectors in real space and $\vec{b}_1$ and $\vec{b}_2$ are reciprocal lattice unit vectors. Energy dispersion relations are obtained for wave vectors within the dotted triangle connecting the high symmetry points, Γ, K and M. (Saito, 1998)

hexagonal lattice are expressed as

$$\vec{a}_1 = \left(\frac{\sqrt{3}}{2}a, \frac{a}{2}\right), \quad \vec{a}_2 = \left(\frac{\sqrt{3}}{2}a, -\frac{a}{2}\right), \tag{1}$$

where $a = |\vec{a}_1| = |\vec{a}_2| = 1.42 \times \sqrt{3} = 2.46\text{Å}$ is the lattice constant of 2D graphite. Correspondingly the unit vectors $\vec{b}_1$ and $\vec{b}_2$ of the reciprocal lattice are given by:

$$\vec{b}_1 = \left(\frac{2\pi}{\sqrt{3}a}, \frac{2\pi}{a}\right), \quad \vec{b}_2 = \left(\frac{2\pi}{\sqrt{3}a}, -\frac{2\pi}{a}\right) \tag{2}$$

corresponding to a lattice constant of $4\pi/(\sqrt{3}a)$ in reciprocal space. The direction of the unit vectors $\vec{b}_1$ and $\vec{b}_2$ of the reciprocal hexagonal lattice are rotated by $90°$ from the unit vectors $\vec{a}_1$ and $\vec{a}_2$ of the hexagonal lattice in real space, as shown in Figure 1. Using the first Brillouin zone of 2D graphite, shown by the shaded hexagon in Figure 1(b), the three high symmetry points, Γ, K and M are defined as the zone center, zone corner, and center of the edge, respectively. The energy dispersion relations are calculated for wave vectors within the dotted lines in Figure 1(b).

Each carbon atom in the graphene layer makes three σ bonds to its three nearest neighbors in a sp^2 configuration, while, and the other $2p_z$ orbital, which is perpendicular to the graphene plane, makes covalent π bonds. In studying the transport properties and Fermi surface characteristics, it is necessary to consider only the π energy bands for 2D graphite, because the π energy bands are covalent and are located close to the Fermi level, thereby determining the solid state properties of graphite. For this reason we first review the electronic structure for the π-bands, and then discuss the electronic structure for the σ-bands which are located some distance from the Fermi level. The 2D model for the σ-bands gives a good first approximation for the σ-bands for 3D graphite.

The two Bloch functions, that are constructed from atomic orbitals for the two inequivalent carbon atoms at A and B in Figure 1, provide the basis functions that describe the energy dispersion relations for the π-bands for 2D graphite (Bassani, 1975). Considering only nearest-neighbor interactions and following the tight-binding approach (Bassani, 1975), we obtain the matrix elements of the tight binding Hamiltonian for the atoms A and B:

$$\mathcal{H}_{AB} = t(e^{i\vec{k}\cdot\vec{R}_1} + e^{i\vec{k}\cdot\vec{R}_2} + e^{i\vec{k}\cdot\vec{R}_3}) = t f(k) \tag{3}$$

where $\vec{R}_1$, $\vec{R}_2$, and $\vec{R}_3$ denote the appropriate lattice vectors for the trigonal bonding on a graphene sheet and t is the two-center integral given by

$$t = \langle \varphi_A(r - R_A)|\mathcal{H}|\varphi_B(r - R_B)\rangle, \tag{4}$$

in which φ_A and φ_B denote the wave functions for carbon atoms on A and B sites in Figure 1(a). Then using the x, y coordinates of Figure 1, $f(k)$ is expressed by the complex function

$$f(k) = e^{ik_x a/\sqrt{3}} + 2e^{-ik_x a/2\sqrt{3}} \cos\left(\frac{k_y a}{2}\right). \tag{5}$$

Using Eq. (5), the overlap integral matrix is given by $\mathcal{S}_{AA} = \mathcal{S}_{BB} = 1$, and $\mathcal{S}_{AB} = sf(k) = \mathcal{S}_{BA}^*$ where the asterisk denotes the complex conjugate and s denotes the overlap integral of the wavefunctions on sites A and B

$$s = \langle \varphi_A(r - R_A)|\varphi_B(r - R_B)\rangle. \tag{6}$$

The explicit form for the tight binding matrix Hamiltonian $\mathcal{H}$ and the overlap integral matrix $\mathcal{S}$ for the π-bands is then written as:

$$\mathcal{H} = \begin{pmatrix} \epsilon_{2p} & tf(k) \\ tf(k)^* & \epsilon_{2p} \end{pmatrix}, \quad \mathcal{S} = \begin{pmatrix} 1 & sf(k) \\ sf(k)^* & 1 \end{pmatrix}. \tag{7}$$

The tight-binding secular equation

$$\det(\mathcal{H} - E\mathcal{S}) = 0 \tag{8}$$

is then solved using Eq. (7) to obtain the energy eigenvalues $E(\vec{k})$ for 2D graphite as a function $w(\vec{k})$, k_x and k_y:

$$E_{g2D}(\vec{k}) = \frac{\epsilon_{2p} \pm t w(\vec{k})}{1 \pm s w(\vec{k})}, \tag{9}$$

where the + signs in the numerator and denominator go together, thereby giving the dispersion relation for the bonding π-energy band. Likewise the dispersion relation for the anti-bonding π^* band are found from Eq. (9) using the − signs in the numerator and denominator, where the function $w(\vec{k})$ is given by:

$$w(\vec{k}) = \sqrt{|f(\vec{k})|^2} = \sqrt{1 + 4\cos\frac{\sqrt{3}k_x a}{2}\cos\frac{k_y a}{2} + 4\cos^2\frac{k_y a}{2}}. \tag{10}$$

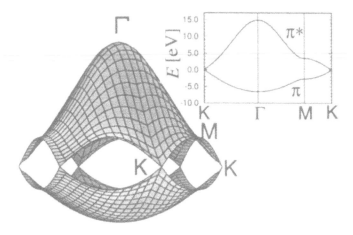

FIGURE 2. On the left the energy dispersion relations for 2D graphite are shown for k points throughout the Brillouin zone. On the right the energy dispersion relations are shown along the high symmetry directions of the triangle $\Gamma M K$ shown in Figure 1(b). (Saito, 1998)

This is one of the simplest applications of the tight binding approximation for more than one atom per unit cell, and is often used as a text book example (Bassani, 1975).

The shaded surfaces of Figure 2 show the energy dispersion relations for the π-bands in two-dimensional graphite throughout the Brillouin zone. The inset to Figure 2 shows the energy dispersion relations along the high symmetry axes along the perimeter of the triangle shown in Figure 1(b). For these dispersion relations, the values of $\epsilon_{2p} = 0$, $t = -3.033$ eV, and $s = 0.129$ are used for the 2D band parameters in order to reproduce the first principles calculations of the graphite energy bands (Painter, 1970; Dresselhaus, 1988; Saito, 1998). The upper half of the energy dispersion curves describes the π^*-energy anti-bonding band, and the lower half describes the π-energy bonding band. The upper π^* band and the lower π band are degenerate at the K points through which the Fermi energy passes. Since there are two π electrons per unit cell in 2D graphite, these two π electrons fully occupy the lower π band. Since a detailed calculation of the density of electronic states shows that the density of states at the Fermi level is zero, two-dimensional graphite is a zero-gap semiconductor. The existence of a zero gap at the K points comes from the symmetry requirement that the two carbon sites A and B in the hexagonal lattice are equivalent to each other. If the A and B sites were occupied by different atoms such as B and N, the site energy ϵ_{2p} would be different for B and N, and therefore the calculated energy dispersion would show an energy gap between the π and π^* bands.

When the overlap integral s becomes zero, the π and π^* bands become symmetrical around $E = \epsilon_{2p}$ which can be understood from Eq. (9). The energy dispersion relations in the case of $s = 0$ (i.e., in the Slater–Koster scheme) are commonly used as a simple approximation for the electronic structure of the π bands for a graphene layer:

$$E_{g2D}(k_x, k_y) = \pm t \left\{ 1 + 4\cos\left(\frac{\sqrt{3}k_x a}{2}\right)\cos\left(\frac{k_y a}{2}\right) + 4\cos^2\left(\frac{k_y a}{2}\right) \right\}^{1/2}. \quad (11)$$

In this case, the energies at the high symmetry points, Γ, M and K in the Brillouin zone have the values of $\pm 3t$, $\pm t$ and 0, respectively, and the band width shown in Figure 2 is $(6|t|)$ when $s = 0$, and $6|t - \epsilon_{2p}s|/(1 - 9s^2)$ when $s \neq 0$. In considering carbon nanotubes as a rolled up piece of a graphene sheet, a zone folding of the tight binding dispersion relations for the π-bands given by Eq. (11) and the periodic boundary conditions imposed by the formation of the cylinder form the basis for determining the dispersion relations for 1D carbon nanotubes (Saito, 1992).

By expanding Eq. (11) about the K point $(2\pi/3a)(\sqrt{3}, 1)$ where the Fermi level is located, we obtain an energy dispersion relation that is linear in κ:

$$E(\kappa) = \pm\frac{\sqrt{3}}{2}a|t|\kappa \quad (12)$$

in which the $+$ and $-$ signs, respectively, denote the conduction and valence bands, and the electron energy $E(\kappa)$ and wave vector $\kappa = \sqrt{\kappa_x^2 + \kappa_y^2}$ are measured from the zone corner (the K point), a is the lattice constant and t is the transfer integral given by Eq. (4). The electronic density of states, $g(E)$ for 2D graphite is then given by the linear E dependence

$$g(E) = \frac{8|E|}{3\pi a^2 t^2} \quad (13)$$

where the spin degeneracy has been included. Equations (12) and (13) hold only for small κ, or for energies which are close (e.g., < 1 eV) to the K-point energy ($E \sim 0$). From the linear dispersion relation of Eq. 12, the corresponding energy-independent electron velocity is obtained:

$$v(\kappa) = \frac{1}{\hbar}\frac{\partial E}{\partial \kappa} = \frac{\sqrt{3}}{2\hbar}a|t| \sim 10^6 \text{ m/s.} \quad (14)$$

In applying the electronic structure of 2D graphite to explaining the electronic properties of turbostratic graphite, the full dispersion relations given by Eq. (11) are used when states over a broad range of E and k are important (e.g., optical properties), while Eqs. (12–14) are used when only states very close to $E = 0$ are needed (e.g., transport properties). When a broad range of E is included, it is also necessary to consider the σ-bands near the center of the Brillouin zone (the Γ point in Figure 2).

In considering the σ bands of 2D graphite, we use the three atomic orbitals $2s$, $2p_x$ and $2p_y$, with sp^2 covalent bonding, yielding six Bloch orbitals in the 2 atom unit cell of Figure 1(a), and giving rise to six σ bands. The dispersion relations at each $\vec{k}$ point for these six σ bands are derived from a 6×6 Hamiltonian $\mathcal{H}$ and a 6×6 overlap matrix S. For the eigenvalues thus obtained, three of the six σ bands are bonding σ bands which appear ~ 10 eV below the Fermi energy at the K point, and the other three σ bands are antibonding σ^* bands, appearing ~ 10 eV above the Fermi energy at the K point.

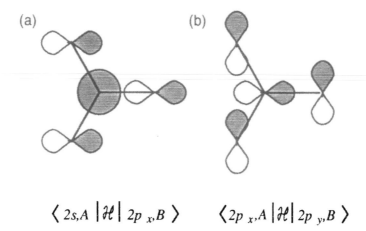

$$\langle \, 2s,A \, |\mathcal{H}| \, 2p_{x},B \, \rangle \qquad \langle 2p_{x},A \, |\mathcal{H}| \, 2p_{y},B \, \rangle$$

FIGURE 3. Examples of the Hamiltonian matrix elements of σ orbitals, (a) $\langle 2s^{A}|\mathcal{H}|2p_{x}^{B}\rangle$ and (b) $\langle 2p_{x}^{A}|\mathcal{H}|2p_{y}^{B}\rangle$. By rotating the $2p$ orbitals, we get the matrix elements in Eq. (17) and Eq. (18), respectively. (Saito, 1998)

The matrix elements of the (6×6) Hamiltonian for the σ bands are expressed in accordance with the basis set for the free atom: $2s^{A}, 2p_{x}^{A}, 2p_{y}^{A}, 2s^{B}, 2p_{x}^{B}, 2p_{y}^{B}$ and break up into 3×3 blocks for the A atom, the B atom and for the coupling between them. Within the nearest neighbor site approximation, the (3×3) diagonal blocks of the Hamiltonian and overlap matrices are given by

$$\mathcal{H}_{AA} = \begin{pmatrix} \epsilon_{2s} & 0 & 0 \\ 0 & \epsilon_{2p} & 0 \\ 0 & 0 & \epsilon_{2p} \end{pmatrix}, \quad \mathcal{S}_{AA} = \begin{pmatrix} 1 & 0 & 0 \\ 0 & 1 & 0 \\ 0 & 0 & 1 \end{pmatrix}, \tag{15}$$

where ϵ_{2p} and ϵ_{2s} are the $2s$ and $2p$ unperturbed levels, and likewise for $\mathcal{H}_{BB}$. The matrix elements for the Bloch orbitals coupling the A and B atoms are be obtained by taking the components of $2p_{x}$ and $2p_{y}$ in the directions parallel and perpendicular to the σ bond. The decomposition of the $2p_{x}$ orbital for the B atom at $(-a/2\sqrt{3}, a/2)$

$$|2p_{x}\rangle = \cos\frac{\pi}{3}|2p_{\sigma}\rangle + \sin\frac{\pi}{3}|2p_{\pi}\rangle, \tag{16}$$

while the decomposition of the $2p_{x}$ orbital for the B atom at $(a/\sqrt{3}, 0)$ is $|2p_{x}\rangle = -|2p_{\sigma}\rangle$, and the corresponding results are obtained for the $2p_{y}$ orbital. In Figures 3(a) and (b) we show examples of the interacting orbitals that are needed to calculate the matrix elements for the σ-bands $\langle 2s^{A}|\mathcal{H}|2p_{x}^{B}\rangle$ and $\langle 2p_{x}^{A}|\mathcal{H}|2p_{y}^{B}\rangle$, respectively (Saito, 1998).

Multiplying the phase factor with the matrix elements, yields the results:

$$\langle 2s^{A}|\mathcal{H}|2p_{x}^{B}\rangle = \mathcal{H}_{sp}\left(-e^{ik_{x}a/\sqrt{3}} + e^{-ik_{x}a/2\sqrt{3}}\cos\frac{k_{y}a}{2}\right) \tag{17}$$

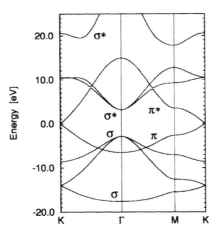

FIGURE 4. The energy dispersion relations for σ and π bands of two-dimensional graphite, using the parameters listed in Table 1. (Saito, 1998)

and

$$\langle 2p_x^A|\mathcal{H}|2p_y^B\rangle = \frac{\sqrt{3}}{4}(\mathcal{H}_\sigma + \mathcal{H}_\pi)e^{-ik_xa/2\sqrt{3}}e^{ik_ya/2} - \frac{\sqrt{3}}{4}(\mathcal{H}_\sigma + \mathcal{H}_\pi)e^{-ik_xa/2\sqrt{3}}e^{-ik_ya/2}$$

$$= \frac{\sqrt{3}i}{2}(\mathcal{H}_\sigma + \mathcal{H}_\pi)e^{-ik_xa/2\sqrt{3}}\sin\frac{k_ya}{2}. \qquad (18)$$

When all the matrix elements of the 6×6 Hamiltonian and overlap matrices are calculated as outlined above, the 6×6 Hamiltonian matrix for the σ-bands is obtained as a function of k_x and k_y. For given $\vec{k}$ points the energy dispersion relations for the σ bands are calculated from the secular equation Eq. (8).

The results thus obtained for the calculated σ and π energy bands for 2D graphite are shown in Figure 4, using the values for the band parameters of 2D graphite listed in Table 1 (Painter, 1970). These results can be used as a first approximation for describing the electronic band structure of most sp^2 bonded graphites for which the nearest neighbor carbon-carbon distance is close to that of graphite, 1.42 Å. Since the three antibonding σ^* bands lie far above the π^* bands near the K-point, and since the three bonding σ bands lie far below, the dispersion relations for the π bands are in most cases sufficient for describing the transport properties of graphites.

As is shown in Figure 4, the π band and two of the σ bands cross each other (i.e., have different symmetries), as does the π^* band and the two σ^* bands. However, because of the different group theoretical symmetries between σ and π bands, no band separation occurs at the crossing points. The relative position of these crossings is known to be important when considering: (1) dipole allowed electronic photo-transitions taking an electron from

TABLE 1. Values for the coupling parameters for carbon atoms in the Hamiltonian for π and σ bands in 2D graphite.

$\mathcal{H}$	value (eV)	$\mathcal{S}$	value
$\mathcal{H}_{ss}$	-6.769	$\mathcal{S}_{ss}$	0.212
$\mathcal{H}_{sp}$	-5.580	$\mathcal{S}_{sp}$	0.102
$\mathcal{H}_{\sigma}$	-5.037	$\mathcal{S}_{\sigma}$	0.146
$\mathcal{H}_{\pi} \equiv t$	-3.033	$\mathcal{S}_{\pi} \equiv s$	0.129
$\epsilon_{2s}{}^{(a)}$	-8.868		

[a] The value for ϵ_{2s} is given relative to setting $\epsilon_{2p} = 0$.

the σ band to the π^* band, and likewise from the π band to the σ^* band, and (2) charge transfer from alkali metal ions to graphene sheets in graphite intercalation compounds.

2.3 THREE DIMENSIONAL GRAPHITE

The ideal graphite crystal is a three-dimensional solid with ABAB... stacking of the graphite layers, giving rise to four carbon atoms per unit cell and therefore four π-bands near the Fermi surface, six bonding σ-bands well below E_F and six antibonding σ-bands well above E_F. Although the interlayer interaction is weak, this interaction has a profound effect on the four π-bands near the 3D Brillouin zone edges, causing a band overlap that is responsible for the semimetallic properties of graphite, in contrast to two-dimensional graphite which is a zero gap semiconductor, as discussed in Section 2.2. Since the bonding and antibonding σ-bands lie far from the Fermi level, the transport properties depend almost entirely on the electronic dispersion relations for the π-bands. Starting from the basic approach presented in Section 2.2 for 2D graphite, detailed phenomenological models for the dispersion relations for the four π-bands have been developed by Slonczewski and Weiss (Slonczewski, 1958), and by McClure (McClure, 1957, 1960) and are known as the Slonczewski–Weiss–McClure (SWMcC) model. This SWMcC model has become the standard model used to describe the dispersion relations for 3D graphite near the Fermi level E_F. This model has been applied extensively to explain the transport, diamagnetic, quantum oscillatory, optical and magneto-optical properties dependent on the electronic structure of bulk crystalline graphite near the Fermi level. Comprehensive summaries of the electronic properties of graphite in this context have been given by many authors (Spain, 1973, 1981; Dresselhaus, 1988).

The SWMcC model gives a phenomenological treatment of the electronic structure for the four π-bands based on crystal symmetry, and focuses on k values in the vicinity of the 3D Brillouin zone edges (see Figure 5), close to the location of the Fermi surface for electrons and holes which are shown schematically in Figure 5. In the k_z direction, a Fourier expansion based on a tight-binding approach is made, and rapid convergence of the expansion is obtained because of the weak interplanar binding. In the layer planes, a $\vec{k} \cdot \vec{p}$ expansion is made with respect to the Brillouin zone edge, since the extent of the graphite Fermi surface is small compared with Brillouin zone dimensions in the basal planes.

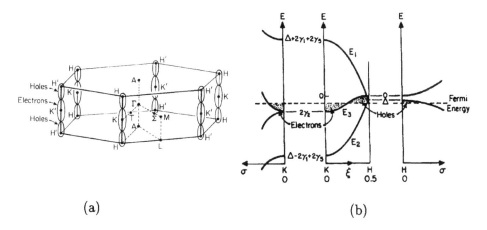

FIGURE 5. (a) 3D graphite Brillouin zone showing several high symmetry points and a schematic version of the 3D graphite electron and hole Fermi surfaces located along the HK axes. (b) Electronic energy bands near the HK axis in three-dimensional graphite as obtained from the SWMcC band model (Dresselhaus, 1981).

The SWMcC model is commonly written in terms of the (4×4) Hamiltonian for the π-bands

$$
H = \begin{pmatrix} E_1 & 0 & H_{13} & H_{13}^* \\ 0 & E_2 & H_{23} & -H_{23}^* \\ H_{13}^* & H_{23}^* & E_3 & H_{33} \\ H_{13} & -H_{23} & H_{33}^* & E_3 \end{pmatrix} \tag{19}
$$

where the band edge energies are given by

$$
\begin{aligned}
E_1 &= \Delta + \gamma_1 \Gamma + \frac{1}{2}\gamma_5 \Gamma^2 \\
E_2 &= \Delta - \gamma_1 \Gamma + \frac{1}{2}\gamma_5 \Gamma^2 \\
E_3 &= \frac{1}{2}\gamma_2 \Gamma^2
\end{aligned} \tag{20}
$$

and the interaction terms are

$$
\begin{aligned}
H_{13} &= (-\gamma_0 + \gamma_4\Gamma)\sigma \exp(i\alpha)/\sqrt{2} \\
H_{23} &= (\gamma_0 + \gamma_4\Gamma)\sigma \exp(i\alpha)/\sqrt{2} \\
H_{33} &= \gamma_3\Gamma\sigma \exp(i\alpha)/\sqrt{2}
\end{aligned} \tag{21}
$$

in which κ is the in-plane wave vector measured from the HKH Brillouin zone edge [see Figure 5(a)], α is the angle between $\vec{\kappa}$ and the $K-\Gamma$ direction, and

$$
\Gamma = 2\cos\pi\xi. \tag{22}
$$

TABLE 2. Magnitude of the Slonczewski–Weiss–McClure band parameters for 3D graphite (Dresselhaus, 1988).

Band parameter	Orbital interaction	Magnitude [eV]
γ_0	$A - B \ (a/\sqrt{3}, 0, 0)$	3.16 ± 0.05
γ_1	$A - A \ (0, 0, c)$	0.39 ± 0.01
γ_2	$B - B \ (0, 0, 2c)$	-0.020 ± 0.002
γ_3	$B - B \ (a/\sqrt{3}, 0, c)$	0.315 ± 0.015
γ_4	$A - B \ (a/\sqrt{3}, 0, c)$	$\sim 0.044 \pm 0.024$
γ_5	$A - A \ (0, 0, 2c)$	0.038 ± 0.005
Δ	$\epsilon_\pi^A - \epsilon_\pi^B$	-0.008 ± 0.002
E_F	–	-0.024 ± 0.002

We note that the labels of the high symmetry points in the $k_z = 0$ plane of the 3D Brillouin zone [see Figure 5(a)] are the same as those for 2D graphite [see Figure 1(b)]. The dimensionless wave vectors ξ along the k_z direction and σ in the basal plane are, respectively, given by

$$\xi = k_z \tilde{c}_0 / \pi \tag{23}$$

and

$$\sigma = \frac{\sqrt{3}}{2} a \kappa, \tag{24}$$

where $\tilde{c}_0 = 3.35$ Å is the interlayer distance between graphene sheets and $a = 2.46$ Å is the in-plane lattice constant, as in the case of 2D graphite. Each of the seven parameters $(\gamma_0, \ldots, \gamma_5, \Delta)$ of the SWMcC model can be identified with overlap and transfer integrals within the framework of the tight binding approximation, but in practice they are evaluated experimentally (see Table 2).

The eigenvalues of the SWMcC Hamiltonian given by Eq. (19) yield the energy dispersion relations which are schematically illustrated in Figure 5(b). Along the Brillouin zone edge HKH, two of the four solutions are doubly degenerate and are labeled by E_3. The remaining two solutions are non-degenerate and are denoted in this figure by E_1 and E_2. The degeneracy of the two E_3 levels is lifted as we move away from the zone edge, and this is indicated on the left-hand side of the figure with reference to the plane defined by $\xi = 0$. At the H point ($\xi = 1/2$), the levels E_1 and E_2 are degenerate, and the double degeneracy of these levels is maintained throughout the planes $\xi = \pm 1/2$, as shown on the right-hand side of Figure 5(b). The SWMcC model for 3D graphite is analogous to the simple model given by Eq. (12) for 2D graphite close to the K-point in the Brillouin zone.

The 3D SWMcC Hamiltonian has simple solutions in certain special cases. If trigonal warping of the Fermi surface is neglected (i.e., $\gamma_3 = 0$), the four solutions to Eq. (19) become

$$E(\xi, \sigma) = \frac{1}{2}(E_1 + E_3) \pm [\frac{1}{4}(E_1 - E_3)^2 + (\gamma_0 - \gamma_4 \Gamma)^2 \sigma^2]^{1/2}, \tag{25}$$

$$E(\xi, \sigma) = \frac{1}{2}(E_2 + E_3) \pm [\frac{1}{4}(E_2 - E_3)^2 + (\gamma_0 + \gamma_4 \Gamma)^2 \sigma^2]^{1/2}. \tag{26}$$

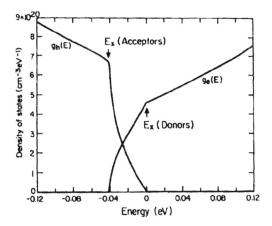

FIGURE 6. Energy dependence of the density of states for electrons and holes for three-dimensional graphite (Dresselhaus, 1981).

The solutions given by Eqs. (25) and (26) as well as the more complete solutions to the SWMcC model have been applied to the interpretation of a large variety of experiments relevant to the electronic structure and Fermi surface of graphite (McClure, 1971; Spain, 1973, 1981; Dresselhaus, 1988). The solutions to Eq. (19) show that three-dimensional graphite is semimetallic with a band overlap of $2\gamma_2$ (~ 0.040 eV) and a bandwidth along the Brillouin zone edge of $4\gamma_1$ (~ 1.56 eV). In the two-dimensional limit, the only non-vanishing band parameter in Table 2 is γ_0, so that in this limit Eqs. (25) and (26) yield the linear κ relation (Eq. 12) that is characteristic of 2D graphite π-bands for small κ, where we identify the transfer integral $-t$ with the band parameter γ_0 of the SWMcC model.

In three-dimensional graphite the density of states assumes a finite value at $E = 0$ as shown in Figure 6, where the densities of states for electrons $g_e(E)$ and for holes $g_h(E)$ are plotted as a function of energy. This result is in contrast to the behavior of the density of states for 2D graphite, where from Eq. (13), we see that $g(E) = 0$ at $E = 0$ and that $g(E)$ is linear in E away from $E = 0$. For the 3D graphite density of states, $g_e(E)$ and $g_h(E)$ in Figure 6 become approximately linear in E for energies below the K point minimum or above the H point maximum in the E_3-band. The singularities in $g_e(E)$ and $g_h(E)$ at these K and H point extrema are denoted by E_x in Figure 6. For energies between E_x(Acceptors) and E_x(Donors) both holes and electrons are present. However, if the Fermi level falls below E_x(Acceptors) in Figure 6, only holes are present, or above E_x(Donors), only electrons are present.

While the SWMcC model has been used successfully to describe many physical properties of 3D graphite (such as the transport properties (Issi, 2000), and the magnetic properties (Flandrois, 2000)), the SWMcC model for the π-bands cannot be used to describe the optical properties of graphite because of the limited energy and wave vector range of the model. The SWMcC model was therefore generalized (Johnson, 1973) to yield dispersion relations for the π-bands throughout the Brillouin zone. Using symmetry requirements to

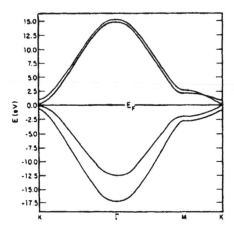

FIGURE 7. Graphite π-bands along several high symmetry directions of the 3D Brillouin zone using a generalization of the SWMcC model based on a 3D Fourier expansion of the electronic energy dispersion relations for graphite (Johnson, 1973).

specify the form of the Hamiltonian, a three-dimensional Fourier expansion was used to obtain the basis functions of the extended SWMcC model throughout the Brillouin zone. The band parameters of the extended SWMcC model were evaluated using (1) Fermi surface data in the vicinity of the Brillouin zone edges HKH and $H'K'H'$ [see Figure 5(a)], (2) fits to experimental optical data below 6 eV, and (3) the requirement that the dispersion relations reduce to those of the SWMcC model in the vicinity of the Brillouin zone edges. The resulting dispersion relations for the π-bands along several high symmetry directions are shown in Figure 2. These results are analogous to the 2D results given by Eq. (11) and shown in Figure 2.

More complete 3D models for the electronic structure including contributions from both the π and σ bands are also available to describe the optical properties, plasmon studies and other properties sensitive to states several eV away from E_F where contributions from the σ-bands become important. One example of a first principles calculation of the 3D graphite electronic band structure is shown in Figure 8 (Tatar, 1982). Near the Fermi level these first principles calculations (Haeringen, 1969; Nagayoshi, 1973; Zunger, 1978; Tatar, 1982; Fretigny, 1989; Dresselhaus, 1988) generally yield good agreement with the widely used phenomenological SWMcC model. These first principles calculations have also been extended to consider the effect of pressure on the graphite electronic structure (Nagayoshi, 1977).

If the interlayer site correlation is weak due to disorder, then the simpler 2D ab initio calculations (Painter, 1970; Nagayoshi, 1973; Zunger, 1978) should be sufficient for the interpretation of optical data. It should be mentioned that the electronic structure for diamond (which is an allotropic crystalline form of carbon with a free energy that is only slightly higher than that of graphite) corresponds to a wide band gap semiconductor ($E_g = 5.4$ eV), with very different electronic structure from that of graphite. Consequently

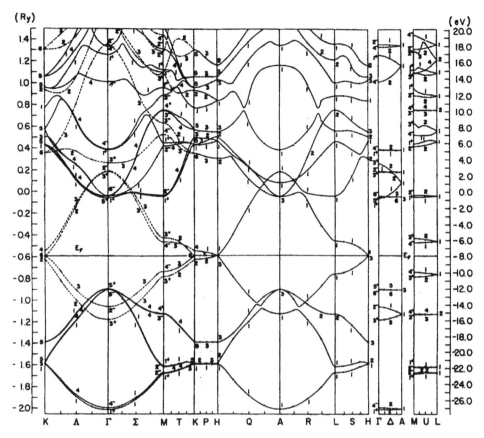

FIGURE 8. Ab initio electronic energy band structure for 3D graphite including both the π-bands (dashed) and the σ-bands (solid curve) (Tatar 1982). The ab initio results for the π-bands are in good agreement with those from the phenomenological Fourier expansion method shown in Figure 7.

diamond has a very low intrinsic carrier concentration at room temperature, and therefore is a good electrical insulator. As a result, the optical properties of diamond are much more sensitive to impurities and defects than is the case of graphite. The thermal properties of both diamond and graphite are, however, dominated by phonons, so that both graphite (in-plane) and diamond have very high thermal conductivities (Issi, 1998). Both diamond and graphite (in-plane) have a very high Young's modulus, consistent with the strong in-plane carbon-carbon bonding.

2.4 GRAPHENE RIBBONS

Interest in graphitic nanostructures and the new synthesis capabilities of thin graphite films has stimulated study of the electronic structure of *graphene ribbons*. A graphene ribbon is

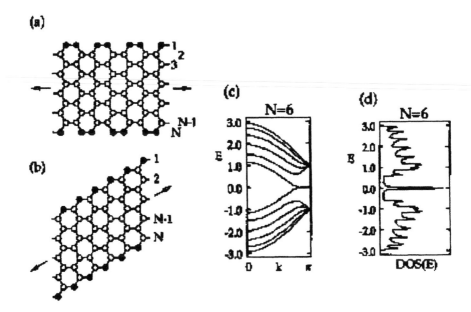

FIGURE 9. Graphene ribbons terminated by: (a) armchair edges and (b) zigzag edges, indicated by filled circles. The indices N denote the atomic rows for each ribbon. The model for the electronic structure considers hydrogen termination of the dangling bonds. (c) Calculated $E(k)$ for a zigzag ribbon ($N = 6$) showing an edge state at $E = 0$ extending out to the zone boundary. The energy E is measured with respect to the intrinsic Fermi level and is normalized to t, the transfer integral, while the dimensionless wave vector is normalized to the Brillouin zone dimension 2π. (d) The density of electronic states [DOS(E)] for a zigzag ribbon for $N = 6$ showing a sharp peak in the DOS at the Fermi level $E = 0$ (Nakata, 1996).

a strip taken from a graphene plane which has the honeycomb structure, as shown in Figure 1. The graphene ribbon is very long (or effectively infinite in length) along one direction, but has only a few unit cells in the transverse direction (see Figure 9). It is believed that many commercially used graphites contain structures somewhat connected to graphene ribbons (McClure, 1982; Hoarau, 1976).

Of the many possible edge structures, the two shown in Figure 9 have high symmetry (armchair and zigzag edges), while other edge structures can be related to these fundamental structures (Nakada, 1996; Dresselhaus, 1988). Many important graphites and commercial carbon materials, such as carbon fibers and glassy carbon, contain carbon networks that can be approximated by graphene ribbons (Nakada, 1996; Dresselhaus, 1988). These carbon networks play an important role in the transport and mechanical properties of these disordered materials.

It has been found recently that zigzag terminated ribbons [Figure 9(b)] possess a unique edge state at $E = 0$ [Figure 9(c)] with an unusually high density of states close to the Fermi level E_F [Figure 9(d)]. As the width of the graphene ribbon increases,

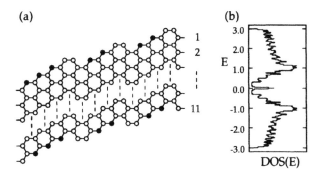

FIGURE 10. (a) The unit cell of a general graphene ribbon of $N = 11$. The zigzag sites are indicated by full closed circles. (b) The corresponding density of states (DOS) showing the contribution of the special zigzag "edge state" to the DOS at the Fermi level (Nakada, 1996).

the intensity of the contribution of this special edge state to the density of states decreases. This special state is not present for armchair-terminated ribbons [Figure 9(a)], so that there is no special contribution made to the density of states at $E = 0$ in the case of armchair-terminated ribbons. The findings concerning the special edge state have stimulated interest in the electronic structure of graphene ribbons (Nakada, 1996; Dresselhaus, 1988). If the ribbon width of a random ribbon is narrow, and if the ribbon has a sufficient fraction of zigzag terminations, this feature in the density of states near E_F persists, as shown in Figure 10 where the unit cell of a ribbon containing a relatively high fraction of zigzag edge terminations is shown on the left. The figure on the right shows that if the fraction of edge states is large enough, then a significant contribution is made to the density of states at the Fermi level ($E = 0$). This result suggests that a collection of narrow ribbons would have enough random zigzag segments to give rise to the feature in the density of states shown in Figure 10(b). This finding might also help with the tailoring of commercial graphites to possess certain desired properties, by controlling the graphene ribbon widths. Studies of edge states could also have relevance to the electronic properties of porous carbons which have a very high density of edge states. Because of the availability of dangling bonds, graphene ribbon edge sites are more reactive than bulk sites. Thus carbon clusters have been investigated in terms of their potential for the uptake of active species (Dresselhaus, 1997).

2.5 CONCLUSIONS

In subsequent chapters of this volume, the electronic structure of graphites is used for the interpretation of a variety of experiments related to their transport (Issi, 1998), magnetic (Flandrois, 1998), and other properties. For graphites showing the greatest structural ordering, the 3D electronic structure is most useful, while for graphites which have well-developed in-plane ordering but little or no interlayer site correlations, electronic structure models for a 2D graphene sheet are most useful. However, for more disordered graphites, where the ordered portion of a generally disordered graphite is in the form of graphene ribbons, a model for the density of states of a 2D graphene ribbon should be used, at least for gaining physical insights.

Acknowledgments

The authors wish to acknowledge valuable conversations with Professors M. Endo and J.P. Issi. The MIT authors thank NSF Grant DMR-95-10093 and the Lawrence Livermore National Laboratory DOE Sub-Contract #B287707 for support of this work. This work is funded in part by International Joint Research Program of the New Energy and Industrial Technology Organization (NEDO), Japan. Part of the work by RS is supported by a Grant-in Aid for Scientific Research (No. 08454079 and 9243211) from the Ministry of Education and Science of Japan.

References

Bassani, F. and Pastori Parravicini, G. (1967) *Nuovo Cim.*, **B50**, 95.

Bassani, F. and Pastori-Parravicini, G. (1975) *Electronic States and Optical Transitions in Solids.* Pergamon Press, Oxford.

Corbató, F.J. (1957) In Mrozowski, S., Studebaker, M.L. and Walker, P.L. Jr., editors, *Proceedings of the Third Conference on Carbon*, page 173. Pergamon Press, New York.

Coulson, C.A., Schaad, L.J. and Burnelle, L. (1957) In Mrozowski, S., Studebaker, M.L. and Walker, P.L. Jr., editors, *Proceedings of the Third Conference on Carbon*, page 27. Pergamon Press, New York.

Delhaès, P. and Carmona, F. (1981) Carbon fibers from rayon precursors. In Walker, P.L. Jr. and Thrower, P.A., editors, *Chemistry and Physics of Carbon*, volume 17, page 89. Marcel Dekker, New York, Vol. 17.

Doni, E. and Pastori Parravicini, G. (1969) *Nuovo Cim.*, **B64**, 117, 1969.

Dresselhaus, M.S. and Dresselhaus, G. (1981) Intercalation compounds of graphite. *Advances in Phys.*, **30**, 139–326.

Dresselhaus, M.S., Dresselhaus, G. and Fischer, J.E. (1977) Graphite intercalation compounds. Electronic properties in the dilute limit. *Phys. Rev. B*, **15**, 3180.

Dresselhaus, M.S., Dresselhaus, G. Sugihara, K., Spain, I.L. and Goldberg, H.A. (1988) *Graphite Fibers and Filaments*, volume 5 of *Springer Series in Materials Science*. Springer-Verlag, Berlin.

Dresselhaus, M.S. (1997) Future directions in carbon science. In Elton N. Kaufmann, editor, *Annual Reviews of Materials Science*, volume 27, pages 1–34, Palo Alto, CA. Annual Reviews Press. Vol. 27.

Flandrois, S. (2000) In *Graphite and Precursors*, Pierre Delhaès, editor, volume 1, page 71, Paris, France. Gordon and Breach. Series World of Carbon.

Fretigny, C. and Kamimura, H. (1989) Electronic structures of unoccupied bands in graphite. *J. Phys. Soc. Jpn.*, **58**, 2098–2108.

Hoarau, J. and Volpilhac, G. (1976) *Phys. Rev. B*, **14**, 4045.

Issi, J.P. (2000) In *Graphite and Precursors*, volume 1, Pierre Delhaès, editor, page 45, Paris, France. Gordon and Breach. Series World of Carbon.

Johnson, L.G. and Dresselhaus, G. (1973) Optical properties of graphite. *Phys. Rev. B*, **B7**, 2275–2285.

McClure, J.W. (1957) *Phys. Rev.*, **108**, 612.

McClure, J.W. (1960) Theory of 3D diamagnetism in graphite. *Phys. Rev.*, **119**, 606.

McClure, J.W. (1969) *Carbon*, **7**, 425.

McClure, J.W. (1971) In Carter, D.L. and Bate, R.T., editors, *Proceedings of the International Conf. on Semimetals and Narrow Gap Semiconductors*, page 127. Pergamon Press, (New York).

McClure, J.W. and Hickman, B.B. (1982) *Carbon*, **20**, 373.

Mrozowski, S. (1971) *Carbon*, **9**, 97.

Nagayoshi, H. (1977) *J. Phys. Soc. Jpn.*, **43**, 760.

Nagayoshi, H., Tsukada, M., Nakao, K. and Uemura, Y. (1973) *J. Phys. Soc. Jpn.*, **35**, 396.

Nakada, K., Fujita, M., Dresselhaus, G. and Dresselhaus, M.S. (1996) Edge state in graphene ribbons, Nanometer size effect and edge shape dependence. *Phys. Rev. B*, **54**, 17954–17961.

Oberlin, A. (1984) *Carbon*, **22**, 521.

Ohno, T., Nakao, K. and Kamimura, H. (1979) *J. Phys. Soc. Jpn.*, **47**, 1125.

Painter, G.S. and Ellis, D.E. (1970) *Phys. Rev. B*, **1**, 4747.

Saito, R., Dresselhaus, G. and Dresselhaus, M.S. (1998) *Physical Properties of Carbon Nanotubes*. Imperial College Press, London.

Saito, R., Fujita, M., Dresselhaus, G. and Dresselhaus, M.S. (1992) Electronic structures of carbon fibers based on c_{60}. *Phys. Rev. B*, **46**, 1804–1811.

Samuelson, L., Batra, I.P. and Roetti, C. (1980) *Solid State Commun.*, **33**, 817.

Slonczewski, J.C. and Weiss. P.R. (1958) *Phys. Rev.*, **109**, 272.

Spain, I.L. (1973) The electronic properties of graphite. In Walker, P.L. Jr. and Thrower, P.A., editors, *Chemistry and Physics of Carbon*, volume 8, page 1. Marcel Dekker, Inc., New York, Vol. 8.

Spain, I.L. (1981) Electronic transport properties of graphite, carbons, and related materials. In Walker, P.L. Jr. and Thrower, P.A., editors, *Chemistry and Physics of Carbon*, volume 16, page 119. Marcel Dekker, New York, Vol. 16.

Tatar, R.C. and Rabii, S. (1982) Electronic properties of graphite, a unified theoretical study. *Phys. Rev.*, **B25**, 4126.

Van Haeringen, W. and Junginger, H.G. (1969) *Solid State Commun.*, **7**, 1723.

Wallace. H.R. (1947) *Phys. Rev.*, **71**, 622.

Wyckoff, R.W.G. (1964) In *Crystal Structures, Interscience*, New York, volume 1.

Zunger, A. (1978) *Phys. Rev. B*, **17**, 626.

3. Electronic Conduction

J-P. ISSI

Unité de Physico-chimie et de Physique des matériaux, Université Catholique de Louvain, 1, place Croix du Sud, B-1348 Louvain-la-Neuve, Belgique

3.1 INTRODUCTION

3.1.1 Graphites and Transport

We could have chosen another more general title to this chapter: Electronic conduction mechanisms in solids: from semiclassical to quantum transport. This is to say that the story of transport in carbons and graphites follows closely that of transport in general, and that, because of the various aspects of conduction which are covered by these materials, whether electrical or thermal, one may cover the whole spectrum of solid state conduction. In some instances, one has the possibility to tailor these conductivities, mainly through intercalation and, to a lesser extent, by doping.

Indeed, carbon in its various forms is a fascinating material for solid state physicists. With these forms are associated a wide range of properties on which they can test their models. Transport is no exception to this rule. By transport properties we mean those concerned with the movement of a quasi particle system, which is initially isotropic and which is directed in a preferred orientation under the action of external forces. The charge carriers may be either electrons or holes or both, which may either carry the electrical current (electrical conductivity) or the heat energy (electronic thermal conductivity).

In metals and semimetals the charge carrier distributions are described by their Fermi surfaces, while in semiconductors the knowledge of the electronic band structure is essential for the discussion of electronic transport effects. Collisions tend to bring the particle system back into equilibrium and the relaxation time is generally the essential parameter describing the process. Around and below room temperature, the charge carriers are mainly scattered by lattice vibrations (phonons), impurities and crystal boundaries. At low and ultralow temperatures, i.e. in the milliKelvin region, specific effects show up in the scattering of charge carriers, revealing quantum effects at the macroscopic scale. These are the *weak localization* effects, *Coulomb interaction* effects and the *universal conductance fluctuations* in a quasi-balistic regime.

In order to discuss electron transport properties we thus need to know about two essential ingredients:

- the electronic distribution of the quasi-particle system considered. In graphites this means that we must have a model for the Fermi surface and for the phonon spectrum. The electronic structure is discussed in Chapter 2 of this book (Dresselhaus *et al.*).
- some characteristic lengths, one of them being the mean free path, but for the case of charge carriers, some other characteristic lengths should also be considered, especially at low temperatures (cfr. Section 3.4.5).

Macroscopically, the electrical conductivity is expressed by Ohm's law.

In a few pages we will be reviewing a subject which could be covered in an entire book. We will thus concentrate in the following on what are the most prominent features of conduction in carbons and graphites, whether pristine or in deliberately modified forms, referring for more details to reviews of earlier works by pioneers in the field (Mrozowski, 1971; Pacault, 1974; Kelly, 1981; Dresselhaus and Dresselhaus, 1981; Spain, 1981) and more recent monographies (Charlier and Issi, 1996; Delhaes, 1997; Dresselhaus *et al.*, 1988; Ebbesen, 1996; Issi and Nysten, 1998; Zabel and Solin, 1992) and the references therein. We will discuss the physics of electronic conduction related to various forms of carbon leaning heavily on four essential aspects:

- the *semimetallic behavior* due to the presence of a small density of charged particles, electrons and holes, which has a drastic effect on the band structure and on the scattering mechanisms (cfr. Section 3.3)
- the effects of *reduced dimensionality* or very high anisotropy
- the dramatic effect of disorder, even when relatively weak, in relation to *quantum aspects of conduction*
- the possibility of *intercalation* and its effects.

We believe that the effect of a small Fermi surface on the scattering mechanisms, though already identified for semimetals in the early fifties, has been overlooked in previous reviews, or at least not been given proper emphasis for the case of graphites. With regard to quantum aspects they were mostly observed in the mid-eighties on carbons and graphites, and very recently on nanotubes.

The ability of carbon to form either single or multiple homopolar bonding, leading to various coordination numbers, results in a rich polymorphism for this element (Delhaès, 1997), from crystalline sp^3-bonded diamond to planar sp^2-bonded graphite, hybridized ball- and tube-shaped fullerene clusters. These different kinds of chemical bondings allow the existence of a large set of pure solid state crystalline phases of carbon. We will show below

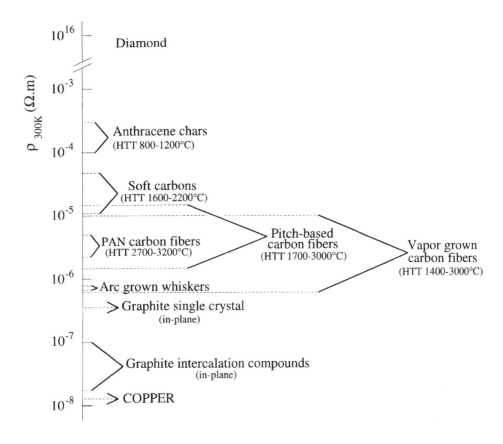

FIGURE 1. Room temperature electrical resistivities of various forms of carbons.

how this wide variety of carbon forms provides a large spectrum of *types* and *ranges* of electrical conductivities whether in the pristine or in the intercalated material (Charlier and Issi, 1996) (Figure 1). In the latter case one may obtain the highest room temperature electrical conductivities ever observed in synthetic conductors. Conduction in these systems is also dependent on the degree of atomic disorder and on their *low dimensionality*.

The fact that various geometries are available: bulk, fibers, particles, nanotubes is also a remarkable feature of these materials.

3.1.2 Solid State Physics and Transport in Carbons and Graphites

When solid state physicists started in the fifties to investigate quantitatively the transport properties of typical semiconductors such as germanium and silicon, or that of almost perfect metals, they usually started by growing the most perfect possible single crystals.

They were aiming at determining the electronic properties of ideal systems. They were able to use the tools of quantum theory, together with the periodicity associated with perfect lattices. Then, to describe more realistic situations, they studied the various types of lattice defects to determine their effects on these properties, assuming that local perturbations were introduced to an otherwise periodic system. Once the electronic structure was established with increasing accuracy, they investigated the interaction of the electronic system with static lattice defects, such as point defects, dislocations, sample boundaries,... and temperature dependent perturbations, mainly phonons. The interpretation of the results were generally made using semiclassical models.

By contrast, in the early eighties, measurements performed on "dirty systems", such as disordered metallic thin films and heavily doped semiconductors, revealed new features in these transport properties in the form of pure quantum effects originating from the quantum interference of the electron waves. Among these are those generating weak localization effects and universal conductance fluctuations (UCF). We will show below that these effects, which require the presence of defects, modified to some extent the concept of electron scattering (cfr. Sections 3.2.4 and 3.4.5). Indeed, these dirty systems, which were initially held in contempt by physicists, proved to be ideal candidates to test the quantum models for conduction. The emergence of powerful tools to characterize these systems helped giving credibility to the results.

The story of transport properties of carbons and graphites started differently, since no nearly perfect single crystals were available, except natural ones. So the electronic systems were "dirty" at the start, and it required a lot of ingenuity from researchers in the field to try to fit reasonable models explaining their experimental findings (see for example: Klein, 1964; Delhaès and Marchand, 1965; Mrozowski, 1971; Pacault, 1974; Bright, 1979; Spain, 1981).

Because of the large number of lattice defects carbons and graphites usually contain, even for the best quality HOPG, they recently revealed to be exceptionally interesting systems on which transport properties, including their fascinating quantum aspects, could be investigated with success. In addition, the fact that they may be profoundly modified by *intercalation* has lead to interesting findings, which could not be made on any other single class of solids. Besides, since *disorder* can be modified by heat treatment and the Fermi energy by varying the charge transfer in intercalated materials, the effects could be studied under different controlled conditions.

Their variety of structures, their particular geometry and dimensionality, helped the experimentalist in his investigations. For example, it is the large length to cross section ratio in fibers that made possible the discovery of quantum transport effects on pristine and intercalated fibers, which was at a later stage confirmed in bulk carbons and graphites.

In the following, we shall only consider conduction in graphitic planes. This can be done in bulk graphites, carbon fibers and multiwall nanotubes. Thus, we shall not discuss here the particular case of fullerenes and other carbon particles. Bulk HOPG presents the advantage that the anisotropy can be investigated since the resistivity can be measured in-plane and in the c-axis direction, while in carbon fibers high resolution resistivity measurements can be performed (Issi and Nysten, 1998).

3.2 ELECTRICAL RESISTIVITY

3.2.1 Boltzmann Zero-Field Resistivity

The expression for the electrical conductivity for a given group of charge carriers is given by:

$$\sigma = qN\mu = \frac{q^2 N \tau}{m^*} \qquad (1)$$

where q is the electronic charge, N the charge carrier density, μ the mobility, τ the relaxation time and m^* the carrier effective mass.

The relaxation time, which is related to the mean free path, $l = v_F \tau$, is defined as the time elapsed between two collisions, and its inverse $1/\tau$ reflects the probability for a carrier to experience a scattering event. v_F is the carrier velocity at the Fermi level. The mean free path is the distance between two scattering centers. In a naive way, one may imagine the mobility as expressing the ease with which the carriers move in the crystal lattice.

Expression (1) was derived for a single type of charge carrier. If there is more than one type of carrier, i.e. electrons and positive holes, as in pristine graphite fibers, or many bands for electrons or holes as in intercalation compounds for stages higher than 1, the contribution of each type of carrier should be taken into account. In that case, the total electrical conductivity is given by the sum of the partial conductivities, σ_j, of each group of charge carrier, j:

$$\sigma = \sum \sigma_j \qquad (2)$$

3.2.2 Ideal and Residual Resistivities

Usually, the main contributions to the electrical resistivity of metals, ρ, consists of an intrinsic temperature-sensitive ideal term, ρ_i, which is mainly due to electron-phonon interactions and an extrinsic temperature independent residual term, ρ_r, due to static lattice defects:

$$\rho = \rho_r + \rho_i \qquad (3)$$

In graphites the *residual resistivity* is determined by the defect structure of the graphitic layers, which may vary widely according to the heat treatment temperature and, to a lesser extent, to the type of carbon and the quality of the precursor. In Figure 2 we present a typical illustration of the validity of Matthiessen's rule for graphite intercalation compounds.

Expression (2) shows that the contributions to the conductivity from different carrier groups add, while expression (3) shows that it is the resistivities due to various scattering mechanisms that add.

In the presence of weak disorder, one should consider an additional contribution to the residual resistivity (curves *d* and *e* in Figure 2) due to *weak localization* resulting from quantum interference effects and/or that due to *Coulomb interaction* effects. These quantum effects, though they do not generally affect significantly the magnitude of the resistivity, introduce new features in our understanding of low temperature transport effects. So, in addition to the semiclassical ideal and residual resistivities of carbons discussed above, we must take into account the contributions due to quantum localization and interaction

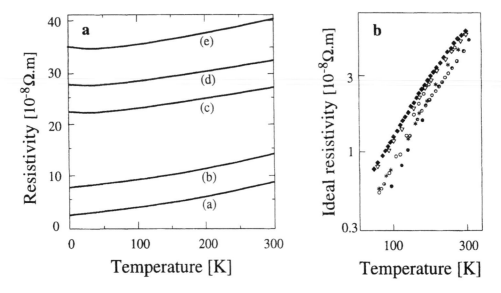

FIGURE 2. Illustration of the validity of Matthiessen's rule for GICs. Temperature dependence of the (a) total resistivity (as measured) and (b) ideal electrical resistivities of various low stage fibrous acceptor GICs. The resistivities are given as the equivalent of the 3D electrical resistivity. The data in Figure 2a are relative to a $CuCl_2$ intercalate for different host materials: (a) BDF, $n = 2$ (b) PX5, $n = 4$ (c) VSC25, $n = 1 - 2$ (mainly $n = 2$) and (d) P55 ($n = 3$) (e) P100-4, $n = 1 - 2$ (mainly $n = 2$). The ideal resistivities are almost the same in spite of the large differences in total resistivities, i.e., whatever the stage or host material. Note that the temperature dependent part of the resistivity is a tiny fraction of the total resistivity in most of the compounds. For the ideal contribution (b) an almost linear variation is observed at low temperature and an almost T^2 behavior around room temperature. For curve e, the arrow indicates a region where the resistivity increases with decreasing temperature due to weak localization effects (From Issi, 1992).

effects (Section 3.2.4). These localization effects were found to confirm the 2D character of conduction in turbostratic carbons (Bayot et al., 1989) and acceptor GICs (Piraux et al., 1985, 1987; Meschi et al., 1986).

In the same way, experiments performed at the mesoscopic scale revealed quantum oscillations of the electrical conductance as a function of magnetic field, the so-called universal conductance fluctuations (Section 3.4.5).

3.2.3 Typical Results

As is the case for most conductors, the temperature dependence of the electrical resistivity of carbon-based materials is very sensitive to lattice perfection. The higher the structural perfection, the lower the resistivity. Oddly enough, the differences observed between samples are more pronounced when the heat treatment temperature (HTT) varies within one class of carbon-based materials than they are between various classes for samples heat treated at approximately the same HTT. This is rather obvious when we compare the results presented in Figure 3 and Figure 4a. Whether we consider bulk or fibrous materials, the general trend of the electrical resistivity is as follows (Dresselhaus *et al.*, 1988).

Samples of *high structural perfection* exhibit resistivities below a few 10^{-6} Ωm. For these samples, one can describe the resistivity results using the semimetallic graphite band model. Around room temperature, electrons and holes are scattered by both static defects and phonons. Since the carrier mobilities and densities are very sensitive to defects and temperature and have opposite temperature dependences, the exact temperature variation cannot be predicted with confidence. Indeed, contrary to metals, where the highly degenerate electron system is temperature insensitive and only the mobility varies with temperature, in a semimetal like graphite, because of the very small carrier densities, these are very sensitive to defects and temperature. At the other end of the scale, *partially carbonized* samples exhibit resistivities higher than 10^{-4} Ωm which generally increase with decreasing temperature. An *intermediate behavior* between these two extremes is represented by curves which depend less on the HTT and does not show significant temperature variations.

In Figure 4a we present the room temperature resistivity of various experimental pitch-based carbon fibers versus in-plane coherence length (Issi and Nysten, 1998). We may see that there is a direct correlation between the two parameters, and, as expected, the electrical resistivity decreases with increasing structural perfection.

3.2.4 Quantum Interference and 2-D Localization and Interaction Effects

High resolution electrical resistivity measurements performed during the eighties at very low temperatures on thin films and wires of very small diameters have shown deviations from Matthiessen's rule. Instead of observing the plateau of the residual resistivity, a small increase of the resistivity of the order of a percent was observed with decreasing temperature.

This effect has been observed in a variety of quasi-two dimensional (2D) systems, including graphites and its intercalation compounds (curves d and e in Figure 2a), in the form of a logarithmic increase of resistivity with decreasing temperature. This behavior, which is more pronounced in the presence of significant defect scattering, is interpreted in terms of two mechanisms:

- a single-carrier weak localization effect produced by constructive quantum interference between elastically back-scattered partial-carrier-waves,
- charge carrier many-body Coulomb interaction. Disorder attenuates the screening between charge carriers, thus increasing their Coulomb interaction.

So, both effects are enhanced in the presence of weak disorder, or, in other words, by static defect scattering. These effects modify our concept of classical transport at low temperatures. It suggests that the general tendency for metallic systems is an increase in resistivity as the temperature is lowered, instead of levelling-off.

The particular geometry and the varying defect structures of carbon fibers have allowed the observation of *quantum transport effects* in these materials. These were first observed by Piraux *et al.* (1985) in intercalated acceptor carbon fibers, then on pristine fibers (Bayot *et al.*, 1989) and more recently on nanotubes (Langer *et al.*, 1996). The observed negative

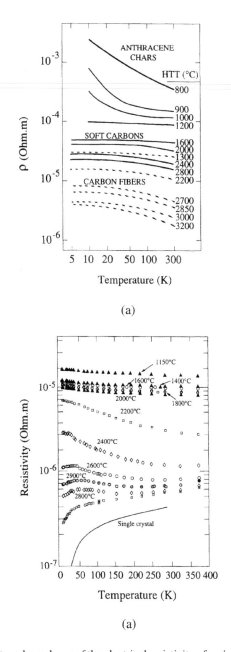

(a)

(a)

FIGURE 3. Temperature dependence of the electrical resistivity of various forms of pristine carbons. (a) Comparison of bulk heat treated carbons to that of ex-PAN fibers heat treated at various temperatures (From Dresselhaus *et al.*, 1988). (b) Benzene-derived carbon fibers (BDF) heat treated at various temperatures, as indicated. The data are compared to those for a graphite single crystal in-plane (From J. Heremans, *Carbon*, **23**, 431 (1985)).

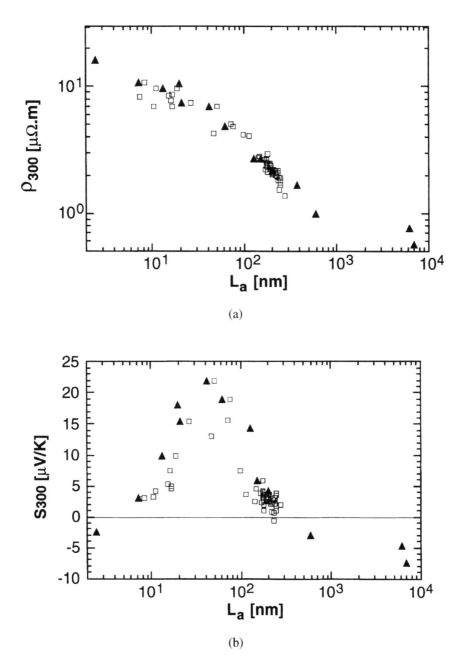

(a)

(b)

FIGURE 4. Room temperature resistivity (a) and thermoelectric power (b) of various experimental pitch-based carbon fibers versus in-plane coherence length, L_a (From B. Nysten, Ph.D. Thesis, Université Catholique de Louvain, Louvain-la-Neuve (1991)).

magnetoresistance (see below) in pristine carbon fibers, which is one of the signatures of the two-dimensionnal (2D) weak localization, was attributed to the turbostratic layers.

One can explain these findings in terms of *quantum interference* phenomena. When the time of flight, τ_r, between two elastic collisions is much smaller than the time, τ_{in}, between two inelastic collisions, the electron does not lose memory of the phase of its wave function for a time of the order of τ_{in}. As may be seen from Figure 5, an electron in the state $\mathbf{k}$ may be scattered to the state $-\mathbf{k}$ through successive scattering events: $\mathbf{k} \rightarrow \mathbf{k}'_1 \rightarrow \mathbf{k}'_2 \rightarrow \mathbf{k}'_{3...} \rightarrow -\mathbf{k}$, and $\mathbf{k} \rightarrow \mathbf{k}''_1 \rightarrow \mathbf{k}''_2 \rightarrow \mathbf{k}''_{3...} \rightarrow -\mathbf{k}$, thus experiencing changes in momentum $\mathbf{g}_i$. As shown in Figure 5, the momenta exchanged in these successive collisions are the same and thus their magnitudes are also the same. The quantum interference phenomenon between the two sequences described above leads to an increased backscattering probability $\mathbf{k} \rightarrow -\mathbf{k}$ with respect to the semiclassical value and thus to a larger resistivity. This effect which tends to localize electrons through increasing backscattering is called *weak localization*. It gives rise to characteristic resistivity variations with temperature and magnetic field.

The amplitude of the back scattering $\mathbf{k} \rightarrow -\mathbf{k}$ is the sum of a large number of contributions from sequences like those presented in Figure 5. During these successive scattering events the electron explores a region in space whose size depends on *temperature*. When the temperature decreases, the inelastic lifetime τ_{in} due to electron-phonon or electron-electron interactions increases. As a result, the number of sequences leading to constructive interference increases and backscattering is more efficient. This results in an increased resistivity as the temperature is lowered. When the temperature increases, the inverse situation occurs and when electron-phonon or electron-electron interactions become so numerous that the inelastic lifetime, τ_{in}, is of the order of τ_r, the electron loses its phase memory after each collision and we are back in the semiclassical situation.

It was demonstrated that when a *magnetic field* is applied to a sample it decreases the effect of the constructive interferences described in Figure 5. This effect is more pronounced for longer sequences. The magnetic field reduces the backscattering probability leading to a decrease in resistivity. This *negative magnetoresistance*, which tends to restore the semiclassical resistivity has nothing to do with the conventional *positive magnetoresistance* which is generally observed in electrical conductors and is discussed in Section 3.5. In the case of positive magnetoresistance, the Lorentz force due to the magnetic field deviates electrons from their initial path and increases the resistivity in the applied electric field direction (Section 3.5).

A weak external magnetic field suppresses the phase coherence of the backscattered waves, but does not influence the Coulomb interaction phenomenon. Thus magnetoresistance measurements allow one to distinguish between the two effects.

In order that the above considerations apply, we need elastic scattering, i.e; the presence of static defects. Though these defects do not destroy the translational periodicity of the lattice, provided that they are not too numerous, they introduce some disorder, *weak disorder*, as opposed to strong disorder which leads to *strong localization*, as is the case for amorphous solids. Thus in a subtle way, the concept of weak disorder, introduced in the early eighties, modifies significantly our way of dealing with the electron gas. Previously, the electron gas was treated independently of the scattering mechanism as discussed above (Sections 3.1.2 and 3.2.1). One was in that case considering how a given electron distribution was interacting

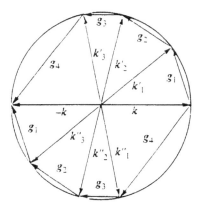

FIGURE 5. Schematic diagram showing how the electron is scattered via two parallel series of intermediary scattering states (From G. Bergmann, *Phys. Rev.*, **B28**, 2914 (1983)). An electron in the state $\mathbf{k}$ may be scattered to the state $-\mathbf{k}$ through the successive scattering events $\mathbf{k} \rightarrow \mathbf{k}'_1 \rightarrow \mathbf{k}'_2 \rightarrow \mathbf{k}'_3 \ldots \rightarrow -\mathbf{k}$ and $\mathbf{k} \rightarrow \mathbf{k}''_1 \rightarrow \mathbf{k}''_2 \rightarrow \mathbf{k}''_3 \ldots \rightarrow -\mathbf{k}$, thus experiencing changes in momentum $\mathbf{g}_i$. The momenta exchanged in these successive collisions are the same and thus their magnitudes are also the same.

with scattering centers. Now, these scattering centers, provided that they lead to elastic interactions, modify the electron gas in that sense that they lead to quantum interference effects within its distribution.

The limit of weak disorder, which is also the condition for the Boltzmann approximation to hold, implies that the electron mean free path, l, should be large compared to its wavelength, i.e.:

$$k_F l \gg 1 \qquad (4)$$

where k_F is the Fermi wave number. When these two terms are of the same magnitude, we are in a situation of strong disorder and strong localization.

As shown above, the weak localization generates an additional contribution to the low temperature electrical resistivity which adds to the classical Boltzmann resistivity. A magnetic field destroys this extra contribution and restores the classical temperature variation predicted by the simple two band (STB) model (Bayot *et al.*, 1989). This results in an apparent negative magnetoresistance. Weak localization is also destroyed when the carriers are scattered by magnetic impurities or by phonons since the phase coherence is not conserved in these processes. Weak localization effects thus take place at low temperatures where elastic scattering by lattice defects is the main mechanism limiting electrical conduction. They occur whatever the dimensionality of the system but the effect is more pronounced for quasi 2D electronic systems than for 3D systems.

As a typical example, we present in Figure 6 the electrical resistivity at 4.2 K for a fluorine graphite intercalation compound showing the effect of the transition from weak to strong disorder by varying the fluorine content. This is a unique situation where by varying the relative amount of the two constituents, one may at a given pressure and temperature, starting

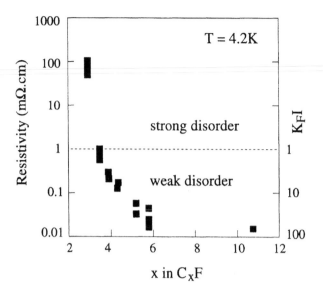

FIGURE 6.   Electrical resistivity at 4.2 K for various fluorine GICs showing the effect of the transition from weak to strong disorder by varying the fluorine content (From L. Piraux, V. Bayot, J-P. Issi, M.S. Dresselhaus, M. Endo and T. Nakajima, *Phys. Rev.*, **B41**, 4961 (1990)).

from a a semimetal, obtain a metal, then a bad conductor, and end up with an insulator for high fluorine concentration. While the electronic structure is drastically modified, the degree of disorder and localization is also continuously modified in a dramatic way.

Weak localization effects were also observed on individual multiwall carbon nanotubes (MWCN). In Figure 7 we present the temperature dependence of the conductance for MWCN measured, respectively, in zero magnetic field, 7 T and 14 T (Langer *et al.*, 1996). Electrical gold contacts have been deposited to the carbon nanotube via local electron beam lithography with a scanning tunneling microscope (STM). In order to eliminate contact resistances, three contacts were attached. The measured individual MWCN had a diameter of about 20 nm and a total length of the order of 1 μm. In zero magnetic field, a logarithmic decrease of the conductance at higher temperature, followed by a saturation of the conductance at very low temperature, was observed. The saturation occured at a critical temperature which shifts to higher temperatures in the presence of a magnetic field. The data in Figure 7 show that there is an additional contribution to the magnetoconductance of the carbon nanotube which is temperature independent up to the highest temperature investigated, including in the ln T variation range. This magnetoconductance was ascribed to the formation of Landau states predicted by Ajiki and Ando. 2D weak localization and "Landau level" contributions to the magnetoconductance can be separated as illustrated in Figure 7.

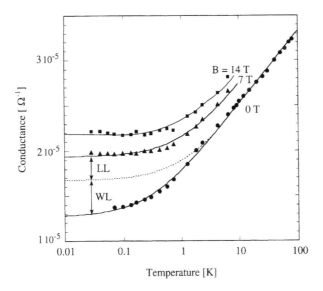

FIGURE 7. Electrical conductance as a function of temperature of a SWCN at the indicated magnetic fields. The dashed line separates the contributions to the magnetoconductance due to both the Landau levels (LL) and weak localization (WL). The solid lines are theoretical fits to the data points (From Langer *et al.*, 1996).

3.3 SEMIMETALLIC BEHAVIOR

Pristine HOPG is a semimetal with an equal density of electrons and holes, while some carbons are narrow-gap semiconductors (Dresselhaus *et al.*, 1998). This means that in any case these materials have very few charge carriers compared to metals. Since carbons and graphites have generally more than one type of charge carriers, this complicates the analysis of electronic transport properties data. Intercalation increases the carrier density leading to metallic behavior.

The semimetallic behavior of graphites is due to their small Fermi surfaces, i.e. the presence of very small densities of electrons and holes, which have a dramatic effect on the band structure and on the scattering mechanisms:

– contrary to metals, the charge carrier system is fully degenerate only at low temperatures. When the temperature increases, above say 50 K, the carrier distributions become partially degenerate and the carrier densities vary with temperature. In fact, all the parameters of the band structure and the Fermi surface are very sensitive to temperature. Thus the models, which are derived theoretically at 0 K (Dresselhaus *et al.*, 1998), and the corresponding parameters, which are measured in the liquid helium range, are no longer valid at higher temperature. In order to study transport properties at high temperature, one must rely on inspired guesses rather than on a sound model with quantitative data for the electronic parameters.

- associated with small semimetal Fermi surfaces is weak electron-phonon scattering leading to very high carrier mobilities (cfr. Section 3.3.2). We will introduce the concept of the *ineffectiveness of phonon scattering* in graphites, which associated with weak disorder allows the observation of scattering mechanisms, such as electron-electron scattering, which can hardly be observed in other types of solids. Also, the unusually high inelastic relaxation times allow the observation of quantum interference effects and the mesoscopic regime at much higher temperatures than in metals.
- weak electron-point defect scattering associated with large Fermi wavelengths. From Raleigh's law it is known that point defect scattering depends dramatically on the scattered wave wavelength. This is the case for both electron and phonons.

Thus, one of the salient features of semimetals with very small Fermi surfaces is the huge mobilities of their charge carriers at all temperatures, which are due to:

- ineffective electron-phonon interactions in order to satisfy momentum and energy conservation requirements, and, for the particular case of carbons and graphites, the very high Debye temperatures (cfr. Section 3.4.3),
- very low effective masses in some directions,
- low dimensionality of the electronic system.

3.4 SCATTERING AND RELAXATION TIME

3.4.1 *General*

Once the electronic distribution in the solid is known, one usually turns to classical concepts to describe in a phenomenological way the response of this distribution to external forces. This is the semiclassical approach which has the great advantage of leading to naive physical pictures. As indicated above, this is not always applicable.

The starting assumption is that all collision processes may be described in terms of a relaxation time, τ. This means that, if we apply an external force, e.g. that caused by an electric field, and we remove it rapidly, the distribution will tend to equilibrium in an exponential way with a time constant τ.

We suppose instantaneous collisions, i.e. that the time during which the electron interacts with a lattice imperfection, is negligible compared to that elapsed between two collisions. If, as it is generally the case, the energy exchanged during the collision is negligible compared to the initial electron energy, the electron experiences a change of momentum but its energy remains almost unchanged. In *classical terms* this means that after a collision the magnitude of the electron velocity remains almost unchanged but that its direction is altered.

In *quantum terms*, we must consider the wave-like motion of electrons. Since a wave may propagate in a perfect periodic array without being scattered, so will be the electronic waves in a perfect lattice. Thus, the positive ions in the lattice do not impede the wave propagation, except in the particular case of Bragg reflection, i.e. when the wavelengths of the electronic waves are close to that of the lattice periodicity, i.e. twice the interatomic spacing. If we exclude this particular situation, the electrons in an infinite perfect crystal at 0K will have infinite relaxation times. However, any deviation from such an idealized situation will cause a scattering of the electronic waves, so that any perturbation destroying the lattice periodicity limits τ.

3.4.2 Electron-Phonon Interactions

One may think in a naive way that one of these perturbations occurs when the temperature is increased since the atomic vibrations destroy the periodicity. This means that electrons are scattered by phonons, leading to a temperature-dependent relaxation time. For electron-phonon scattering the Bloch-Grüneisen relation predicts for metals a T^n law with $n = 5$ at very low temperature, followed by a gradual decrease of n until it is equal to 1 around and above the Debye temperature.

In a collision process between two quasi-particle systems, such as electrons and phonons, there is an exchange of momentum and energy and the laws of momentum and energy conservation must be obeyed. Thus, the only carriers which participate effectively in transport are those which can exchange momentum and energy. In metals, because of the Pauli principle, such carriers are only those which are near the Fermi surface and electrons are always scattered from one point to another on the Fermi surface.

The effectiveness of an electron-phonon scattering event for a degenerate electron gas depends on the *angle* through which the charge carriers are scattered on the Fermi surface and on the *number of phonons* which are available for a given scattering angle. This number of phonons depends on temperature. The scattering angle ϕ is determined by the magnitude of the interacting phonon wave vector (Figure 8a), $\mathbf{q}$. If its magnitude q is comparable to k_F, the resulting scattering is through a large angle, while when q decreases with temperature, small angle scattering results. So, the scattering angle is also temperature dependent. For ordinary metals k_F is generally comparable to q_D, the radius of the Debye sphere, i.e. the magnitude of the phonon wave vector on the Debye sphere.

Thus, we have made a distinction between two sorts of electron-phonon collisions:

- *large angle* collisions. These collisions, which are schematically represented in Figure 8a, are very efficient since the electron velocity component along the electric field is reversed after collision. They thus lead to an important contribution to the resistance. For such a situation to occur there must be sufficient phonons with large q-vectors, i.e. the temperature must be high. This situation will prevail in metals around and above the Debye temperature.
- *small angle collisions*. It may be seen from Figure 8a that these collisions are less efficient in producing resistivity than large angle collisions. Here, there is no reversal of the electron velocity component along the electric field, but instead a small decrease in the magnitude of this component along the field. We are here in the presence of a resistive process, but much less dramatic than for the preceeding case. Since at low temperatures only phonons with small wave numbers are available, small angle collisions will be dominant in metals.

Though the classification of the relaxation time into two distinct categories is somewhat artificial, it is justified physically. Indeed, though small angle collisions occur at high temperatures as well, they can be totally neglected with respect to the more frequent and more efficient large angle collisions. Thus in the limit of high temperatures only large angle scattering is taken into account, while at very low temperatures only small angle scattering is operative.

In metals, the electrons are scattered by the Debye phonons around and above the Debye temperature and by phonons of energy $\approx k_B T$ below the Debye temperature. This leads to small relaxations times around room temperature and large ones at low temperatures (cfr. below). We will show below that, for semimetals like graphites, the charge carriers are scattered by low energy phonons — which are not the dominant ones at the temperature considered — through large angles around room temperature and below. Only at very low temperatures are they scattered by phonons of energy $\sim k_B T$ through small angles.

Before discussing the case of graphites we must make an important remark. Though we have stated above that the energy of the phonon is always negligible with respect to that of electrons at the Fermi surface, the electron still absorbs a small amount of energy in the collision process. As a result, it loses memory of its phase after collision. This is an inelastic process and the resulting scattering time is called *inelastic*. This is in contrast to what happens for collisions with lattice static defects, which are *elastic* and will be discussed in the next paragraph.

Also, the fact that we have used the mobility concept so far, means implicitly that we have considered electronic *diffusive motion*. This means that we assume that between two successive collisions, an electron is accelerated by the electric field, experiences a scattering event, is accelerated again, experiences another collision, and so on.

3.4.3 Ineffectiveness of Electron-Phonon Interaction in Graphites

We have previously discussed the electron-phonon interaction in the particular case of pristine graphite (Issi and Nysten, 1998) and its intercalation compounds (Issi, 1992). We have shown that, contrary to a 3D metallic system, the quasi-2D electron and hole systems in semimetallic graphites do not interact with the thermal phonons. For small Fermi surfaces, like those characterizing semimetals, in order to satisfy energy and momentum conservation requirements, electrons and holes are scattered in-plane through large angles by low energy *subthermal phonons*, except at very low temperatures (Figure 8b). The limit is determined by a characteristic temperature:

$$\theta^* = \frac{2k_F v_s \hbar}{k_B} \tag{5}$$

The subthermal phonons are of lower frequencies than the thermal phonons which dominate at the temperature considered and are thus less numerous and the resulting electron-phonon scattering is less effective, even for large angle scattering.

For graphite, the velocity of sound in the graphene planes, $v_s \sim 2.1 \ 10^6$ cm/s, is one order of magnitude higher than in metals, while k_F, the Fermi wave vector, is orders of magnitudes smaller than in ordinary metals. The result is that θ^* is much smaller than the Debye temperature of 3D metals and, *a fortiori*, of the in-plane Debye temperature of graphite.

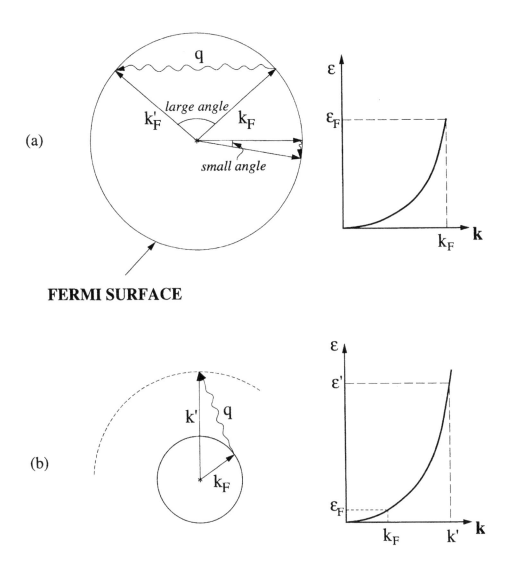

FIGURE 8. Schematic representation of electron-phonon interactions. In (a) allowed transitions in metals are shown for the case of large and small angle scattering. In both cases, electrons are scattered on the Fermi surface. In (b) transitions which are not allowed because of energy and momentum conservation requirements. This is the case for semimetals, where interactions cannot take place between electrons at the Fermi surface and thermal phonons at high temperatures. In order to interact with a thermal phonon, the electron would need an energy of the order of 1 eV which an acoustic phonon cannot provide. This illustrates the concept of the ineffectiveness of electron-phonon interactions in semimetals.

As a result, the electron-phonon interaction is much weaker in pristine carbons and graphites than in 3D metals at a given temperature because, though scattering is through large angles above θ^*, the number of interacting phonons is very small. This is due to the fact that, as stated above, the interaction is with subthermal phonons, and that in addition we are well below the in-plane Debye temperature, even at room temperature. Thus, for pristine carbons and graphites, at low and high temperatures, phonon scattering should be much less effective than in 3D metals leading to unusually high relaxation times. This, associated with the fact that the effective masses, m^*, on which the mobility depends (relation 1), are much smaller than the free electron mass, explains why we observe very high mobilities in these materials. It explains also why, despite the very small densities of electrons and holes, which are about 4 orders of magnitudes lower than in metals at room temperature, well graphitized samples display resistivities less than two orders of magnitudes higher than in metals (cfr. Figure 1).

When the temperature is decreased, large angle scattering persists until we reach the temperature $T = \theta^*$, where the scattering angle begins to decrease.

3.4.4 Scattering by Lattice Defects

Lattice defects, such as impurities, vacancies, dislocations,... destroy the periodicity locally. The fact that a crystal has finite dimensions will cause reflections at the boundaries too. Electron or hole scattering by such so-called static defects lead to a relaxation time, τ_r, which is temperature independent.

When the temperature decreases, the relaxation time for intrinsic scattering increases. If only electron-phonon interactions are considered, it should follow a T^5 law at very low temperatures. If the sample were ideally free from all lattice static defects, one should thus expect that τ becomes infinite in the limit of 0 K. However, this is never the case, since any sample will have definite geometrical dimensions. When the sample dimensions become comparable to the electron mean free path, reflections will occur at the sample boundaries for a single crystal, or at the grain boundaries for a polycrystalline sample. This is the so-called size-effect or *boundary scattering*. The scattering may be either specular or diffuse. In the latter case, the mean free path will be equal, within a factor of the order of unity, to the smallest dimension of the sample or to the grain size. This is the situation which prevails in HOPG at low temperatures.

Contrary to phonon scattering, scattering by static defects is *elastic* and electron waves do not lose their phase memory after collision (see below).

3.4.5 Scattering in Mesoscopic Systems

In a metal sample of very small dimensions, though larger than that of the atom, it may happen that the mean diffusion length between inelastic collisions, L_{in},

$$L_{in} = \sqrt{D\tau_{in}} \tag{6}$$

D being the diffusion constant, is larger than the dimensions of the sample (Figure 9). This is likely to occur for samples of very small dimensions at very low temperatures when the

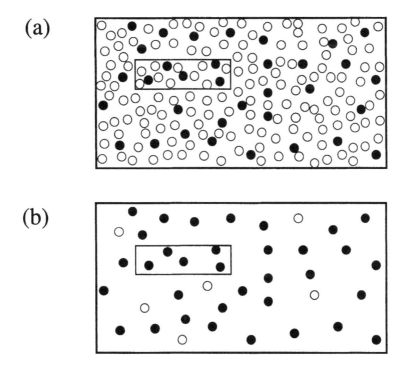

FIGURE 9. Schematic spatial representation of scattering events in mesoscopic systems. The white circles represent inelastic scattering events, while the dark ones are relative to elastic scattering. In (a) the room temperature situation is represented where inelastic electron-phonon scattering dominates. The motion of electrons is diffusive. In (b) the low temperature situation is shown and the small rectangle is the image of a mesoscopic system where only elastic scattering occurs, i.e where the inelastic diffusion length is larger than the sample dimensions. In such a system, electrons propagate without losing memory of their phase, and the regime is quasi-ballistic.

probability for electron-phonon inelastic collisions becomes very low. Indeed, below 1 K, L_{in} may attain a few micrometers in a metallic sample. If there are no lattice defects, the electrons will propagate balistically from one end of the sample to the other and we are no longer in a resistive regime. In the *ballistic regime*, the laws of conductivity discussed above no longer apply. The propagation of an electron is then directly related to the quantum probability of transmission across the global potential of the sample.

In a real crystal, static effects will be present and the only scattering processes will be elastic collisions. Then we are in a particular situation. In fact, it was shown in the early fifties that after these sort of collisions the electrons do not lose memory of the phase contained in their wave functions and thus propagate through the sample in a coherent way. Electrons in a given state interacting with the same impurity experience the same phase shift. As a corolary, the geometrical configuration of the impurities will have a direct effect on electron propagation. This is what happens in *mesoscopic systems*. Also, the fact that electrons do not lose their phase memory after an elastic collision gives rise to interference effects which cause weak localization effects, as discussed above (Section 3.2.4).

This is the situation which was demonstrated to prevail in multiwall carbon nanotubes (MWCN). Langer *et al.* (1996) recently performed electrical resistance measurements on *individual MWCNs* down to very low temperatures and with a magnetic field applied perpendicular to the tube axis (cfr. Section 3.2.4). They found a significant *positive magnetoconductance*. The temperature and field dependences of the nanotube conductance were interpreted consistently in the framework of the theory for 2D weak localization that we previously discussed (Section 2.4). However, for the particular case of nanotubes, owing to the very small dimensions of the sample, we are close to the mesoscopic regime in the lowest temperature range. This situation is responsible for the conductance fluctuations that we will now discuss.

At lower temperature, reproducible aperiodic fluctuations appear in the magnetoconductance (Figure 10a). The positions of the peaks and the valleys with respect to magnetic field are temperature independent. In Figure 10b we present the temperature dependence of the peak-to-peak amplitude of the conductance fluctuations for three selected peaks as well as the rms amplitude of the fluctuations, $rms[\Delta G]$. It may be seen that the fluctuations have a constant amplitude at low temperature, which decreases slowly with increasing temperature with a weak power law at higher temperature. The turnover in the temperature dependence of the conductance fluctuations occurs at a critical temperature $T_c^* \approx 0.3$ K which, in contrast to the T_c values presented in Figure 7, is independent of the magnetic field. This behavior was found to be consistent with a quantum transport effect of universal character, the *universal conductance fluctuations* (UCF). UCFs were previously observed in mesoscopic weakly disordered metals and semiconductors of various dimensionalities. In such systems, where the size of the sample, L, is smaller or comparable to both L_ϕ, the phase coherence length, and the thermal diffusion length:

$$L_T = \sqrt{\frac{\hbar D}{k_B T}} \tag{7}$$

elastic scattering of electron wave functions generate an interference pattern which gives rise to a sample-specific, time-independent correction to the classical conductance. The interference pattern, and hence the correction to the conductance, can be modified by applying a magnetic field in order to tune the phase of the electrons. The resulting phenomenon is called UCFs, because the amplitude of the fluctuations ΔG has a universal value: $rms[\Delta G] \approx q^2/h$ as long as the sample size $L < L_\phi$, L_T. When the relevant length scale, L_ϕ or L_T, becomes smaller than L, the amplitude of the observed fluctuations decreases due to self-averaging of the UCF in phase-coherent subunits. When the relevant length scale decreases with increasing temperature, the amplitude of the fluctuations decreases as a weak power law: $T^{-\alpha}$, where α depends on dimensionality and limiting diffusion length, L_ϕ or L_T. $\alpha = 1/2$ for a 2D system with $L_\phi \ll L$, L_T. The observed fluctuations have therefore been interpreted in terms of universal conductance fluctuations for mesoscopic 2D systems.

So, in spite of the small diameter of the MWCN compared to the de Broglie wavelengths of the charge carriers, the cylindrical structure of the honeycomb lattice gives rise to a 2D electron system for both weak localization and universal conductance fluctuation effects. Both the amplitude and the temperature dependence of the conductance fluctuations were found to be consistent with the universal conductance fluctuations models applied to the particular cylindrical structure of nanotubes (Langer *et al.*, 1996). At high temperature a behavior similar to that known for bulk graphites was observed. This is in contrast with what was very recently observed on single-wall nanotubes.

Experiments performed in Delft on individual single-wall nanotubes (Tans *et al.*, 1997) have confirmed their one-dimensional nature. They appear to behave as genuine coherent quantum wires, or quantum dots if the length of the tube is reduced.

Oddly enough, it is found that to a given dimensionality is associated a specific quantum transport behavior at low temperature: while some multiwall nanotubes behave as 2D systems, single-wall nanotubes exhibit a lower dimensionality in their properties.

3.4.6 *Intercalation Compounds*

The work on the transport properties of graphite intercalation compounds (GIC) was initially stimulated by the promise of realizing electrical conductors with conductivities that could reach or even exceed that of copper. Indeed, as a result of charge transfer, one can increase significantly the electrical conductivity of graphitic materials by intercalation of various species (Dresselhaus and Dresselhaus, 1981). The resulting intercalation compounds exhibit typical metallic behaviors with room temperature electrical resistivities 3 to 4 times that of pure copper (Figure 1 and Figure 2). As a result of the anisotropic band structure, the electrical resistivity of acceptor GICs is also highly anisotropic. For some acceptor GICs, the ratio of the in-plane conductivity to that along the *c*-axis may reach six orders of magnitude at room temperature. This allows one to consider the hole system in-plane as a quasi two-dimensional (2D) gas.

At low temperature, the ideal electrical resistivity of acceptor GICs was found to be mainly due to hole-hole interaction in the presence of weak disorder, while electron-phonon scattering dominates the scene at higher temperature. A simple model of a 2D hole gas, introduced by the intercalate through charge transfer and interacting with the phonons and defects of the host layers, gives a good picture of electronic transport in the Boltzmann conductivity range (Issi, 1992).

Because of the inherent 2D nature of the electronic structures of acceptor GICs, weak localization effects and hole-hole interaction effects were found to be particularly interesting in these materials (Figure 2 and Figure 7). The possibility of varying the Fermi level of the compound through charge transfer and the defect structure of the host material over wide ranges provided a large range of experimental conditions for the investigation of these phenomena (Issi, 1992).

3.5 MAGNETORESISTANCE

The effect of the Lorentz force on the charge carriers, resulting from the application of a magnetic field on a conductor carrying an electrical current is to increase its resistance.

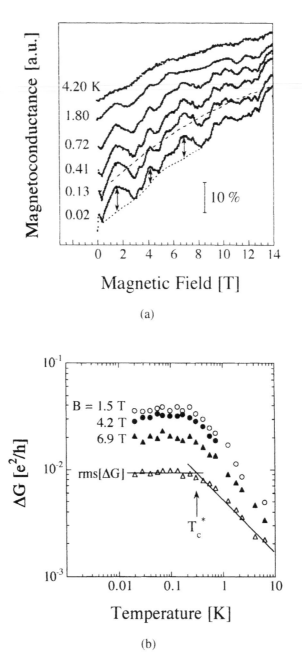

(a)

(b)

FIGURE 10. (a) Magnetic field dependence of the conductance of a single MWCN at different temperatures showing the aperiodic universal conductance fluctuations. (b) Temperature dependence of the amplitude of the aperiodic conductance fluctuations (δG) for the first three peaks selected in (a) as well as $rms[\delta G]$. (From Langer et al., 1996).

This effect is called *"positive magnetoresistance"*. The fractional change in the resistance caused by the application of an external magnetic field is expressed by:

$$\frac{\Delta\rho}{\rho_o} = \frac{\rho_H - \rho_o}{\rho_o} \tag{8}$$

where ρ_H and ρ_o are the electrical resistivities with and without a magnetic field, respectively. Solid state theory predicts that the classical positive magnetoresistance at low magnetic fields depends essentially on the carrier mobilities. However, *negative magnetoresistances*, i.e. decreases in resistivities with increasing magnetic fields, have been observed in pregraphitic carbons by Mrozowski and Chaberski (1956) as well as in other forms of carbons. These include poorly graphitized bulk carbons, PAN-based fibers, pitch-derived fibers and vapor-grown fibers. As is the case for the zero-field resistivity, these observations are consistent with the weak localization theory for two dimensional systems that we discussed in Section 3.2.4.

It was also observed that the sign and the magnitude of the magnetoresistance was closely related to the microstructure of the sample (Issi and Nysten, 1998). Highly graphitized samples present large positive magnetoresistances at all temperatures, as expected from high mobility charge carriers. With increasing disorder, a negative magnetoresistance appears at low temperature, where the magnitude and the temperature range at which it shows up increase as the relative fraction of turbostratic planes increases in the material.

The Bright model (1979) has been invoked to explain the negative magnetoresistance. In this model it is assumed that the electronic structure for turbostratic graphite is nearly two-dimensional (2D). The magnetic field induces changes in the electronic density of states which lead to an increase in carrier concentration. A critical analysis of this model is given by Dresselhaus *et al.* (1988).

An alternative explanation to the negative magnetoresistance has been proposed by Bayot *et al.* (1989). This was based on weak localization effects resulting from weak disorder in a 2D electronic system, which were previously observed in intercalated graphite by Piraux and coworkers (1985). Bayot *et al.* suggested that the 2D weak localization effects observed in pristine carbons occurs in the quasi 2D turbostratic phase. Later on, it was confirmed by Nysten and co-workers that the magnitude of the negative magnetoresistance increases with increasing structural perfection.

The analysis of the field dependence of the magnetoresistance, in the framework of the 2D weak localization theory, allowed the determination of two essential parameters: the 2D resistance and a characteristic magnetic field associated with scattering by magnetic impurities. Scattering by magnetic impurities was found to be the main mechanism destroying the phase coherence at low temperatures.

3.6 THERMOELECTRIC POWER

3.6.1 Introduction

The thermoelectric power (TEP) or Seebeck coefficient, S, is defined as the potential difference, ΔV, resulting from an applied unit temperature difference, ΔT, across an electrical conductor. There are essentially two mechanisms for thermoelectric power generation:

- the *diffusion thermoelectric power*, which is generated by the spontaneous diffusion of the charge carriers from hot to cold caused by the redistribution of their energies due to the difference in temperature. Charge carriers tend to accumulate at the cold end of the sample giving rise to an electric field. This field acts to counterbalance the stream of diffusing carriers until a steady state is reached.
- the *phonon drag thermoelectric power* which consists in an anisotropic transfer of momentum from the phonon system to the electron system when the coupling between the two systems is strong. This results in a drag on the charge carriers causing an extra electronic drift with an additional electric field to counterbalance it.

When, as it is the case for graphites, there is more than one type of carriers, the total thermoelectric power is obtained by considering the different groups of carriers with partial thermoelectric powers that contribute to the total thermoelectric power, S, as emf's in parallel. The total thermoelectric power is thus expressed as:

$$S = \frac{\sum \sigma_j S_j}{\sum \sigma_j} \tag{9}$$

The general expression for the diffusion thermoelectric power for a given group of charge carriers is given by the Mott formula:

$$S_d = \frac{\pi^2 k_B^2}{3q} T \left[\frac{\delta \ln \sigma}{\delta \varepsilon} \right]_{\varepsilon_F} \tag{10}$$

where ε_F is the Fermi energy and s the electrical conductivity. The derivative is taken at the Fermi level. Expression (10) shows that, for a given scattering mechanism, the diffusion thermoelectric power for a degenerate electron gas depends mainly on the Fermi energy, or on the carrier density. The smaller the Fermi energy, the higher the relative fraction of the charge carriers concerned with respect to the total Fermi gas. This is why semimetals like graphites exhibit higher partial diffusion thermoelectric powers than metallic GICs. In GICs the diffusion thermoelectric power depends on the charge transfer which determines the magnitude of the Fermi energy.

3.6.2 *Typical Results*

In Figure 11 we present the temperature variation of the TEP of a graphite single crystal (Takezawa *et al.* 1969). It is generally agreed that above 100 K the TEP is due to the diffusion of electrons and holes. The low temperature peaks were attributed to a phonon-drag mechanism. It may be seen from Figure 4c that, as for other transport effects, the TEP is sensitive to lattice perfection. For disordered carbons, the thermoelectric power is low and does not vary significantly with temperature. For highly disordered carbons, the room temperature thermoelectric power may even be negative. Decreasing disorder leads to a more marked temperature dependence and higher magnitudes for the thermoelectric power (Figure 4b). Graphitic fibers behave very much like bulk graphites.

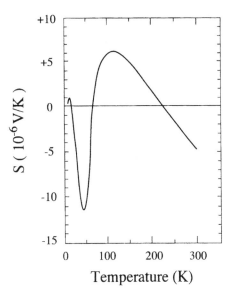

FIGURE 11. Temperature variation of the thermoelectric power of a graphite single crystal (From Takezawa *et al.*, 1969).

Observations concerning the in-plane thermoelectric power of GICs were also found particularily interesting. While at high temperature almost all compounds show the same temperature variation, in the low temperature region the behavior varies according to the stage of the compound (Issi, 1992). At low temperatures, the diffusion thermoelectric power dominates in non magnetic compounds. For stages higher than stage-1, a stage-dependent linear function of the temperature is observed for the diffusion thermoelectric power. A more complex behavior is found for stage-1 compounds.

3.7 CONCLUSION

In conclusion, in many aspects the electrical and thermal transport properties of carbons and graphites are quite interesting. This is mainly due to their variety of structures and the ability to engineer their properties at the atomic and microstructural levels, mainly through heat treatment and through intercalation.

We have shown that, because of the various aspects of conduction which are covered by carbons and graphites, whether electrical or thermal, one may observe in the different forms of this element the whole spectrum of solid state conduction phenomena.

In this review, we have focused on four important aspects to describe the physics of electronic conduction: (1) the semimetallic behavior and its effect on the band structure and on the scattering mechanisms, (2) the effects of reduced dimensionality, and (3) that of disorder leading to quantum aspects of conduction, and (4) the effects of intercalation.

References

Bayot, V., Piraux, L., Michenaud, J-P. and Issi J-P. (1989) *Phys Rev B*, **40**, 3514.

Bright, A.A. (1979) *Phys. Rev. B*, **20**, 5142.

Charlier, J-C. and Issi, J-P. (1996) Electrical conductivity of novel forms of carbon, *Physics and Chemistry of Solids*,, **57**, 957.

Delhaès, P. and Marchand, A. (1965) *Carbon*, **3**, 115, *Ibid.*, **3**, 125.

Delhaès, P., *Le carbone dans tous ses états*, (ed. P. Bernier and S. Lefrant, Gordon and Breach, in press).

Dresselhaus, M.S. and Dresselhaus, G. (1981) *Adv. Phys.*, **30**, 139.

Dresselhaus, M.S., Dresselhaus, G., Sugihara, K., Spain, I.L. and Goldberg, H.A. (1988) *Graphite Fibers and Filaments*, Springer Series in Materials Science 5, Springer-Verlag.

Dresselhaus, M.S., Dresselhaus, G. and Saito, R. (1998) *Electronic Band Structure of Graphites*, Chapter 2, this issue.

Ebbesen, T.W. (1996) *Physics Today*, **49**, 26.

Issi, J-P. (1992) *Transport properies of metal chloride acceptor graphite intercalation compounds* in Zabel, H. and Solin, S.A. eds., *Graphite Intercalation Compounds II*, Springer Series in Materials Science, Vol.18 (Springer-Verlag, Berlin).

Issi, J-P. and Nysten, B. (1998) *Electrical and thermal transport properties in carbon fibers* in *Carbon Fibers*, ed. J-B. Donet.

Kelly, B.T. (1981) *Physics of Graphite*, Applied Science Publishers, London.

Klein, C.A. (1964) *J. Appl. Phys.*, **35**, 2947.

Langer, L., Bayot, V., Grivei, E., Issi, J-P, Heremans, J.P., Olk, C.H., Stockman, L., Van Haesendonck, C. and Bruynseraede, Y. (1996) *Phys. Rev. Letters*, **76**, 479.

Marchand, A. (1977) Electronic Properties of Doped Carbons, *Physics and Chemistry of Carbons*, **7**, 155.

Meschi, C., Manceau, J.P., Flandrois, S., Delhaes, P., Ansart, A. and Deschamps, L. (1986) *Ann. de Physique*,, **11**, 199, colloq. 2, suppl. 2.

Mrozowski, S. and Chaberski, A. (1956) *Phys. Rev.*, **104**, 74.

Mrozowski, S. (1971) *Carbon*, **9**, 97.

Pacault, A. (1974) Along the carbon way, *Carbon*, **12**, 1.

Piraux, L., Issi, J-P., Michenaud, J-P., McRae, E. and Marêché, J-F. (1985) *Solid State. Commun.*, **56**, 567.

Piraux, L., Bayot, V., Gonze, X., Michenaud, J-P. and Issi, J-P. (1987) *Phys. Rev.*, **B36**, 9045.

Spain, I.L. (1981) *Electronic Transport Properties of Graphite, Carbons, and Related Materials*, ed. P.L. Walker Jr. and P.A. Thrower, in *Chemistry and Physics of Carbon*, volume 13, Marcel Dekker, Inc., New York, p. 119.

Tans, S.J., Devoret, M.H., Dai, H., Thess, A., Smaller, R.E., Geerligs, L.J. and Dekker, C. (1997) *Nature*, **386**, 474.

Takezawa, T., Tsuzuku, T., Ono, A. and Hishiyama, Y. (1969) *Phil. Mag.*, **19**, 623.

Zabel, H. and Solin, S.A. (1992) *Graphite Intercalation Compounds II*, Springer Series in Materials Science, Vol. 18 (Springer-Verlag, Berlin).

4. Magnetic Properties of Graphite and Graphitic Carbons

SERGE FLANDROIS

Centre de recherche Paul Pascal, avenue Albert Schweitzer, 33600 Pessac, France

4.1 DEFINITIONS AND METHODS

When a substance is placed in a magnetic field H, the magnetic induction, B, is given by the sum of the applied field plus a contribution $4\pi M$, where M is the magnetization per unit volume due to the substance itself[*]: $B = H + 4\pi M$.

The ratio $\kappa = M/H$ is the magnetic susceptibility per unit volume and is a dimensionless quantity. In practice, susceptibility is more conveniently expressed per unit mass (gram susceptibility) than per unit volume: $\chi = \kappa/\rho$, where ρ is the mass per unit volume of the substance. The susceptibility is a second rank tensor.

In diamagnetic substances, the induced magnetization, M, and hence the susceptibility is negative: a magnetic flux is produced in a direction opposite to the applied field. In simple terms, the application of an external field to the orbital motion of an electron around a nucleus induces a magnetization in the opposite direction, in accordance with Lenz's law. However, in a substance containing unpaired electrons, the permanent moment arising from the unpaired spins gives rise to a positive, paramagnetic susceptibility, generally much larger than the diamagnetic susceptibility: a flux is produced in the same direction as the applied field.

---

[*]We have retained the use of the unrationalised cgs system (whose units are abbreviated as cgsemu or emu) throughout this chapter because magnetochemists are more familiar with this system. A conversion to the rationalized mks SI (Système International) units can be easily carried out; conversion tables are found in textbook on magnetism. For the mass susceptibility, 1 emu/g = $(10^3/4\pi)$ SI Units.

Thus, in a non-homogeneous field, the substance will tend to move to regions of lowest field strength when the susceptibility is negative (diamagnetism) and to regions of highest field strength when χ is positive (paramagnetism). This is the basis of some common methods of susceptibility measurement. The force exerted on a sample is proportional to the susceptibility and to $H(\partial H/\partial x)$, where $\partial H/\partial x$ is the field gradient. Thus, the susceptibility can be obtained from the force measurement using some sensitive weighing technique. In the Faraday method, the sample volume is sufficiently small for $H(\partial H/\partial x)$ to be constant over the whole sample. With this method, samples of only a few milligrams can be used.

Other more recent techniques determine the magnetization itself, when a sample is placed in a homogeneous magnetic field. A detection coil is used to detect the change in magnetic flux due to the presence of the magnetic moment. In a SQUID magnetometer, the sample passes through the detection coil and the output from the SQUID yields the value of the magnetic moment. The susceptibility is then obtained from the ratio M/H, in weak field regime.

4.2 DIAMAGNETISM OF GRAPHITE

One of the most characteristic properties of graphite is its exceptional diamagnetism. After superconductors, graphite exhibits the largest diamagnetic susceptibility. Moreover, a strong anisotropy is observed: the susceptibility at room temperature is about -21×10^{-6} emu/g with the magnetic field applied along the c-axis and -0.4×10^{-6} emu/g along the a-axis (parallel to the carbon layers). Finally, although χ_a is temperature independent, as generally expected for diamagnetic susceptibilities, χ_c and hence the diamagnetic anisotropy increases (in absolute value) with decreasing temperature.

The first reliable measurements were performed by Ganguli and Krishnan (1941) between 90 K and 1000 K. The data were confirmed by Poquet *et al.* (1960) and extended down to 77 K and up to 2000 K. As the in-plane susceptibility χ_a has a value close to that expected for the free-atom susceptibility, it was thought, from the first measurements, that the large diamagnetic χ_c value was due to the π electron delocalization in the graphite layers. Indeed, graphite is a narrow band overlap semimetal (Dresselhaus *et al.*, 2000) and the particular magnetic behaviour, as for other electronic properties, results from the presence of charge carriers. Thus, the large diamagnetic susceptibility of graphite has to be explained in terms of its electron energy band structure. First calculations (Mc Clure, 1956), based on a simplified two-dimensional band structure gave the correct order of magnitude and temperature dependence of the susceptibility at high temperatures ($T > 1000$ K) but not at low temperatures. A full understanding of the magnetic behaviour of graphite was obtained with the help of a three-dimensional band model elaborated by Sharma, Johnson and McClure (1974), taking account of the progress in the energy-band structure of graphite over the years. From the Slonczewski-Weiss model (Slonczewski and Weiss, 1958), they were able to obtain the magnetic energy levels and to derive the expression of the magnetic susceptibility as a function of the band-parameters of the model.

The energy band parameters may be obtained experimentally by a variety of experimental methods, such as magnetoreflection, optical absorption, de Haas-van Halphen effect, etc...

Using the values admitted at that time, Sharma *et al.* showed that a reasonable agreement was obtained between experiment and theory of graphite diamagnetism.

Nevertheless, the experimental behaviour of the diamagnetism of graphite at temperatures below 77 K was still unknown. This gap was filled by Maaroufi *et al.* (1982) who measured the diamagnetism of graphites and of various carbon materials down to liquid helium temperature. In addition, as the band parameters of graphite had been remeasured since Sharma's work and refined (some had changed their sign!), these authors showed that the diamagnetism behaviour could be interpreted using these new accepted values or with small adjustments to these values.

This chapter will show these results after a short presentation of the McClure's theory. Then, we will see how the diamagnetism of graphite is modified by the presence of defects or by doping. The diamagnetism of the rhombohedral form of graphite will be examined. Finally, an overview will be given on the Pauli paramagnetism as measured by electron spin resonance methods.

4.3 THEORETICAL DIAMAGNETISM OF GRAPHITE

The Slonczewski-Weiss-Mc Clure (SWMcC) hamiltonian describes the four-electron energy bands that produce the Fermi surface of graphite near the H-K edge of the Brillouin zone. The corresponding matrix is shown in chapter 1.2 (Eq. 19), with the matrix elements given as a function of the in-plane wave vector and the seven band parameters of the model ($\gamma_0, \gamma_1, \gamma_2, \gamma_3, \gamma_4, \gamma_5, \Delta$). The definition of the energy band parameters is given in Table 1, together with their generally accepted values and the corresponding references (see also Eqs. 20 to 24 of Chapter 1.2).

From this hamiltonian and using Fukuyama's formalism for the diamagnetism of band electrons (Fukuyama, 1971), Sharma *et al.* (1974) calculated the magnetic susceptibility with the magnetic field applied perpendicular to the graphitic layers. They showed that tthe susceptibility χ_{SW} calculated using the SWMcC hamiltonian can be written in two parts:

$$\chi_{SW} = \chi_0 + \partial\chi$$

where χ_0 is independent of γ_3 and $\partial\chi$ is the correction to second order in γ_3. They found:

$$\chi_0 = N_0\gamma_0^2 \int_{-\pi}^{\pi} d\xi(D + I)$$

with:

$$D = -\frac{1}{12}\{\omega_1[f(E_1) - f(E_3)] + \omega_2[f(E_2) - f(E_1)]\}$$

$$I = \int_{E_3}^{E_1} dE\left(-\frac{\partial f}{\partial E}\right)\phi_1 - \int_{E_3}^{E_2} dE\left(-\frac{\partial f}{\partial E}\right)\phi_2$$

where $f(E)$ is the Fermi-Dirac distribution function and E_1, E_2, E_3 the band edge energies (Eq. 20, Chapter 1.2)

$$\xi = C_0 k_z$$
$$N_0 = (e^2/\hbar C)^2 (3a_0^2/2\pi^2 C_0)$$
$$\omega_1 = (1 - v)^2/(E_1 - E_3)$$
$$\omega_2 = (1 + v)^2/(E_2 - E_3)$$
$$v = \gamma_4 \Gamma/\gamma_0$$
$$\phi_1 = \frac{1}{2\Omega_0} \ln \left[\frac{(E - E_3)(E_1 - E_2)}{\Omega(E_1 - E_3)(1 + v)^2} \right]$$
$$\phi_2 = \frac{1}{2\Omega_0} \ln \left[\frac{(E - E_3)(E_1 - E_2)}{\Omega(E_2 - E_3)(1 - v)^2} \right]$$
$$\Omega = -[(E - E_1)/(1 - v)^2] + [(E - E_2)/(1 + v)^2]$$

and

$$\delta\chi = -\frac{N_0 \gamma_3^2}{4} \int_{-\pi}^{\pi} d\xi \Gamma^2 \frac{\partial}{\partial E_3} \left[\int_{E_2}^{E_1} \frac{G(E) - G(E_3)}{E - E_3} dE + G(E_3) \ln \left(\frac{1 - v}{1 + v} \right)^2 \right.$$
$$\left. + G_1(E_3) - \frac{8}{3} f(E_3) \right]$$

where

$$G(E) = f(E)[(x_1 + x_2)/(x_1 - x_2)^3] \left(\frac{7}{2} x_1^2 + 24 x_1 x_2 + \frac{7}{2} x_2^2 \right)$$
$$G_1(E) = \frac{31}{2} f(E)[(x_1 + x_2)/(x_1 - x_2)]^2$$
$$x_1 = (1 + v)^2 (E - E_1)$$
$$x_2 = (1 - v)^2 (E - E_2)$$

To obtain χ_{SW} as a function of temperature and Fermi level, the integrations on E and ξ must be done numerically. Thus, from these expressions, χ_{SW} can be calculated for any set of band parameters.

4.4 EXPERIMENTAL DIAMAGNETISM OF GRAPHITE AND COMPARISON WITH THEORY

Measurements have been carried out on single crystals of natural graphites or on synthetic graphites: HOPG (from Union Carbide) and PGCCL (from Le Carbone-Lorraine, France). The structure and physical properties of these materials are very close to those of natural graphite (Moore, 1973).

χ_a (the magnetic susceptibility with the magnetic field applied parallel to the layers) was generally measured using the Faraday method. It was shown to be temperature independent with a value close to -0.4×10^{-6} emu/g.

The second susceptibility component has been determined by a torque method derived from the critical angle method first proposed by Krishnan *et al.* (1934) and modified by Lumbroso and Pacault (1957). The sample is suspended by a fine quartz fiber ($\phi \sim 10 \, \mu$m) in a rotating homogeneous field. The method, very accurate, gives the difference between the two principal susceptibilities in the plane of rotation, here ($\chi_c - \chi_a$) if the graphite sample is suspended along a-axis.

The results obtained on a single crystal by Poquet *et al.* (1960) and on HOPG and PGCCL by Maaroufi *et al.* (1982) are shown in Figures 1 and 2. These results can be compared with the theoretical susceptibility χ_{SW}. However χ_{SW} is an approximate value of χ_c, since only the contributions from the states close to the Fermi level were taken into account. The rest of the energy bands gives a temperature-independent contribution, so that in fact:

$$\chi_c - \chi_a = \chi_{SW} + C$$

where the constant term C contains χ_a in addition. $C \sim 2 \times 10^{-6}$ emu/g, according to Sharma *et al.* (1974).

On Figures 1 and 2 are shown the fits obtained by Maaroufi *et al.* (1982) using this value of C and adjusting the band parameters around the values previously determined for graphite. A remarkable agreement is obtained for the whole temperature range. The values of the band parameters used for fitting the curves are given in Table 2. There is no significative difference between the values obtained from diamagnetism and from other properties (Table 1).

4.5 INFLUENCE OF STRUCTURE DEFECTS

Due to the peculiarities of the band structure of graphite, a small shift in the Fermi level E_F can induce dramatic changes in the diamagnetic susceptibility χ_c and its temperature dependence. This has been shown by Maaroufi (1986) who examined the influence of each band parameter independently on the theoretical χ_{SW}. Calculation of χ_{SW} in the range $-1 < E_F < 1$ eV, with all other parameters kept constant at the values given in Table 1, showed an almost symmetrical variation around $E_F = 0$. Except for some singularities for negative E_F values close to $E_F = 0$, χ_{SW} increases with the absolute value of E_F. It becomes slightly paramagnetic for $E_F < -0.4$ eV and $E_F > +0.6$ eV. Also, the maximum of diamagnetism observed in the temperature dependence ($T_{\max} = 32$ K for HOPG, Figure 1) is generally shifted to higher temperatures and large changes in E_F lead to small and temperature-independent χ_{SW}.

Structure defects, if they are in small amounts, i.e. act as electron donors or acceptors with only local distortions of the lattice, may produce only changes in the Fermi level position. To observe experimentally this effect on the diamagnetism, Maaroufi *et al.* (1982) measured the magnetic anisotropy of two neutron-irradiated PGCCL samples, whose irradiation doses were 6.5×10^{17} neutrons/cm^2 (26C sample) and 20×10^{17} neutrons/cm^2 (28C sample), respectively. X-ray diffraction showed identical spectra for pristine and irradiated PGCCL samples: the effect of neutron irradiation was weak. The results of magnetic measurements are given in Figure 2 and Table 2. As expected, the diamagnetism becomes smaller with the

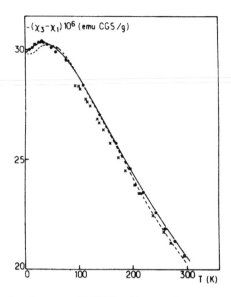

FIGURE 1. Diamagnetic anisotropy of HOPG, with χ_3 and χ_1 standing for χ_c and χ_a, respectively. Dots are experimental data from Maaroufi *et al.* (1982). Crosses are single crystal data from Poquet *et al.* (1960). Solid line is the fit with McClure's theory and values of band parameters given in Table 2. Dotted line is the fit with values of Table 1.

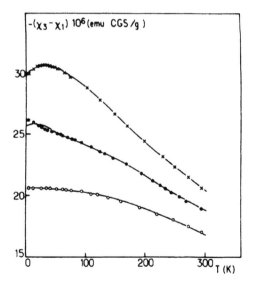

FIGURE 2. Diamagnetic anisotropy of PGCCL samples, with χ_3 and χ_1 standing for χ_c and χ_a, respectively: ×; pristine; •, neutron-irradiated (26C sample); ○, neutron-irradiated (28C sample). Solid lines are fits with McClure's theory and values of band parameters given in Table 2. From Maaroufi *et al.* (1982).

TABLE 1. Energy band parameters of graphite (all in eV).

Parameter	Definition	Accepted value (eV)	Reference
γ_0	In-plane interaction between nearest neighbour atoms	3.16	Toy et al. (1977)
γ_1	Nearest layer interaction between atoms of type α	0.39	Misu et al. (1979)
γ_2	Next nearest layer interaction between atoms of type β	–0.019	Mendez et al. (1980)
γ_3	Nearest layer interaction between atoms of type β producing trigonal warping of bands	0.315 ± 0.015	Doezema et al. (1979)
γ_4	Nearest layer interaction between atoms of type α and β	0.044 ± 0.024	Mendez et al. (1980)
γ_5	Next nearest layer interaction between atoms of type α	0.038	Misu et al. (1979)
Δ	Shift resulting from difference in α and β atom sites	–0.008	Toy et al. (1977)
E_F	Fermi level	-0.024 ± 0.002	Dillon et al. (1977)

TABLE 2. Band parameters (in eV) of SWMcC model obtained from diamagnetism (Maaroufi et al., 1982; Biensan et al., 1991).

Sample	E_F	γ_0	γ_1	γ_2	γ_3	γ_4	γ_5	Δ
HOPG	–0.0248	3.22	0.41	–0.019	0.32	0.06	0.04	–0.005
KISH	+0.006	3.17	0.45	–0.019	0.29	0.06	0.04	–0.005
PGCCL	–0.0243	3.31	0.45	–0.018	0.29	0.06	0.04	–0.011
Irradiated PGCCL (26C)	–0.0314	3.17	0.41	–0.021	0.27	0.06	0.04	–0.011
Irradiated PGCCL (28C)	–0.0339	3.13	0.45	–0.022	0.17	0.06	0.04	–0.011

increase of defects number due to irradiation. There is lowering of the Fermi level, but a good fit could not be obtained with only a change in the Fermi level. The data seemed to show that from PGCCL to 28C sample, in addition to the expected lowering of Fermi level, γ_3 and γ_2 would be lowered by 40% and 22% respectively, whereas γ_0 and γ_1 would be little affected. The authors explained these γ changes by the presence of some interstitial atoms produced by irradiation, which would lower the interplane interaction.

Kish graphite, a highly crystalline graphite generated by carbon recrystalllization during steel-making operations (Liu and Loper, 1991), is another example of the effect of structural defects on the magnetic properties. STM examination of Kish flakes (Biensan et al., 1991) has shown the existence of in-plane defects associating a seven-atom ring (heptagon) with a three-atom ring (triangle). These defects, which can be understood by the presence of extra carbon atoms, produce a shift of the Fermi level in the conduction band. Indeed, measurements of the temperature dependence of χ_c susceptibility showed (Figure 3) a maximum of diamagnetism at a temperature (45 K) higher than for HOPG (32 K). The fit with Mc Clure's theory gave band parameters close to those of HOPG, with the exception of the Fermi level, which is higher (+0.006 eV) than for HOPG (–0.0248 eV) (Table 2). This shift, up to a value slightly above the top of the valence band, corresponds to a gain in charge carriers (electrons) which can be calculated from the SWMcC band model (Dresselhaus et al., 1977) as being equal to 2×10^{19} carriers per cubic centimeter. Assuming that each

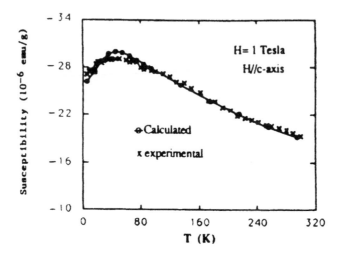

FIGURE 3. Temperature dependence of the magnetic susceptibility χ_c of Kish graphite and fit with McClure's theory using the values of band parameters given in Table 2. From Biensan *et al.* (1991).

defect gives one extra electron, the gain corresponds to a ratio of one defect for about 6000 carbon atoms. This defect concentration was in agreement with STM observations.

4.6 EFFECT OF DOPING ON THE MAGNETIC PROPERTIES OF GRAPHITE

Graphite can be doped by substitution of carbon atoms by boron, which acts as an electron acceptor because it has one less valence electron than carbon. The thermodynamic limit of substitutional boron content has been shown to be 2.35 at. % (Lowell, 1967), although it seems possible to increase the B content up to 20-25 % with the use of non-equilibrium methods such as chemical vapor deposition (CVD) (Derré *et al.*, 1994).

When the boron content is weak, let us say of the order of 1% or less, it is reasonable to deal with electronic properties on the basis of the rigid band model, i.e. the effect of introducing boron in the lattice is only a downward shift of the Fermi level. The first magnetic measurements on boronated graphite samples were performed at room temperature by Soule (1962). He showed that the diamagnetism drops off sharply for very low boron content: it becomes equal to the core diamagnetism when the B/C atom ratio reaches values between 10^{-3} and 10^{-2}. This result and the temperature dependence of the diamagnetism of dilute boronated graphites (Okura *et al.*, 1981), which shows a typical diamagnetism maximum at high temperature, can be ascribed to the Fermi level shift into the rigid valence band and are fully understood within the framework of McClure's theory.

Another means of graphite doping is intercalation, i.e. introduction of a chemical species between the carbon layers. Graphite can be intercalated either with electron donors (e.g. alkali metals) or with electron acceptors (e.g. halides, acids). Intercalation induces an expansion of the lattice along c-axis and a charge transfer between graphite and intercalated species. Clearly, the rigid band model can be considered only with dilute compounds, i.e. high-stage compounds (the stage number is the number of carbon layers separating two intercalated layers).

The magnetic susceptibilities of a series of potassium-graphite compounds from stage 2 to 16 have been studied by Maaroufi et al. (1987). They showed that stages higher than 10 behave as mixtures of stage 10 and graphite, i.e. the charges transferred from potassium to carbon are distributed on only five carbon layers on either side of the potassium layers. Therefore, the magnetic behaviour cannot be described by Mc Clure's theory, even for stage-16 compounds where the K/C atom ratio is of the order of 4×10^{-3}. Similar conclusions were put forward for the diamagnetism of dilute acceptor compounds of graphite with ICl and nitrate ions (Matsubara et al., 1983).

4.7 DIAMAGNETISM OF RHOMBOHEDRAL GRAPHITE

In the previous sections of this chapter, the most stable form of graphite, the hexagonal structure, was considered. Another form, with a rhombohedral structure, can exist, always mixed with the hexagonal form. It has been found in natural graphite flakes (Jagodzinski, 1949) and can be produced by grinding, with contents reaching 30% (Boehm and Hoffman, 1955). The essential difference between the two structures comes from the stacking sequence: ABAB in the hexagonal form and ABCABC in the rhombohedral one.

The electron energy band structure of rhombohedral graphite was first calculated by Haering (1958), who used the nearest-neighbor tight-binding approximation. McClure (1969) extended these calculations by including interaction between next nearest neighboring planes and calculated the carrier concentrations, the density of states and the diamagnetic susceptibility. The main result is an increase in the diamagnetic susceptibility χ_c, whose room temperature value is of the order of -35×10^{-6} emu/g (instead of -21×10^{-6} for the hexagonal form). χ_c would increase (in absolute value) with decreasing temperature, with eventually a slight maximum depending on the Fermi level position. At absolute zero, $\chi_c \simeq -91 \times 10^{-6}$ to -210×10^{-6} emu/g.

These results were in good agreement with measurements performed on samples of Madagascar graphite enriched in rhombohedral form by grinding (Gasparoux, 1967). Assuming addivity of hexagonal and rhombohedral susceptibilities, the value of χ_c was estimated to be -34×10^{-6} emu/g at room temperature. Measurements carried out down to 77 K showed an increase in diamagnetism and χ_c would extrapolate to about -97×10^{-6} emu/g at absolute zero.

4.8 DIAMAGNETISM OF PREGRAPHITIC CARBONS

Various carbon materials are obtained by heat-treatment of organic substances. The first step is the loss of heteroatoms (carbonization step) with the formation of basic structural units (B.S.U.) constituted of two or three stacked aromatic molecules with lateral size of about 1 nm (Oberlin, 1979). At increasing heat-treatment temperatures (HTT) in the range 1000–1500°C, the B.S.U. build columnar structures. Then graphitization can occur with at

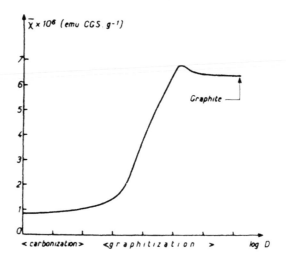

FIGURE 4. Schematic evolution of the average diamagnetic susceptibility of carbons at room temperature (actual values are negative). From Marchand (1986).

first a two-dimensional growth for HTT up to 2100–2200°C and, at higher temperatures, a three-dimensional organization which can eventually lead to a structure close to graphite.

The diamagnetic susceptibility of pregraphitic carbons and its evolution with HTT have been extensively studied. The average susceptibility $\overline{\chi}$ is generally measured, since the structural units in these materials are not oriented parallel to a same direction and, in addition, the samples are generally in powder form. Figure 4 shows the schematic evolution of $\overline{\chi}$ during the carbonization and graphitization steps (Marchand, 1986): the rapid increase of $\overline{\chi}$ corresponds to the two-dimensional growth of carbon layers, which allows for larger and larger orbits of the π electrons. A small maximum at the end of this growth phase is the signature of the beginning of interplane interactions when the three-dimensional ordering sets in.

The increase of $\overline{\chi}$ is accompanied by the appearance of its temperature dependence. Figure 5 shows various curves where $\overline{\chi}$ is plotted vs. the reciprocal temperature. A gradual change from a temperature-independent diamagnetism to graphite-like behavior is observed when graphitization proceeds. All the curves can be fitted above 77 K with the expression:

$$\overline{\chi} = \overline{\chi}_0[1 - \exp(-T_0/T)],$$

where $\overline{\chi}_0$ and T_0 are adjustable parameters which can be used to characterize the progressive two-dimensional ordering. Actually, this expression was derived from theoretical calculations of the diamagnetism of a two-dimensional electron gas (Marchand, 1957) and its use is questionable for pregraphitic carbons. The strong disorder present in these materials, associated with the existence of many various defects, prevents indeed of calculating theoretically their diamagnetism.

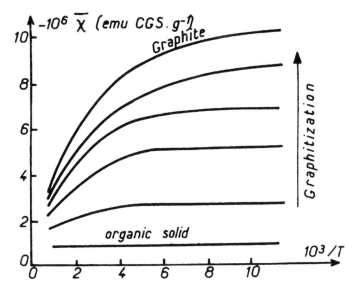

FIGURE 5. Temperature dependence of the average diamagnetic susceptibility of carbons of increasing graphitization stage. From Marchand (1986).

4.9 PARAMAGNETISM OF GRAPHITE AND PREGRAPHITIC CARBONS

In addition to their exceptional diamagnetism resulting from the particular band structure of graphite, the charge carriers have a Pauli-type paramagnetism. The corresponding susceptibility is weak, about three orders of magnitude smaller than the diamagnetic susceptibility. It was discovered independently by Castle (1953) and by Hennig *et al.* (1954) from electron spin resonance (ESR) measurements. A few years later, a thorough investigation of the spin resonance of charge carriers in graphite was made by Wagoner (1960).

ESR measurements provide several kinds of information:

– the integrated intensity of the absorption signal, which is proportional to the paramagnetic susceptibility,
– the line shape and the linewidth, which is inversely proportional to the relaxation time of the spins,
– the *g*-factor (position of the line).

The paramagnetic spin susceptibility of graphite is about 1.3×10^{-8} emu/g (Singer, 1963). It increases slightly above 100 K, in agreement with theoretical calculations (McClure and Smith, 1963). The line shape is of the Dysonian form, characteristic of conduction electron spin resonance in metals: for a single crystal and the field applied parallel to the *c*-axis,

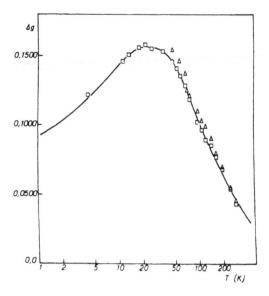

FIGURE 6. Temperature dependence of the g-factor anisotropy, $\Delta g = g_c - g_a$, for HOPG ($\triangle$) and PGCCL ($\square$) samples. From Carmona *et al.* (1973).

the ratio of peak heights in the derivative of the power absorption $A/B = 3.0$ at room temperature (Wagoner, 1960).

The g-factor and the linewidth vary strongly with the orientation of the crystal in the magnetic field, according to the following expressions:

$$g = g_a + (g_c - g_a)\cos^2\theta$$

and

$$\Delta H = \Delta H_a + (\Delta H_c - \Delta H_a)\cos^2\theta$$

where θ is the angle between the c-axis and the magnetic field direction. The value of g_a, independent of temperature, is equal to 2.0026, close to the free electron value (2.0023). At room temperature, $g_c = 2.050$ and is strongly dependent on temperature, like the diamagnetic susceptibility χ_c: at 77 K, $g_c = 2.123$ (Wagoner, 1960). Indeed, the g-factor anisotropy exhibits a temperature dependence similar to the diamagnetism anisotropy, as shown in Figure 6 (Carmona *et al.*, 1973). A similar behavior is observed for the linewidth ΔH: ΔH_a is temperature independent whereas ΔH_c increases with lowering temperature.

McClure and Yafet (1962) have published a theory of the g-factor based on the SWMcC band structure. However, this work seems to be incorrect (Dresselhaus *et al.*, 1988). A satisfactory theory of the g-factor in graphite still remains to be developed.

ESR has been used extensively for the study of pregraphitic carbons (Marchand, 1986). The ESR signal generally contains two contributions: a free electron contribution of Pauli-type, independent of temperature, and a localized radical contribution with a Curie-like temperature dependence (in $1/T$). Although a single resonance line is observed, the two contributions can be evaluated from the temperature dependence of the ESR line intensity. The Curie contribution increases, as expected, for lower HTT and becomes predominant during the carbonization step (HTT < 1000°C), where many localized radicals are created through breaking of C-O, C-H and C-C bonds (Carmona and Delhaes, 1978). A puzzling behaviour is the increase of the Pauli contribution for HTT < 2000°C. Note that for HTT between 1000 and 1400°C the signal may be so broadened that it often seems to disappear completely. The g-factor is anisotropic and its evolutions with HTT and with the measurement temperature parallel those of the diamagnetic susceptibility (Pacault *et al.*, 1965).

4.10 CONCLUSION

Whereas more experimental work is needed for rhombohedral graphite, the diamagnetism of hexagonal graphite is well understood. Mc Clure's theory based on the SWMcC band model accounts perfectly for the measured diamagnetism, with values of band parameters in excellent agreement with other estimates (most notably from optical and magneto-optical effects, Shubnikov-de Haas and de Haas-van Alphen effects).

The same model seems to account for the magnetic anisotropy of irradiated graphites and pyrocarbons. The values of the band parameters seem to be reasonable (Maaroufi *et al.*, 1982). It would be of interest to verify if, as for hexagonal graphite, other physical properties of these materials could be interpreted using these parameters.

In the case of pregraphitic carbons, although the diamagnetic behaviour is qualitatively understood, no quantitative interpretation is available as yet. For example, it is not fully understood why the diamagnetic susceptibility reaches values close to hexagonal graphite for samples whose in-plane coherence length is only of the order of 100–150 Å (samples heat-treated at about 2000°C).

Finally, a new field of research has been opened with the recent discovery of carbon nanotubes. The study of their magnetic properties is of great interest, since carbon nanotubes can be either metallic or semiconducting, depending on their symmetry (Saito *et al.*, 1994). Several susceptibility measurements have been recently reported on multiwall nanotubes produced by the arc-discharge method (Heremans *et al.*, 1994; Chauvet *et al.*, 1995; Kotosonov and Kuvshinnikov, 1997). The data reveal an anisotropic diamagnetic behaviour that could be different from graphite. However, the measurements were carried out on samples containing nanotubes with a large distribution in diameters and containing possibly other carbon particles, and the results are not in quantitative agreement. Clearly, new experiments are needed on well-defined samples, possibly with single-wall nanotubes of calibrated size.

References

Biensan, P., Roux, J.C., Saadaoui, H. and Flandrois, S. (1991) *Microsc. Microanal. Microstruct.*, **2**, 465.

Boehm, H.P. and Hoffmann, H. (1955) *Z. Anorg. Allg. Chem.*, **278**, 58.

Carmona, F., Amiell, J. and Delhaes, P. (1973) *C.R. Acad. Sci. Paris*, **B276**, 765.

Carmona, F., Delhaes, P. (1978) *J. of Applied Physics*, **49**, 618.

Castle, J.G. (1953) *Phys. Rev.*, **92**, 1063.

Chauvet, O., Forro, L., Bacsa, W., Ugarte, D., Doudin, B. and de Heer, W.A. (1995) *Phys. Rev.*, **B52**, 6963.

Derre, A., Filipozzi, L., Bouyer, F. and Marchand, A. (1994) *J. Mater. Sci.*, **29**, 1589.

Dillon, R.O., Spain, I.L. and McClure, J.W. (1977) *J. Phys. Chem. Solids*, **38**, 635.

Doezema, R.E., Datars, W.R., Schaber, H. and Van Schyndel, A. (1979) *Physical Rev. B*, **19**, 4224.

Dresselhaus, M.S., Dresselhaus, G. and Fischer, J.E. (1977) *Phys. Rev. B*, **15**, 3180.

Dresselhaus, M.S., Dresselhaus, G., Sugihara, K., Spain, I.L. and Goldberg, H.A. (1988) *Graphite Fibers and Filaments, Springer series in Materials Science*, vol. 5, Springer-Verlag, Berlin.

Dresselhaus, M.S., Dresselhaus, G. and Saito, R. (2000) in *Graphite and Percursors*, Pierre Delhaus, editor, volume 1, p. 25. Series World of Carbons, Gordon and Breach, Paris, France.

Fukuyama, H. (1971) *Prog. Theor. Phys.*, **45**, 704.

Ganguli, N. and Krishnan, K.S. (1941) *Proc. Roy. Soc. London*, **A117**, 168.

Gasparoux, H. (1967) *Carbon*, **5**, 441.

Haering, R.R. (1958) *Can. J. Phys.*, **36**, 352.

Hennig, G.R., Smaller, B. and Yasaitis, E.L. (1954) *Phys. Rev.*, **95**, 1088.

Heremans, J., Olk, C.H. and Moretti, D.T. (1994) *Phys. Rev.*, **B49**, 1994.

Jagodzinski, H. (1949) *Acta Cryst.*, **2**, 298.

Kotosonov, A.S. and Kuvshinnikov, S.V. (1997) *Phys. Lett.*, **A230**, 377.

Krishnan , K.S. and Bannerjee, S. (1934) *Proc. Roy. Soc. London*, **234**, 265.

Liu, S. and Loper, C.R. (1991) *Carbon*, **29**, 547.

Lowell, C.E. (1967) *J. Am. Ceram. Soc.*, **50**, 142.

Lumbroso, N. and Pacault, A. (1957) *C.R. Acad. Sci. Paris*, **245**, 686.

Maaroufi, A. (1986) PhD. Thesis, University of Bordeaux-I, unpublished.

Maaroufi, A., Flandrois, S., Coulon, C. and Rouillon, J.C. (1982) *J. Phys. Chem. Solids*, **43**, 1103.

Maaroufi, A., Flandrois, S. and Guerard, D. (1987) *J. Chim. Phys.*, **84**, 1443.

Marchand, A. (1986) *in Carbon and Coal Gasification*, NATO ASI series 105, Edited by Figueiredo, J.L. and Moulijn, J.A., Plenum, Dordrech, p. 93.

Matsubara, K., Kawamura, K. and Tsuzuku, T. (1983) *J. Phys. Soc. Japan*, **52**, 3180.

McClure, J.W. (1956) *Phys. Rev.*, **104**, 666.

McClure, J.W. (1969) *Carbon*, **7**, 425.

McClure, J.W. and Smith, L.B. (1963) *Proceedings of the 5th Conference on Carbon*, vol. 2, Pergamon Press, N.Y., p. 3.

McClure, J.W. and Yafet, Y. (1962) *Proceedings of the 5th Conference on Carbon*, vol. 1, Pergamon Press, N.Y., p. 22.

Mendez, E., Misu, A. and Dresselhaus, M.S. (1980) *Physical Rev. B*, **21**, 827.

Misu, A., Mendez, E. and Dresselhaus, M.S. (1979) *J. Phys. Soc., Japan*, **47**, 199.

Moore, A.W. (1973) in *Chemistry and Physics of Carbon*, edited by P.L. Walker, Marcel Dekker, N.Y., vol. 11, 69.

Oberlin, A. (1979) *Carbon*, **17**, 7.

Pacault, A., Uebersfeld, J., Theobald, J.G. and Cerruti, M. (1965) *C.R. Acad. Sci. Paris*, **261**, 3589.

Poquet, E., Lumbroso, N., Hoarau, J., Marchand, A., Pacault, A. and Soule, D.E. (1960) *J. Chim. Phys.*, **57**, 866.

Saito, R., Dresselhaus, G. and Dresselhaus, M.S. (1994) *Physical Rev.*, **B50**, 14698.

Sharma, M.P., Johnson, L.G. and McClure, J.W. (1974) *Physical Rev.*, **B9**, 2467.

Singer, L.S. (1963) *Proceedings of the 5th Conference on Carbon*, vol. 2, Pergamon Press, N.Y., p. 37.

Slonczewski, J.C. and Weiss, P.R. (1958) *Physical Rev.*, **109**, 272.

Soule, D.E. (1962) *Proceedings of the 5th Conference on Carbon*, vol. 1, Pergamon Press, N.Y., p. 13.

Toy, W.W., Dresselhaus, M.S. and Dresselhaus, G. (1977) *Physical Rev. B*, **15**, 4077.

Wagoner, G. (1960) *Phys. Rev.*, **118**, 647.

NOTE

Recently, the magnetic properties of graphite materials containing variable fractional amounts (up to 40%) of the rhombohedral graphite form have been investigated by means of electron spin resonance (Chehab, S., Guerin, K., Amiell, J. and Flandrois, S. (2000) Eur. Phys. J. B, **13**, 235). A narrow parallelism in the behaviors of the g-factor, linewidth and diamagnetic susceptibility has been perceived. From the g-value results, the g-anisotropy of the pure rhombohedral structure was inferred. At room temperature, g_c was found to be ≈ 2.24, compared to 2.05 for the hexagonal graphite.

5. Thermal Properties and Nuclear Energy Applications*

TIMOTHY D. BURCHELL

Oak Ridge National Laboratory, Oak Ridge, Tennessee USA

5.1 INTRODUCTION

When we think of energy applications of carbons it is carbon's use as a fossil fuel that first comes to mind. Carbon is, after all, the major constituent of coal, and other important fossil fuels such as oil and natural gas. However, carbon is the basis of a diverse group of engineering materials which are used, and in some cases, enable energy production, conversion, or storage. The use of graphite in the brushes of electric motors, or carbon anodes in lithium intercalation batteries, are perhaps familiar examples.

Carbon can exist in one of four allotropic forms (see Chapter 1): diamonds, graphites, the Fullerenes, and Carbynes, but is most frequently encountered in nuclear energy applications in the graphitic form, as artificial (manufactured) graphite, carbon fibers, or as carbon fiber-carbon matrix composite materials.

Carbon materials possess unique properties, or property combinations, that make them attractive for energy production and storage applications. Frequently, the unique attributes of carbon materials are related to their thermal behavior and, consequently, the thermal properties of carbon are reviewed here. Other novel properties include, but are not limited to: high temperature strength (in graphite the strength increases with increasing temperature

---

*The submitted manuscripts has been authored by a contractor of the U.S. Government under contract No. DE-AC05-96)R22464. Accordingly, the U.S. Government retains a nonexclusive, royalty-free license to publish or reproduce the published form of this contribution, or allow others to do so, for U.S. Government purposes.

up to ~2200°C); low atomic number — important in nuclear fission and fusion applications; and low density, which is significant in aerospace applications.

Several nuclear energy applications of carbon are reviewed here, including: (1) fission reactor applications, where graphite has been used as the reactor core structural material and nuclear moderator; (2) fusion reactor applications, where graphite and carbon-carbon (C/C) composites are widely used for plasma-facing materials; and (3) space nuclear power applications.

5.2 THERMAL PROPERTIES

The thermal behavior of a solid is controlled by the interatomic forces via the crystal lattice vibrational spectrum. The properties are generally insensitive to the spectrum details because they derive from a wide range of wavelengths in the spectrum. However, the graphite crystal is highly anisotropic because of the in-plane, strong covalent bonds and out-of-plane, weak Van der Walls bonds, so the above generalizations are not necessarily applicable, and deviations from the simple lattice models are required. Moreover, the electronic contributions to the thermal behavior must be considered at low temperatures. A complete and comprehensive review of the thermal properties of graphite has been written (Kelly, 1981).

5.2.1 Specific Heat

Heat energy is stored in the crystal lattice in the form of lattice vibrations. These vibrations are considered to be standing waves, and thus can only have certain permitted frequencies (density of states of waves). These waves produce atomic displacements which can be resolved so as to be parallel to the wave vector (longitudinal waves) and in two directions perpendicular to it (transverse waves). The Debye equation thus gives the specific heat, C, as:

$$C + 9R \left(\frac{T}{\Theta_D} \right)^3 \int_0^{\frac{\Theta}{T}} \frac{z^4 e^z}{(e^z - 1)^2} dz \qquad (1)$$

where,

$$R = \text{gas constant (8.314 J/mol·K)}$$

$$T = \text{temperature}$$

$$\theta_D = \text{Debye temperature}$$

and $z = \hbar\omega/kT$, where

$$\omega = \text{frequency of vibrational oscillations}$$

$$k = \text{Boltzman's constant}$$

and $\hbar = h/2\pi$, where h = Plank's constant.

At low temperatures, where $(T/\theta_D) < 0.1$, z in Equation 1 is large and we can approximate Equation 1 by allowing the upper limit in the integral to go to infinity such that the integral becomes $\sim (\pi^4/15)$, and on differentiating we get

$$C = 1941(T/\theta_D)^3 \quad \text{J/mol.K.} \qquad (2)$$

Thus at low temperatures the specific heat is proportional to T^3. At high temperatures, z is small and the integral in Equation 1 reduces to $z^2 dz$, hence on integrating we get the Dulong-Petit value of 3R, i.e., the theoretical maximum specific heat of 24.94 J/mol.K.

Since we are typically concerned only with the specific heat at temperatures above 10% of the Debye temperature ($0.1\theta_D$), the specific heat should rise exponentially with temperature to a constant value at $T \approx \theta_D$, the Debye temperature. The question then arises, what is the Debye temperature for graphite? Unfortunately, the answer is not straightforward. The situation is somewhat more complicated for graphite because of the anisotropic crystal structure. Therefore, two characteristic Debye temperatures can be identified, one corresponding to in-plane vibrations (parallel to the graphitic layers) and one corresponding to out of plane vibrations (perpendicular to the graphitic layer). Krumhansl and Brooks (1953) used a modified version of the Debye equation which recognized the essentially three-dimensional nature of the lattice at small wave numbers, and the importance of the Brillouin zone boundary in limiting the out-of-plane vibrational mode, and calculated the two characteristic Debye temperatures to be ~2500 K for the in-plane modes and ~950 K for the out-of-plane mode. In comparison, a single value of the Debye temperature may be ascribed to diamond ($\theta_D \sim 2000$ K).

Figure 1 shows the specific heat of graphite over the temperature range 300–3000 K. The data has been shown to be well represented by the equation (ASTM, 1998) in Figure 1, and is applicable to all graphites and carbon-carbon composite materials.

5.2.2 Thermal Expansion

There are two principal thermal expansion coefficients in the hexagonal graphite lattice; α_c, the thermal expansion coefficient parallel to the hexagonal $\langle c \rangle$ axis and α_a, the thermal expansion coefficient parallel to the basal plane $\langle a \rangle$. The thermal expansion coefficient in any direction at an angle ϕ to the $\langle c \rangle$ axis of the crystal is

$$\alpha(\phi) = \alpha_c \cos^2 \phi + \alpha_a \sin^2 \phi \qquad (3)$$

The value of α_c varies linearly with temperature from $\sim 25 \times 10^{-6}$ K^{-1} at 300 K to $\sim 35 \times 10^{-6}$ K^{-1} at 2500 K. In contrast α_a is much smaller and increases rapidly from -1.5×10^{-6} K^{-1} at ~300 K to approximately 1×10^{-6} K^{-1} at 1000 K, and remains relatively constant at temperatures up to 2500 K. The thermal expansion coefficients of engineering carbons such as graphite, fibers, and composites are a function of the (a) crystal anisotropy, (b) the orientation of the crystallites (i.e., textural effects arising during manufacture), and (c) the presence of suitably oriented porosity. A billet of molded or extended graphite would exhibit the same symmetry as the graphite crystal due to alignment of the crystallites during the forming process, with the thermal expansion coefficients α_c and α_a being replaced with $\alpha_\parallel$ (parallel to the molding or extrusion axis) and $\alpha_\perp$ (perpendicular to the molding or extrusion direction), respectively. However, the thermal expansion coefficients of polycrystalline graphites are typically significantly less than that of the graphite crystallites. Mrozowski (1956) associated this phenomenon with the presence of pores and cracks in the polycrystalline graphite that were preferentially aligned with the graphitic basal planes, thereby preventing the high c-axis crystal expansion from contributing fully to the observed

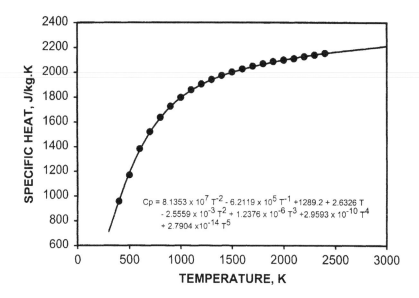

FIGURE 1. Temperature dependence of the specific heat of graphite, a comparison of experimental data for POCO AXM-5Q graphite (Hust, 1984) and calculated values (ASTM, 1989).

bulk expansion. The thermal closure of aligned internal porosity results in an increasing instantaneous coefficient of thermal expansion (CTE) with temperature and, significantly, an increasing strength with temperature up to temperatures of ~2200°C. In high density, isotropic graphites the CTE more closely approaches the graphite crystallite value.

Carbon fibers also exhibit anisotropy resulting from the spinning and stretching processes used in their manufacture. The thermophysical properties of carbon fibers are less well understood than for graphites. Donnet and Bansal (1990) have summarized much of the available data. The coefficients of thermal expansion of carbon fibers are typically small and negative in the parallel to the fiber direction ($\alpha_\parallel \sim -1.0 \times 10^{-6}$ K^{-1}) and in the perpendicular direction are large and positive ($\alpha_\perp = 5 - 18 \times 10^{-6}$ K^{-1}). Carbon fibers exert a dominant influence on the thermal properties of carbon-carbon composites, and thus the fiber fraction and fiber placement (architecture) tend to control the composites properties. Recently, Burchell and Oku (1994) published a comprehensive review of the materials properties data for carbon-carbon composites. The thermal behavior of unidirectional composites mirrors the expansion coefficients of their constituent carbon fibers. Mitsubishi Kasei's MFC-1 (Burchell and Oku, 1994), for example, had an $\alpha_\parallel$ of 0.9×10^{-6} K^{-1} and an $\alpha_\perp$ of 12×10^{-6} K^{-1}. The two- directional CX 2002U (Burchell and Oku, 1994) had in-plane ($\alpha_\parallel$) values of 1.5 and 1.7×10^{-6} K^{-1} and an out-of-plane ($\alpha_\perp$) value of 5.8×10^{-6} K^{-1}. Similarly, the two-directional composite SEPCARB N112 exhibited identical in-plane ($\alpha_\parallel$) values. FMI's three-direction 223 carbon- carbon composite (Burchell and Oku, 1994) exhibited isotropic thermal expansion over the temperature range

25–1400°C, with the expansion coefficients in the composite's x, y, and z directions varying from $\sim -1 \times 10^{-6}$ K^{-1} to $+1 \times 10^{-6}$ K^{-1} over this temperature range.

5.2.3 Thermal Conductivity

Graphite is a phonon conductor of heat. Consequently, the thermal conductivity of a graphite single crystal is highly anisotropic, reflecting the different bond types within, and between, the carbon basal planes. In the crystallographic $\langle a \rangle$ directions (within the basal plane) the atom bonding is of the primary, covalent type, whereas between the basal plane (crystallographic $\langle c \rangle$ direction) the bonding is of the much weaker secondary, or van der Walls, type. Phonons (elastic waves) may thus travel considerably more easily in the $\langle a \rangle$ direction than in the $\langle c \rangle$ direction within a graphite single crystal.

Data for the thermal conductivity of natural and pyrolytic graphite are reviewed by Kelly (1981). The room temperature thermal conductivity parallel to the basal planes is typically > 1000 W/m.K, whereas perpendicular to the basal planes the room temperature thermal conductivity is typically < 10 W/m.K. The thermal conductivity of graphite shows a maximum with temperature at approximately 100 K. Below this maximum the conductivity is dominated by the specific heat and varies as $\sim T^3$. At higher temperatures, above the maxima, the thermal conductivity decreases with increasing temperature due to phonon scattering.

Measurements on single crystals by Smith and Rasor (1956) showed that the maxima in thermal conductivity parallel to the basal plane was located at $\sim$ 80 K at a value of 2800 W/m.K. Nihira and Iwata (1975) reported the maximum thermal conductivity (perpendicular to the basal planes) for a pyrolytic graphite to be located at 75 K with a value of $\sim$ 20 W/m.K. At extremely low temperatures, i.e., $T < 10$ K the thermal conductivity is dominated by an electronic contribution which is proportional to temperature.

The temperature dependence of the in-plane thermal conductivity is shown in Figure 2 for various pyrolytic graphites. The substantial improvements in thermal conductivity caused by thermal annealing and/or compression annealing are attributed to increased crystal perfection and increases in the size of the regions of coherent ordering (crystallites). This minimizes the extent of phonon-defect scattering and results in a larger phonon mean-free path. With increasing temperature, the dominant phonon interaction becomes phonon-phonon scattering (Umklapp processes). Therefore, the observed reduction in thermal conductivity with increasing temperature and the convergence of the curves in Figure 2 are attributed to the dominant effect of Umklapp scattering in reducing phonon mean-free path.

A popular method for determining the thermal conductivity of carbon and graphite is the thermal "flash" technique. A small specimen is exposed to a thermal pulse, usually from a xenon flash lamp or a laser, and the back face of the specimen observed with an infrared detector. The specimen's thermal conductivity is then determined from the back face temperature rise transient. The thermal conductivity at the measurement temperature, T, is calculated from the relationship

$$K_T = \alpha C_p \rho \quad \text{(W/m.K)} \tag{4}$$

where α is the thermal diffusivity (m^{2}s), C_p is the specific heat at temperature T (J/kg.K), and ρ is the density (kg/m^3).

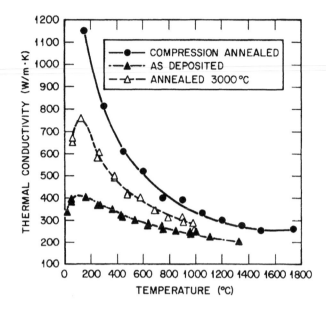

FIGURE 2. The temperature dependence of thermal conductivity for pyrolytic graphite in three different conditions. Adapted from Roth *et al.* (1989).

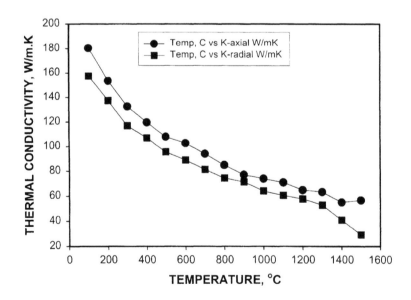

FIGURE 3. The temperature dependence of the thermal conductivity of grade H-451 graphite measured parallel and perpendicular to the extrusion direction.

Figure 3 shows data for the temperature dependence of thermal conductivity of a near-isotropic nuclear graphite (SGL Carbon grade H-451). The data were obtained using the laser-flash method over the temperature range 373–1873 K, and illustrate the reduction of thermal conductivity with increasing temperature and textural effects in an extruded graphite due to filler coke orientation.

5.3 NUCLEAR ENERGY APPLICATIONS

5.3.1 Fission Reactors

Nuclear fission reactors produce useful thermal energy from the fission of isotopes such as $_{92}U^{235}$. The fission of heavy elements, with the release of further neutrons, is usually initiated by an impinging neutron, e.g.,

$$_{92}U^{235} + {_0}n^1 \rightarrow {_{92}}U^{236*} \rightarrow F_1 + F_2 + n + \text{energy} \qquad (5)$$

with an average yield of 2.5 neutrons per fission. The bulk of the energy in a fission reactor is obtained as the kinetic energy of the fission fragments (F_1 and F_2 in Equation 5). The energy is degraded to heat by successive collisions within the body of the uranium fuel mass in which the fission occurred. The fission neutrons give up their energy by elastic collisions, typically within the moderator, and the γ-ray energy is absorbed in the bulk of the reactor materials outside of the fuel (moderator, pressure vessel, and shielding). Finally, some of the energy release occurs slowly, through the radioactive decay of fission products.

The longer a neutron dwells in the vicinity of a nucleus as it passes, the more probable it is that it will react with the nucleus. Hence, it is advantageous to slow the fission neutrons (referred to as "fast" neutrons) to lower (thermal) energies ($\sim$0.025 eV at room temperature), corresponding to a neutron velocity of 2.2×10^5 cm/s. The slowing down (or thermalization) process is also termed "moderation." In thermal reactors the fission neutrons leave the fuel with high energy, i.e., fast neutrons, and are slowed down outside the fuel in a non-absorbing medium called a "moderator."

A good moderator should possess the following properties:

4. The moderator should not react with neutrons because if they are captured in the moderator they are lost to the fission process.
5. Neutrons should be slowed down over a short distance and with few collisions in the moderator.
6. The moderator should be inexpensive, yet must satisfy reactor core structural requirements, and must be compatible with other structural materials used in the construction of the reactor.
7. The moderator should not undergo deleterious physical or chemical changes when bombarded with neutrons.

A good moderator must efficiently thermalize fast neutrons. This thermalization process occurs by neutron-nucleus elastic collisions. The maximum energy loss per collision occurs when the target nucleus has unit mass, and tends to zero for heavy target elements. Therefore,

TABLE 1. Properties of good moderators.

Moderator	Slowing-down power, cm^{-1}	Moderating ratio
Graphite	0.064	170
Be	0.176	159
H_2O	1.530	72
D_2O	0.370	12,000

low atomic number (Z) is a prime requirement of a good moderator. The maximum energy is always lost in a head-on collision. However, elastic collisions occur at many scattering angles, and thermalization takes place over numerous collisions. Additionally, we must take into account the number of atoms per unit volume of moderator, and the chance of a scattering collision taking place. Frequently, when assigning orders of merit to a moderator, the slowing-down power is used, which accounts for the average energy loss per collision, the number of atoms per unit volume, and the scattering cross section of a moderator. However, a complete picture of the nuclear performance of a moderator requires a comparison of its slowing-down power (which should be large) with its tendency to capture neutrons (capture cross-section) which should be small. Thus the ratio of the slowing-down power to the absorption (capture) cross section, or moderating ratio, should also be considered when evaluating potential moderators. The slowing-down power and moderating ratios for several candidate moderators are given in Table 1.

The choice of potential moderators is, in practice, limited to a handful of elements of atomic number less than sixteen. Gases are of little use as moderators because of their low density, but can be used in chemical compounds such a H_2O and D_2O. The choice of potential moderators of practical use rapidly reduces to the four materials shown in Table 1. Water is relatively unaffected by neutron irradiation, is low cost, and easily contained. However, the moderating ratio is reduced by neutron absorption in the hydrogen, necessitating the use of enriched (in U^{235}) fuels to achieve the required neutron economy. Heavy water (deuterium oxide) is a particularly good moderator because $_1H^2$ and $_8O^{16}$ do not absorb neutrons. The slowing-down power is high and the moderating ratio is very large. However, the cost of separating the heavy hydrogen isotope is high. Beryllium or beryllium oxide are good moderators but suffer from toxicity problems and are expensive and difficult to machine. Finally, graphite is an acceptable moderator. It offers an acceptable compromise between nuclear properties, cost, and utility as a structural material for the reactor core. Unfortunately, the properties of graphite are markedly affected by neutron irradiation. The application of graphite in fission reactors has recently been reviewed by Burchell (1999). The earliest experimental reactors or "piles" had cores of stacked blocks of graphite moderator. Commercial energy-producing and dual-use military (materials production) reactors have been developed based on the use of graphite moderators and have utilized both gas (He or CO_2) and water coolants.

5.3.2 *Fusion Reactors*

Carbon materials such as polygranular graphite, pyrolytic graphite, and carbon-carbon composites are widely used in many tokamak fusion devices (Burchell, 1996). As reviewed by Snead (1999), there are numerous nuclear fusion reactions that could be utilized, but the reaction that is easiest to sustain and has an acceptably high energy yield is the deuterium-tritium fusion reaction:

$$1D^2 + 1T^3 \rightarrow 2He^4 + \text{neutron} = 17.6\,\text{MeV} \tag{6}$$

However, the plasma (ionized gas) temperature required to sustain this reaction, which is related to the kinetic energy of the ions, is in excess of 50×10^6 K (> 10 keV). Carbon materials are employed as armor materials within the fusion system to prevent higher atomic number elements of the vessel's structural material from entering the plasma and extinguishing it. The vessel's armor is frequently referred to as a "plasma-facing component" and the materials used are referred to as "plasma-facing materials." The containment of high-temperature, high-density fusion plasmas requires a highly specialized reactor design. The "tokamak" concept utilizes a toriodal magnetic field confinement system in which the plasma ions are confined within, and coupled to, continuous magnetic field lines which travel helically through a toroidal vacuum vessel.

The reactions that occur between the fusion plasma and the plasma-facing materials are severe and can cause melting, sublimation, component failure due to high thermal stress, and excessive surface erosion. Consequently, a good plasma-facing material must exhibit the following desirable properties (Burchell, 1996): high thermal conductivity to minimize the surface temperature; low atomic number to minimize radiative heat losses from the plasma occurring when plasma-facing material enters the plasma; good thermal shock resistance to withstand the enormous heat loads that arise when the plasma collapses (disruption) and the plasma energy is dumped onto the plasma-facing components; low sputtering yield to minimize impurities sputtered into the plasma; low coefficient of thermal expansion; low out gassing and tritium retention; low toxicity, low neutron activation; low irradiation induced swelling and property degradation; good oxidation resistance, and low cost. Beryllium and carbon (graphite, pyrolytic graphite, and carbon-carbon composites) have been used for tokamak plasma-facing materials. They offer many of the desirable attributes listed above. The use of beryllium is, however, complicated by its low melting temperature and toxicity. Consequently, the majority of plasma fusion devices use carbon materials for their plasma-facing components.

Figure 4 illustrates the major plasma-facing components of a tokamak. The limiter, or bumper limiter, is a sacrificial component which intercepts the plasma edge particle flux and thus defines the plasma edge. Other fusion devices magnetically capture and divert the edge plasma onto a "divertor" plate which is removed from the central plasma. The point where the particle flux strikes the "divertor" experiences a significant ion heat loading and can suffer excessive erosion. The cooled plasma gasses are pumped away from beneath the divertor. In a tokamak the majority of the heat and particle flux is intercepted by the limiter or divertor, regardless of which design is employed. However, a significant flux will also strike the balance of the torus vessel lining, which is referred to as the "first wall." For example, the DIII-D tokamak in San Diego, U.S.A., has a first-wall heat loading of 0.6 MW/m^2 and a divertor loading of 5.3 MW/m^2. The JET (U.K.) has a negligible first-wall loading and a divertor/limiter loading of 18 MW/m^2. Both DIII-D and JET have carbon material as their plasma-facing material, the former utilizes POCO ATJ graphite for the first wall and

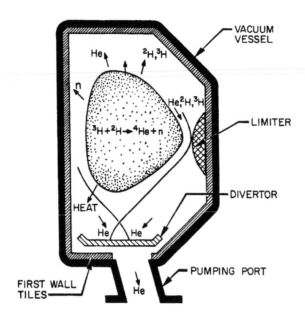

FIGURE 4. The major plasma-facing components of a tokamak.

divertor, whereas the latter utilizes Dunlop DMS-704 C/C composite for the divertor/limiter, and Dunlop DMS-704 and SEPCARB N11 C/C composites for the first wall (Snead, 1999).

The environment inside a tokamak is very harsh and plasma-facing components and materials must withstand numerous damaging phenomena. Full accounts of these can be found elsewhere (Hofer and Roth, 1996; Snead, 1999). Here, the major processes that influence a plasma-facing component's lifetime are briefly reviewed.

Physical sputtering

When sufficient energy is transferred from an impinging particle to a near surface carbon atom to overcome the surface binding energy, or lattice bond energy, the carbon atom may be displaced and can be ejected from the material (Eckstein and Philips, 1996). As the impacting ion energy increases, the sputtering yield decreases because the depth of penetration and interactions increases beyond the back scatter limit for the displaced atoms. In the case of graphite, the majority of the sputtered material comes from the top few atomic layers. However, with the correct combination of incident energy and target mass it is possible for the sputtering yield to exceed unity, i.e., an impacting ion causes the sputtering of more than one target atom. This situation rapidly develops into the catastrophic "carbon bloom," or the self accelerating sputtering of carbon into the plasma. The occurrence of "carbon blooms" during tokamak operation frequently causes a plasma disruption.

Chemical erosion

For intermediate temperatures in the range 400–1000°C the reaction of carbon atoms with energetic plasma particles such as hydrogen isotopes can lead to the formation of volatile molecules (Vietzke and Haasz, 1996). The volatile molecules emitted from the plasma-facing material enter the plasma (e.g. hydrocarbons, carbon dioxide, carbon monoxide) and become ionized. Depending upon the particle transport behavior, they may be redeposited on the tokamak walls or contribute to the impurity concentration in the plasma. Impurities in the plasma can lead to Bremsstrahlung radiation (the electromagnetic radiation produced by the sudden retardation of an electrical particle [electron or position] in an intense electric field), causing reduced plasma temperature. The radiation level depends upon the atomic number, Z of the impurities, ($\sim Z^2$). Hence the widespread use of carbon as a plasma-facing material.

Chemical erosion can be suppressed by doping with substitutional elements such as boron (Roth *et al.*, 1992). Oxygen is the most damaging impurity found in current tokamaks because of its presence in the molecular form, or as water vapor, and its tendancy to be adsorbed by carbon plasma-facing materials. Consequently, oxygen has a significant influence on the plasma performance and the carbon erosion rate. Extraordinary measures are usually taken prior to tokamak operation to remove the oxygen and water vapor from the carbon plasma-facing materials, including prolonged bake-outs and plasma-surface treatments to deposit oxygen gathering elements. The surface treatment of graphite walls has recently been discussed by Winter (1994).

Radiation enhanced sublimation

Neutron irradiation of graphite (see below) leads to the formation of Frenkel pairs (interstitial atom and lattice vacancy). Some fraction of these interstitial are trapped at defect sites, or may recombine with other lattice vacancies. However, some fraction may migrate to the carbon surface where they may be sputtered into the plasma. This phenomenon, termed radiation enhanced sublimation (RES), becomes the dominant sputtering mechanism above a temperature of $\sim 1000°C$ and increases exponentially with temperature. RES has been recently reviewed by Eckstein and Philips (1996).

5.3.3 Neutron Irradiation Damage

The damage mechanism and induced structural and dimensional changes

Displacement damage in graphite and carbon based materials can occur when energetic particles, such as neutrons, ions, or electrons impinge on the crystal lattice. The displacement of carbon atoms from their equilibrium position results in lattice strain, bulk dimensional change, and profound changes in physical properties. The binding energy of a carbon atom in the graphite lattice is about 7 eV (Thrower and Meyer, 1978). Impinging energetic particles such as fast neutrons, electrons, or ions can displace carbon atoms from their equilibrium positions. There have been many studies of the energy required to displace a carbon atom

(E_d), as reviewed by Kelly (1981) and Burchell (1996). The value of E_d is not well defined but lies between 24 and 60 eV. The latter value has gained wide acceptance and use in displacement damage calculations, but a value of 30 eV would be more appropriate.

The primary atomic displacement [primary knock-on carbon atoms (PKAs)] produced by energetic particle collisions produce further carbon atom displacements in a cascade effect. The cascade carbon atoms are referred to as secondary knock-on atoms (SKAs). The displaced SKAs tend to be clustered in small groups of 5–10 atoms and, for most purposes, it is satisfactory to treat the displacements as if they occur randomly. The total number of displaced carbon atoms will depend upon the energy of the PKA, which is itself a function of the neutron energy spectrum, and the neutron flux. For example, a high energy neutron arising from the deuterium-tritium fusion reaction (14.1 MeV) will produce over 500 PKAs. Once displaced the carbon atoms recoil through the graphite lattice, displacing other carbon atoms and leaving vacant lattice sites. However, not all of the carbon atoms remain displaced. The displaced carbon atoms diffuse between the graphite layer planes in two dimensions and a high proportion will recombine with lattice vacancies. Others will coalesce to form C_2, C_3, or C_4 linear molecules. These in turn may form the nucleus of a dislocation loop — essentially a new graphite plane. Interstitial clusters may, on further irradiation, be destroyed by a fast neutron or carbon knock-on atom (irradiation annealing). Adjacent lattice vacancies in the same graphitic layer are believed to collapse parallel to the layers, thereby forming sinks for other vacancies which are increasingly mobile above $600°C$, and hence can no longer recombine and annihilate interstitials.

A principal result of carbon atom displacements is crystallite dimensional change. Interstitial defects will cause crystallite growth perpendicular to the layer planes (c-axis direction), and relaxation in the plane due to coalescence of vacancies will cause a shrinkage parallel to the layer plane (a-axis direction). The damage mechanism and associated dimensional changes are illustrated in Figure 5. Dimensional changes can be very large, as demonstrated in studies on well-ordered graphite materials, such as pryolytic graphite, which has frequently been used to study the neutron-irradiation induced dimensional changes of the graphite crystallite (Engle and Eatherly, 1972; Kelly, 1981). Price (1974) conducted a study of the neutron-irradiation induced dimensional changes in pyrolytic graphite. Figure 6 shows the crystallite shrinkage in the a-direction for neutron doses up to $\sim$12 displacements per atom (dpa) for samples which were graphitized at a temperature of 2200–3300°C prior to being irradiated at 1300–1500°C. The a-axis shrinkage increases linearly with dose for all of the samples, but the magnitude of the shrinkage at any given dose decreases with increasing graphitization temperature. Similar trends were noted for the c-axis expansion. The significant effect of graphitization temperature on irradiation induced dimensional change accumulation can be attributed to thermally induced improvements in crystal perfection, thereby reducing the number of vacancy trapping sites in the lattice.

Polygranular graphites posses a polycrystalline structure, usually with significant texture resulting from the method of forming during manufacture. Consequently, structural and dimensional changes in polygranular graphites are a function of the crystallite dimensional changes and the graphite's texture. In polygranular graphite, thermal shrinkage cracks that occur during manufacture and that are preferentially aligned in the crystallographic a-direction, will initially accommodate the c-direction expansion, so mainly a-direction contraction will be observed. The graphite thus undergoes a net volume shrinkage.

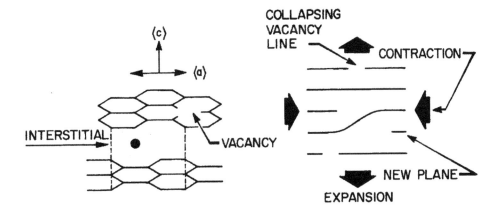

FIGURE 5. The graphite irradiation damage mechanism, illustrating the induced dimensional changes.

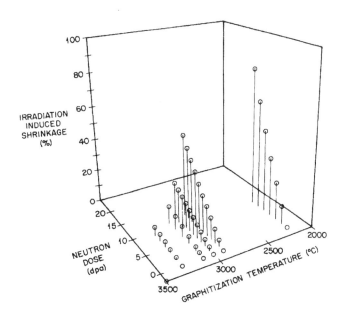

FIGURE 6. High temperature neutron irradiation induced a-axis shrinkage of pyrolytic graphite showing the influence of graphitization temperature on the extent of dimensional change. Adapted from Price (1974).

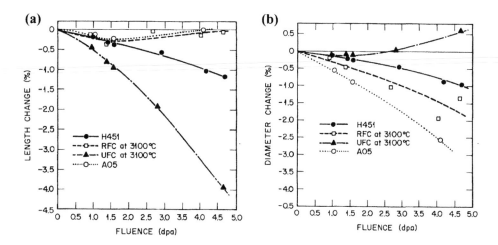

FIGURE 7. Neutron irradiation induced dimensional changes in specimens of unidirectional and two-directional C/C composites. (a) length change, (b) diameter change. RFC = random fiber composite. UFC = unidirectional fiber composite.

With increasing neutron dose (displacements) the incompatibility of crystallite dimensional changes leads to the generation of new porosity oriented parallel to the basal planes, and the volume shrinkage rate falls, eventually reaching zero. The graphite now begins to swell at an increasing rate with increasing neutron dose. The graphite thus undergoes a volume change "turnaround" into net growth which continues until the generation of cracks and pores in the graphite, due to differential crystal strain, eventually causes total disintegration of the graphite. A general theory of dimensional change in graphite due to Simmons (1965) has been extended by Brocklehurst and Kelly (1993). For a detailed account of the treatment of dimensional changes in graphite, the reader is directed to Kelly and Burchell's (1994) analysis of H-451 graphite irradiation behavior.

In contrast to graphite, little work has been performed on radiation damage effects in C/C composites. Much of the literature on this subject has recently been comprehensively reviewed (Burchell, 1996; Burchell and Oku, 1994). Moreover, recent accounts of dimensional changes in C/C composites due to neutron damage are given by Burchell (1996a) and Bonal and Wu (1996). Processing variables such as fiber type, architecture, and graphitization temperature, are known to influence the dimensional change behavior of C/C composites when irradiated. Figures 7(a) and (b) show the length and diameter changes, respectively, of several C/C composites and grade H-451 graphite irradiated to ∼5 dpa. The behavior of the composites can be explained through the application of the graphite single crystal behavior (Figure 5) and appropriate fiber microstructure and composite macrostructure models (Burchell, 1996). H-451 graphite behaves isotropically with respect to irradiation induced dimensional changes. However, the C/C composites are generally extremely anisotropic in their behavior (Figure 7). For example, the unidirectional composite (UFC) shrinks markedly in the specimen length direction (Figure 7(a)), whereas in the specimen diameter direction it shrinks only very slightly before turning around and beginning to swell (Figure 7(b)). The behavior of the random fiber composite (RFC) and A05 composite are similarly anisotropic, although for these composites the shrinkage occurs in the specimen diameter direction and the turn-around behavior is observed in the specimen length direction. The anisotropic behavior of a C/C composite can be attributed to the dominant effect of the fiber's irradiation dimensional changes. Composite A05 and RFC have their carbon fibers aligned in the specimen diameter direction, whereas composite UFC has the carbon fibers aligned in the length direction only. Fiber axial shrinkage and fiber diametral shrinkage, followed by turnaround to growth, are thus mirrored in the composite's behavior.

Physical property changes

The mechanism of thermal conductivity and the degradation of thermal conductivity have been extensively reviewed (Engle and Eatherly, 1972; Burchell, 1996; Kelly, 1981). The increase of thermal resistance due to irradiation damage has been ascribed to the formation of (Taylor *et al.*, 1969) : (1) submicroscopic interstitial clusters, containing 4 ± 2 carbon atoms; (2) vacant lattice sites, existing as singles, pairs, or small groups; and (3) vacancy loops, which exist in the graphite crystal basal plane and are too small to have collapsed parallel to the hexagonal axis. The contributions of collapsed lines of vacant lattice sites and interstitial loops, to the increased thermal resistance, is negligible. The influence of radiation damage on the thermal conductivity of a C/C composite is shown in Figure 8. The upper curve in Figure 8 shows the temperature dependence of the composite in the unirradiated condition. Note that the conductivity is substantially less than that shown for pyrolytic graphites (Figure 2). The lower curve in Figure 8 shows the temperature dependance of the composite after irradiation. At the irradiation temperature (600°C) the thermal conductivity has been reduced from >80 W/m·K to ~40 W/m·K. Some fraction of the irradiation induced damage may be annealed from the material by heating it above its irradiation temperature. This effect may be seen in the "cooling" curve in Figure 8. On cooling to 600°C from an annealing temperature of 1200°C, the thermal conductivity recovers to 67 W/m·K. More substantial recovery of thermal conductivity can be expected as the annealing temperature is increased. The reduction in thermal conductivity due to irradiation damage is temperature and dose sensitive. At any irradiation temperature, the thermal conductivity will reach a "saturation" limit. This limit is not exceeded until the graphite undergoes gross structural changes at high fluences. The saturated conductivity will be attained more rapidly, and will be at a lower conductivity level, at lower irradiation temperatures. Thus, higher irradiation (operating) temperatures are beneficial for carbon based materials under conditions of irradiation damage.

The thermal expansion of polygranular graphites and C/C composites are controlled by the thermal closure of aligned internal porosity. Irradiation induced changes in the pore structure (see earlier discussion of structural changes) can therefore be expected to modify the thermal expansion behavior of carbon materials. The behavior of GraphNOL N3M is typical of many fine-textured graphites (Burchell and Eatherly, 1991), which undergo an initial increase in the coefficient of thermal expansion followed by a steady reduction to a value less than half the unirradiated value of $\sim 5 \times 10^{-6}\ °C^{-1}$. The electrical resistivity of

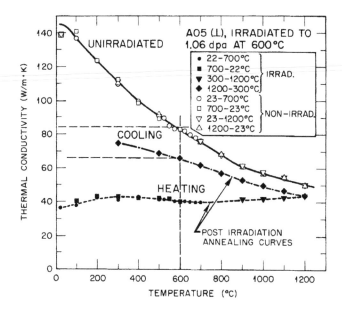

FIGURE 8. The variation of thermal conductivity of a two-directional C/C composite with temperature for three conditions: unirradiated, irradiated, and irradiated and annealed to 1200°C.

graphite and C/C composites will also be affected by radiation damage. The mean free path of the conduction electron in an unirradiated graphite or C/C composite is relatively large, being limited only by crystallite boundary scattering. Neutron irradiation introduces: (1) scattering centers, which reduces charge carrier mobility; (2) electron traps, which decreases the charge carrier density; and (3) additional spin resonance. The net effect of these changes is to increase the electrical resistivity on irradiation, initially very rapidly, with little or no subsequent change to relatively high fluence (Burchell, 1996; Burchell and Eatherly, 1991).

The mechanical properties of graphites and C/C composites are substantially altered by radiation damage. In the unirradiated condition polygranular graphites behave in a brittle fashion and fail at relatively low strains. The stress-strain curve is non-linear, and the fracture process occurs via the formation of sub-critical cracks, which coalesce to produce a critical flaw (Tucker *et al.*, 1986; Burchell, 1996b). C/C composites are frequently more flaw tolerant (tougher). When graphite and C/C composites are irradiated, the stress-strain curves become more linear, the strain to failure is reduced, and the strength and elastic modulus increased. The rapid rise in strength is due to dislocation pinning at irradiation induced lattice defect sites. This effect has largely saturated at doses > 1 dpa. Above ~1 dpa a more gradual increase in strength occurs due to structural changes within the graphite. For polygranular graphites the dose at which the maximum strength is attained loosely corresponds with the volume change turnaround dose, indicating the importance of pore generation in controlling the high-dose strength behavior.

The strain behavior of polygranular graphite subjected to an externally applied load is largely controlled by shear of the component crystallites. As with strength, irradiation induced changes in Young's modulus are the combined result of in-crystallite effects, due to low fluence dislocation pinning, and superimposed structural changes external to the crystallite. The effects of these two mechanisms are generally considered separable, and related by:

$$(E/E_o)_i - (E/E_o)_{\text{pinning}}(E/E_o)_{\text{structure}} \qquad (7)$$

Where E and E_o are the Young's modulus of irradiated and unirradiated graphite, respectively. The pinning contribution to the modulus change, due to relatively mobile small defects, is thermally annealable at $\sim 2000°C$.

The elastic modulus and strength are related by a Griffiths' theory type relationship (Griffiths, 1920).

$$\text{strength, } \sigma = \left(\frac{GE}{\pi c}\right)^{1/2} \qquad (8)$$

where G is the fracture toughness or strain energy release rate (J/m^2), E is the elastic modulus (Pa), and c is the flaw size (m). Thus, irradiation induced changes in σ and E (in the absence of changes in $[G/c]$) should follow $\sigma \propto E^{1/2}$.

Graphite will creep under neutron irradiation and stress at temperatures where thermal creep is negligible. The phenomena of irradiation creep has been widely studied because of its significance to the operation of graphite moderated fission reactors. Indeed, if irradiation induced stresses in graphite moderators could not relax via radiation creep, reactor core disintegration would rapidly result. The reader is directed toward the ample literature for detailed accounts of the phenomenon of irradiation creep (Engle and Eatherly, 1972; Kelly, 1994; Kelly and Burchell, 1994a, see also B. Rand's chapter in this book).

5.3.4 Space Nuclear Power

The National Aeronautics and Space Administration (NASA) has successfully deployed several deep space exploratory probes, including the Galileo, Ulysses, and Cassini missions. The electrical equipment on these probes are powered by converting the heat generated by a radioisotopic heat source called the general purpose heat source (GPHS) into electrical energy through the thermoelectric effects of silicon-germanium alloys (Wei and Robbins, 1985). The GPHS (Figure 9) utilizes the heat produced by the self-absorption of alpha particles in the $^{238}PuO_2$ fuel. A more detailed description of the GPHS is given by Schock (1980). Several stringent requirements must be met by the thermal insulation of the GPHS, chief amongst these are light- weight, long-term thermal stability and an ability to withstand thermal transients to $\sim 2500°C$. A carbon based thermal insulation material, named carbon-bonded carbon fiber (CBCF), was developed at ORNL for this purpose (Wei and Robbins, 1985). The CBCF is made from chopped rayon fibers about 10-μm diameter and 250-μm long, which are carbonized and bonded with phenolic resin. The CBCF is an excellent lightweight insulation material with a nominal density of 0.2 g/cm^3 and a thermal conductivity of 0.24 W/m·K in vacuum at $2000°C$. The mechanical strength of CBCF is satisfactory for the application.

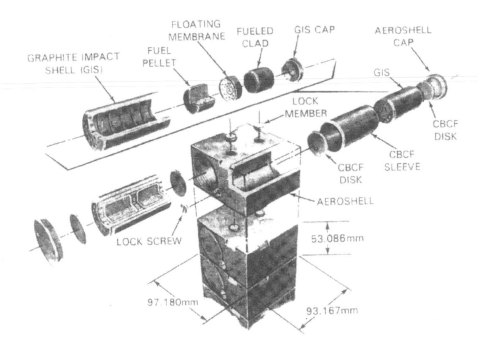

FIGURE 9.   The general purpose heat source.

The response of CBCF to short high temperature, thermal transients has recently been measured and modeled (Dinwiddie *et al.*, 1996). The thermal conductivity increases with increasing measurement temperature due to the contribution of radiation within the insulation (Figure 10). To determine the effect of high temperature transients on the thermal conductivity of CBCF, samples were exposed to temperatures from 2000–3000°C for times ranging from 5.7 to 10.7 seconds. Figure 11 shows typical thermal conductivity data for CBCF at measurement temperatures from 673 to 2273 K after exposure at 2673, 2873, 3073, and 3273 K for 5.7 seconds. Clearly, short exposure to 3000°C temperatures increases the high temperature thermal conductivity of CBCF. The data in Figure 11 have been fitted to an equation of the form:

$$\lambda = X A_1 T^{(\varepsilon + z)} + 2 \cdot 1599 \times 10^{-11} T^3 \qquad (9)$$

where the first term accounts for the conductivity of the carbon and the second term accounts for in-pore radiation (Dinwiddie *et al.*, 1996). In Equation (9) λ is the thermal conductivity, T is the temperature, X and Z are empirical parameters for the CBCF material, and are functions of the transient conditions (time and temperature), and ε and A_1, are data fitting parameters.

The GPHS contains several other carbon components. The graphite aeroshell and graphite impact shell (Figure 9) are both manufactured from Textron 3D Fineweave™ carbon-carbon composite (FWPF). The aeroshell and impact shell provide thermal and impact protection

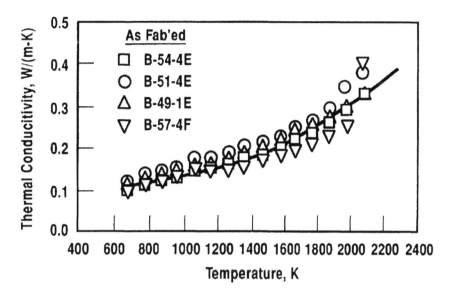

FIGURE 10. The temperature dependence of CBCF thermal conductivity measured in vacuum.

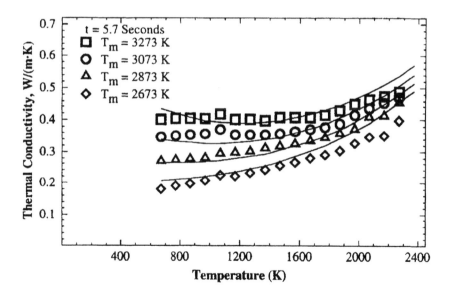

FIGURE 11. Experimental and predicted temperature dependence of thermal conductivity for CBCF.

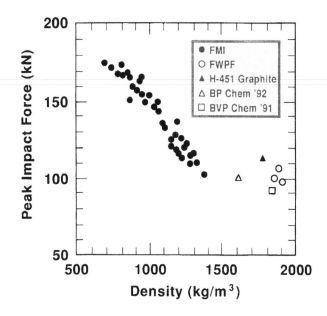

FIGURE 12. Peak impact force versus density for various candidate C/C composite impact shells. (Adapted from Burchell, 1996c).

to the GPHS in the event of reentry. The ability of the GPHS modules to withstand impact without significant fuel clad damage was demonstrated in a series of tests (Pavone *et al.*, 1985) which simulated reentry of the modules into earth's atmosphere and impact at terminal velocity (55 m/s). Postimpact analysis showed that impact shells typically fractured parallel to their longitudinal axis at four locations where the fiber bundles intersected at 45° to the circumferential direction. Although the current configuration and material has proven adequate, increased margins of impact performance are of interest. A research program was conducted at ORNL to develop and characterize alternative carbon-carbon composite materials for the impact shell having greater circumferential strengths and higher energy absorption (Romanoski and Pih, 1995). Carbon-carbon composite materials of cylindrical architecture were fabricated by commercial vendors and evaluated as alternative impact shell materials. Characterization included gas gun impact tests where cylindrical specimens, containing a mass simulant, were fired at 55 m/s to impact a target instrumented to measure force. The force versus time output was analyzed to determine: peak force, acceleration, velocity, and displacement. The impact test data are shown in Figure 12 for the candidate composites and Textron FWPF. Low peak impact force is indicative of good impact protection. Several candidate carbon-carbon composites exhibit impact energy absorption comparable to the FWPF. However in the case of the FMI material, a series of different circumferential (braided) architecture, this equivalence is achieved at significantly lower density. In addition to a potential weight saving, the FMI braided architecture offers a six-fold increase in circumferential strength and a radial thermal conductivity two orders of magnitude less than FWPF. A potential disadvantage of the braided architecture is its inability to retain threads when the desired means of connection or closure is a threaded joint. This issue was addressed in the development of a filament-wound, carbon-carbon composite which incorporated a number of radial elements of triangular geometry around the circumference of the thread for the full length of thread engagement (Romanoski and Burchell, 1996). The radial element significantly increases the shear strength of the threaded joint by transmitting the applied force to the balance of the composite structure. Consequently, future GPHS modules can utilize circumferential architecture (filament wound or braided) carbon-carbon composites which offer improved strength, reduced thermal conductivity, and potential weight savings. Increased margins of impact and thermal protection may thus be provided for future NASA missions.

5.4 SUMMARY

Carbon materials play a vital role in several nuclear energy systems, including graphite moderated fission reactors, fusion reactor plasma-facing components, and space nuclear power thermal and impact protection systems. Their flexibility of product forms, tailorable thermal properties, and in many instances, unique combinations of physical and chemical attributes make carbon an obvious choice for a variety of energy applications. The applications discussed here are examples. Graphite offers an acceptable compromise in the selection of a fission reactor moderator. It exhibits acceptable nuclear properties and can be used as the reactor core structural material. Unfortunately, graphite physical properties are markedly affected by neutron irradiation induced damage. Graphite and C/C composites have become the material of choice for plasma-facing components in tokamak fusion reactors. Again, they exhibit suitable properties such as high thermal shock resistance, low Z, and high thermal conductivity. However, they can suffer from excessive sputtering and erosion. Future tokamaks designs will additionally have to accommodate irradiation damage in their plasma-facing materials. Carbon-bonded carbon fiber offers a light-weight and highly insulative material that can withstand high temperature (in vacuum), and is thus ideally suited to its space nuclear thermal protection role. Dense C/C composites similarly have the ability to withstand the high temperatures associated with an inadvertent earth atmosphere reentry. Moreover, they provide adequate impact protection to assure safety margins.

Acknowledgment

Research sponsored by Oak Ridge National Laboratory, managed by Lockheed Martin Energy Research Corp. for the U.S. Department of Energy under contract number DE-AC05-96OR22464.

References

ASTM Standard C781 (1998) *Annual Book of ASTM Standards*, Vol. **15.01**, p. 199. Pub. American Society for Testing and Materials, West Conshohocken, Pennsylvania.

Bonal, J.P. and Wu, C.H. (1996) *Physica Scripta*, **T64**, 26–31.

Broklehurst, J.E. and Kelly, B.T. (1993) *Carbon*, **31**, 155.

Burchell, T.D. (1996) Radiation Damage in Carbon Materials, in *Physical Processes of the Interaction of Fusion Plasmas with Solids*, edited by Roth and Hoffer, Academic Press, San Diego, pp. 341–384.

Burchell, T.D. (1996a) *Physica Scripta*, **T64**, 17–25.

Burchell, T.D. (1996b) *Carbon*, **34**, 279–316.

Burchell, T. D. (1996c) in *Proc. CARBON '96*, Pub. The British Carbon Group, pp. 185–188.

Burchell, T. D. (1999) Fission Reactor Applications of Carbon. In *Carbon Materials for Advanced Technologies*. Ed. T.D. Burchell. Pub. Elsevier Science, Oxford, UK (1999).

Burchell, T.D. and Eatherly, W.P. (1991) *Journal of Nuclear Materials*, **179-181**, pp. 205–208.

Burchell, T.D. and Oku, T. (1994). *Atomic and Plasma-Materials Interactions Data for Fusion*, Vol. 5, Supplement to *Nuclear Fusion*, pp. 77–128. Pub. IAEA, Vienna.

Dinwiddie, R.B., Nelson, G.E. and Weaver, C.E. (1996) in *Proc. 23rd Int. Thermal Conductivity Conference*, Pub. Technomic Pub. Co., Lancaster, PA., pp. 466–477.

Donnet, J-B. and Bansal, R.C. (1990) *Carbon Fibers*, Second Edition. Pub. Marcel Dekker, Inc., New York.

Eckstein, W. and Philips, V. (1996) Physical Sputtering and Radiation-Enhanced Sublimation, in *Physical Processes of the Interaction of Fusion Plasmas with Solids*, Hofer, W.O. and Roth, J., eds. Pub. Academic Press, Inc., San Diego, U.S.A., pp. 93–133.

Engle, G.B. and Eatherly, W.P. (1972) *High Temperatures–High Pressures*, **4**, 119–158.

Griffiths, A.A. (1920). *Phil. Trans. Roy. Soc.*, **211**, 163.

Hofer, W.O. and Roth, J. (1996) *Physical Processes of the Interaction of Fusion Plasmas with Solids*, Pub. Academic Press, Inc., San Diego, USA.

Hust, J.G. (1984) NBS Special Publication 260–89, Pub. U.S. Department of Commerce, National Bureau of Standards. p. 59.

Kelly, B. T. (1981) *Physics of Graphite*, Applied Science Publishers, London.

Kelly B.T. (1994) in *Materials Science and Technology: Nuclear Materials*, Part 1 (VCH Weinheim) pp. 365–417.

Kelly, B.T. and Burchell, T.D. (1994) *Carbon*, **32**, 499.

Kelly, B.T. and Burchell, T.D. (1994a), *Carbon*, **32**, 119–125.

Krumhansl, J. and Brooks, H. (1953) *J. Chem. Phys.*, **21**, 1663.

Mrozowski, S. (1956) in *Proc. 1st and 2nd Carbon Conference*, p. 31.

Nihira, T. and Iwata, T. (1975) *Japanese J. Applied Phys.*, **14**, 1099.

Pavone, D., George, T.G. and Frantz, C. E. (1985) Los Alamos National Laboratory, LA-10353.

Price, R.J. (1974) *CARBON*, **12**, 159.

Romanoski, G.R. and Burchell, T.D. (1996) *Ceramic Engineering and Science Proceedings*, **17**(4-5), 90–97.

Romanoski, G. R. and Pih, H. (1995) in *Proc. ICCM-10 Whistler, B.C., Canada*, Vol. V, pp. 575–582.

Roth, E.P., Watson, R.D., Moss, M. and Drotning, W.D. (1989) Sandia National Laboratory Report No. SAND-88-2057, UC-423.

Roth, J., Garcia-Rosales, C., Behrisch, R. and Eckstein, W. (1992) *J. Nucl. Mat.*, **191–194**, 45–49.

Schock, A. (1980) In *Proc. of 15th Intersociety Energy Conv. and Eng. Conf.*, **2**, 1032–1042.

Simmons, J.H.W. (1965) *Radiation Damage in Graphite*, Pergamon Press, Oxford.

Smith, A.W. and Rasor, N.S. (1956) *Phys. Rev.*, **104**, 885.

Snead, L.L. (1999) Fusion Energy Applications, in *Carbon Materials for Advanced Energy Applications*, Editor T.D. Burchell. Pub. Elsevier Science, Oxford, UK (1999).

Taylor, R., Kelly, B.T. and Gilchrist, K.E. (1969) *J. Phys. Chem. Solids*, **130**, 2251–2267.

Thrower, P.A. and Mayer R.M. (1978) *Phys. Status Solidi*, **47**, 11.

Tucker, M.O., Rose, A.P.G. and Burchell, T.D. (1996) *Carbon*, **34**, 581–602.

Vietzke, E. and Haasz, A.A. (1996) Chemical Erosion, in *Physical processes of the Interaction of Fusion Plasmas with Solids*, Hofer, W.O. and Roth, J., eds. Pub. Academic Press, Inc., San Diego, U.S.A., pp. 135–176.

Wei, G.C. and Robbins, JM (1985) *Ceramic Bulletin*, **64**(5), 691–699.

Winter, J. (1994) *Plasma Phys. Contr., Fus.*, **36**, B263.

6. Mechanical Properties

BRIAN RAND

Department of Materials, School of Process, Environmental and Materials Engineering, University of Leeds, Leeds, LS2 9JT, UK

6.1 HISTORICAL ASPECTS

The mechanical properties of graphites are important and often are critical in their engineering applications, but usually it is a unique combination of properties that leads to the selection of graphite for a particular application. These other characteristics are low friction coefficient, high electrical and thermal conductivity, resistance to corrosive chemicals, low neutron capture cross section, and high thermal stability which results in exceptional mechanical properties at high temperatures relative to other materials.

In this chapter the emphasis will be placed on materials that have significant graphitic character, although there will be some reference to the mechanical properties of the vast range of quasi-crystalline carbon materials that exists. The first uses of graphite probably involved natural graphite crystals and made use of the easy basal plane shear which led to the pre-eminence of graphite as a writing material. Thus, in a sense, the first class of mechanical properties to be exploited were the wear properties. In the 18th century, with the development of the metallurgical industry, graphite flakes were combined with clays to create so-called 'Plumbago' refractory materials for crucibles to contain molten steel. In these early composites, the thermal stability of graphite arising from the exceptionally strong, 'in-plane' covalent bonding was exploited. Such materials are still used today, but have evolved into a large array of oxide-graphite refractories with quite variable compositions developed to meet specific requirements in the metallurgical industry. In terms of mechanical properties it is the ability of the flakes to deform and to deflect cracks that is significant in enabling the graphite to greatly enhance the thermal shock resistance of the brittle ceramic phase.

The development of electricity and the metallurgical industry were the driving forces for the next major advances in graphite technology, which were the use of graphite as an electrode for the melting of steel and as the preferred material for conducting electricity to moving components. Here the combination of electrical conductivity and high temperature stability (relative strength, thermal shock resistance, dimensional stability) and of conductivity and low frictional coefficient were the significant property combinations. These applications, which continue to the present day, depended on the development of a satisfactory method of producing synthetic graphite components of large size and followed the introduction of the Acheson electric furnace in 1896. The understanding of the factors controlling the mechanical properties at this early stage was poor of course. The availability of synthetic graphite led to its use in other industries, notably in the chemical industry for containing corrosive fluids and especially as seals for pumps used in transporting such liquids, where again the frictional and wear characteristics were relevant. A wide variety of graphitic powders and colloidal dispersions had also been developed as wet and dry and high temperature lubricants as the middle of the 20th century approached.

The second world war brought major advances in the use of graphite materials for what we today call high technology applications and they together led to a burst of scientific activity to provide the basic understanding of structure-property relationships for this type of material. They were the use of graphite as a rocket nozzle material (thermomechanical properties) and as a moderator and reflector material in nuclear reactors. In the latter case, the key factors were that the material, whilst having a low neutron capture cross section could also be used as a structural component in the reactor. Improved types of synthetic graphites of very high purity and improved structure were required as the industry moved from military reactors to large-scale reactors for the production of electricity. It became necessary to predict the behaviour of the graphite core under irradiation conditions, in oxidising gases and over a time period of at least 25 years. It is largely to this need, for controlling the safety of reactors, that we owe our present understanding of graphite properties.

In parallel with the development of nuclear graphite, graphite for rocketry also advanced significantly in the post war years. The polycrystalline graphites available from the electrode and nuclear graphite producers were quite porous and this porosity has a controlling effect on the mechanical properties and on the oxidation characteristics, both factors critical to the lifetime and stability of the rocket nozzle. Rocket nose cones were also being constructed from graphite at this time. There followed a programme of research to produce relatively pore-free graphites by chemical vapour deposition, CVD, of hydrocarbon gases onto substrates. These materials are completely different in structure and usually show high degrees of preferred orientation of the graphite basal planes to the substrate surface. The development of methods of thermally treating such materials under pressure led to the production of highly oriented pyrolytic graphite, HOPG. This material has frequently been used in basic studies of the physical properties of graphite and has been of considerable significance in providing a baseline from which to develop the understanding of the porous polycrystalline graphites, which are in more widespread use.

The development of aerospace materials was also the spur to the invention of strong, stiff carbon fibres. These initially were not graphitic being produced from the

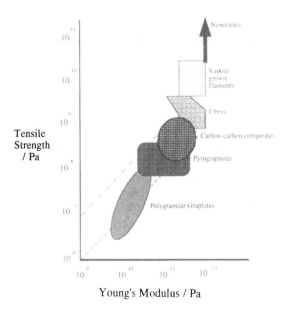

FIGURE 1. Variation of tensile strength and Young's modulus for the range of graphitic materials. Lines relate to relationships at .... 1% and – – – 0.1% strain.

non-graphitising precursor, polyacrylonitrile, PAN. However, the second generation of carbon fibres was based on the polyaromatic liquid crystalline precursor, known as the carbonaceous mesophase, which does produce graphitisable carbons and from this source, given the appropriate heat treatment, highly oriented fibres are now available which are substantially graphitic in character. Highly oriented graphitic filaments are also available from gas phase pyrolytic processes using metallic catalyst particles, vapour grown carbon fibres, VGCF. More recently, following the discovery of the Fullerene forms of carbon molecules, a wide range of single and multiple wall nanotubes have been produced experimentally in small quantities and methods are being devised to elucidate their mechanical and other properties. Since these are some of the most perfect forms of graphitic carbon available the information resulting from these studies will be of the greatest importance in furthering our understanding of this class of materials.

At the present time, a wide range of materials is available, which displays a tremendous variation in the degree of structural perfection of the lamellar structure, in the size distribution and orientation of the lamellar domains and in the presence, size and shape distribution of the variety of pores and cracks that exist. Hence, a wide range of mechanical properties is displayed by engineering graphites, as summarised in Figure 1. This review will attempt to illustrate as far as is known at the present time the effects of these various structural factors, the microstructure, on the mechanical properties of industrial significance. One important factor to consider is the effect on some of these properties of irradiation damage.

6.2 STRUCTURE

The crystal structure of graphite and the nature of the interatomic bonding are described in part 6.1.1, whilst in parts 6.3.1 and 6.3.2 the defective nature of synthetic carbons and how this structure develops during carbonisation is considered. In the synthetic polycrystalline materials, the property controlling factors are the degree of crystallographic perfection combined with the microstructural characteristics of the material. Hence, it is necessary to briefly review these features before examining their influence on properties.

6.2.1 Crystal Structure and Defects in Graphite

The defect and dislocation structure of graphites are reviewed by Thrower (1969), Kelly (1981), Amelinckx *et al.* (1966) and Thomas and Roscoe (1967). The dislocations in graphite have been the subject of considerable investigation. Both basal and non-basal dislocations have been observed, comprising edge and screw dislocations. Basal plane dislocations split into partials which bound a region of stacking faults in which there is rhombohedral stacking. Vacancy and interstitial loops are common non-basal dislocations. Twinning is common in crystalline graphite and can be readily induced by stress. The rhombohedral form of graphite is sometimes present in small amounts in natural flake crystals. However, the proportion is very substantially increased when the crystal is subject to extensive damage, such as takes place in grinding and severe deformation. Indeed, high-energy milling can destroy the crystal structure altogether resulting in a turbostratic form of graphite. The easy mobility of the dislocations at room temperature results in graphite being highly susceptible to stress induced deformation.

The basic types of point defect found in graphite are interstitials, vacancies and interstitial-vacancy pairs. The subject has been reviewed by Kelly (1981). The formation energies have been estimated as 7.0 eV for both types of defect, which is sufficiently large to ensure that only small equilibrium concentrations should exist at room temperature in highly crystalline samples. However, since many synthetic graphites are not highly crystalline then large concentrations of metastable defects can be expected. The migration energy for vacancies is considerably larger (≈ 3 eV) than that for interstitials (< 0.1 eV). Thus, interstitial atoms can readily diffuse along the 'a' directions of the crystals, even at room temperature. Both vacancies and interstitials tend to group into clusters, or loops, as depicted in Figure 2. These loops correspond to regions displaying stacking faults of the ABAABA type for vacancies and ABCAB type for interstitials and are examples of edge dislocations.

6.2.2 Structure of Synthetic Carbons and Graphites

Synthetic carbons and graphites are polycrystalline materials of variable degrees of crystallite perfection, crystallite dimensions, preferred orientation of the crystallites and porosity. It is necessary to understand the relative contributions of each of these factors to the mechanical properties and behaviour under irradiation, both separately and in combination. Unfortunately, this understanding is incomplete. The reader is referred to parts 6.3.1 and 6.3.2 for discussion of the crystallography of imperfect synthetic graphites.

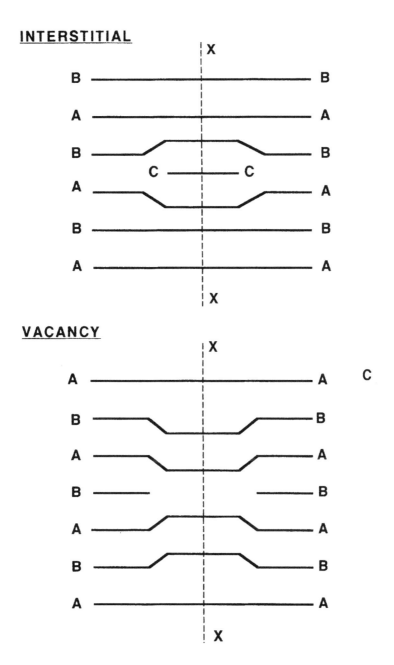

FIGURE 2. Models of interstitial and vacancy loops.

The most dense materials are produced by pyrolytic deposition and can be graphitised at temperatures approaching 3000°C. However, the density and preferred orientation of the deposits is strongly influenced by the deposition conditions. Densities can be as low as 1.4 g cm^{-3} in isotropic deposits or close to the crystal value (2.25 g cm^{-3}) for highly anisotropic materials, especially when severely annealed thermally or annealed under stress. The latter materials are highly oriented graphites (HOPG), with preferred orientation and properties close to those of the single crystal. They have thus been widely used to establish intrinsic structure-property relationships, since they can be produced in bulk, whereas property measurement on natural graphite crystals is experimentally very difficult. The production of pyrolytic and highly oriented pyrolytic graphite and their structure-property relationships are reviewed by respectively by Bokros (1969) and Moore (1973). In HOPG the c-axes are oriented parallel to the deposition direction, whereas there is random distribution of the a-axes within the plane perpendicular to this direction. However, the more usual pyrolytic deposits have variable degrees of orientation and the degree of graphitisation may vary significantly. Methods of characterising the preferred orientation have been reviewed by Ruland (1968).

Fibres can also be produced relatively free of porosity and those from mesophase pitch and catalytically grown from hydrocarbon vapours can also be graphitised. Thus, they may also show extremely high degrees of preferred orientation of the c-axes perpendicular to the fibre axes, when subjected to extremely high temperatures. At lower heat treatment temperatures they, of course, exhibit typical turbostratic non-graphitic structure. The degree of alignment of the lamellar structure along the fibre axis increases steadily as the a-direction x-ray coherence length, L_a, increases and the average interlayer spacing increases. Fibres produced from polyacrylonitrile, PAN and rayon also may have high degrees of orientation of the lamellae along the fibre axis, but are generally not graphitising and do not develop the degrees of structural perfection exhibited by mesophase based fibres. At the other extreme, completely isotropic carbon fibres can be produced from isotropic pitch and from various polymeric precursors.

The major engineering carbon and graphite materials have microstructures that are vastly different from the oriented forms. They are produced by bonding coke particles (grains) with pitch. The mixture is extruded, or otherwise moulded into shape, and the block is then carbonised at about 1000°C and may be further graphitised at temperatures above 2700°C to produce synthetic graphite. The product is porous and for special applications like nuclear graphite it is re-impregnated with molten pitch and heat treated again. A number of such cycles are required to achieve densities of the order of 1.9 g cm^{-3}. However, most industrial products are not so extensively densified and have considerable levels of porosity. The manufacturing process has been described by Hutcheon (1970). In order to distinguish this important class of materials from the pyrographites, which are also polycrystalline, they will be referred to in this chapter as polygranular carbons and graphites. Their crystallinity is dependent upon the graphitisability of the cokes and pitches used and the heat treatment conditions. Some coke particles have within them a preferred orientation of the layer planes and break up into elongated grains during grinding. These 'needle' coke grains tend to align with the extrusion axis and result in a measure of preferred orientation of the crystallite c-axes perpendicular to this axis. The preferred orientation is, however, quite small compared with the pyrolytic deposits and fibres. By appropriate selection of the coke microstructure,

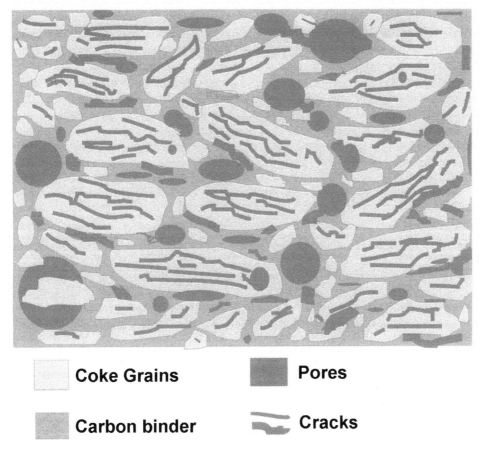

FIGURE 3. Schematic representation of the structure of polygranular graphite to illustrate the presence of pores and cracks within the coke grains as well as pores within the binder phase.

however, it is possible to produce synthetic graphites that have random c-axis orientation of the crystallites, a characteristic particularly desirable for nuclear applications (Kelly, 1981). In addition to the initial crystallinity of the coke and its graphitisability (part 6.3.1), the grain size and pore size distribution can vary enormously from product to product and thus polygranular carbons and graphites of this type can show a wide variation in properties. The porosity arises from the following sources:

1. pores and cracks within the coke particle (grain)
2. pores due to incomplete filling of the intergranular space by the pitch
3. pores arising from gas evolution in the binder phase when in the fluid state

Figure 3 illustrates these structural features, schematically. A very important microstructural feature in synthetic materials of this type is the presence of many lamellar cracks which tend to align with the layer planes. During carbonisation and graphitisation, the true solid density changes from 1.35 g cm^{-3} in the pitch to 2.2 g cm^{-3} in the graphite. Also because of the high c-axis expansion coefficient (approximately 27×10^{-6} at room temperature) a large contraction takes place on cooling from the production temperature. The resultant stresses are partly relieved by the formation of these cracks. Those due to thermal contraction are known as 'Mrozowski' cracks (Mrozowski, 1956) and are of critical importance in controlling the thermal expansion characteristics of the bulk graphite. They also have a significant role to play in the mechanical properties. The 'Mrozowski' cracks are reversible in the sense that they can close on reheating and so change the response of the material to imposed stresses at elevated temperatures, which influences the values of the major properties and can lead to an usual temperature dependence of mechanical properties as described below.

6.2.3 *Effect of Irradiation on the Structure of Graphite*

The use of graphite as a moderator and reflector material has led to an extensive study of the effects of irradiation on its structure and properties. The interstitial and vacancy defects and their clusters are the prime defects produced by irradiation damage. Most of the studies in the literature concern the effects of fast neutrons, but there has also been significant investigation of high-energy electron irradiation and that by other particles. The subject has been extensively reviewed by Kelly (1981) and is discussed by Burchell in Chapter 5. Major contributions to this subject were made by Reynolds and Thrower (1965), Reynolds (1966), Simmons (1965), Thrower (1969).

The extent of damage is very sensitive to the irradiation temperature. This is because the displaced carbon atoms can recombine with vacancies, the vacancies and interstitials can nucleate, homogeneously or heterogeneously, to form loops, or be annihilated at defects in the structure, e.g. at crystallite boundaries, and these effects are influenced by temperature. Since the interstitial atoms are more mobile than the vacancies, they are assumed to dominate the behaviour at all but the highest temperatures. High-energy irradiation alters the physical properties of graphite, but the changes can be largely reversed by thermal annealing. Changes in lattice parameters are also observed, accompanied by anisotropic dimensional changes in the crystals. The effects of these structural changes on the bulk dimensions depend upon the microstructure of the graphite. The understanding of these processes owes much to careful investigation using well characterised pyrolytic graphites and HOPG, which are almost pore free. Polycrystalline materials have a characteristic limit to the crystallite size. At high temperatures, as the interloop spacing increases, it becomes commensurate with this crystallite size at a certain temperature. At this point the population of heterogeneous nucleation sites, at boundaries, becomes equal to that of homogeneously nucleated clusters and above this critical temperature, nucleation of interstitial loops becomes predominantly heterogeneous resulting in a dramatic increase in distortion rate. The critical temperature depends on the crystallite dimensions of the graphite. Also, at high radiation doses and at temperatures above 300°C, the movement of vacancies to crystallite boundaries leads to the generation of porosity and there is an additional increase in bulk volume due to this effect.

Polygranular nuclear graphites on irradiation show many similar characteristics to the dense anisotropic materials and some significant differences. Swelling into the pores can

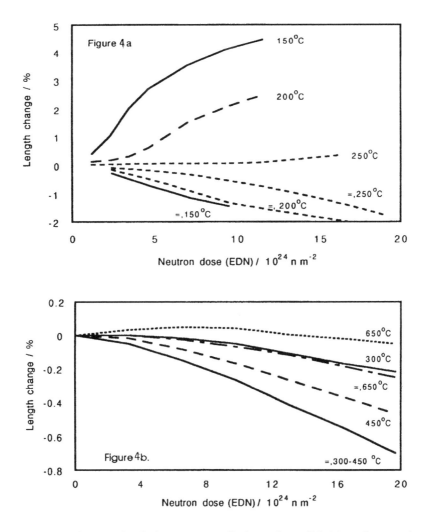

FIGURE 4. Typical dimensional changes perpendicular and parallel (=) to the extrusion axes of nuclear graphite, (PGA) after neutron irradiation, (a) at temperatures, 150–250°C and (b) at 300–650°C.

lead to densification of the structure and these effects are not always reversible. The anisotropic polygranular graphites show anisotropic dimensional changes at low irradiation temperatures, contracting in the direction parallel to the extrusion axis and expanding in the perpendicular direction, as shown in Figure 4. However, in the temperature range 400–800°C, contraction in both principal directions has been observed. This is not fully understood, but appears to be associated with the expansion of crystallites into pores and the local densification of the material. The grossly defective nature of the polygranular reactor

TABLE 1. Elastic constants for single crystal graphite (taken from Kelly, 1981).

Elastic moduli / GPa		Elastic compliance / $10^{-12} Pa^{-1}$	
c_{11}	1060 ± 20	s_{11}	0.98 ± 0.03
c_{12}	180 ± 20	s_{12}	-0.16 ± 0.06
c_{13}	15 ± 5	s_{13}	-0.33 ± 0.08
c_{33}	36.5 ± 1	s_{33}	27.5 ± 1.0
c_{44}	4.5 ± 0.5	s_{44}	240 ± 30

graphites means that they provide many more heterogeneous nucleation sites for mobile interstitials and for the removal of vacancies when they become mobile. Consequently, at temperatures higher than $800°C$ with high neutron doses, the situation is reversed and a dramatic volumetric expansion is observed due to the generation of significant porosity. Ultimately the graphite may disintegrate.

6.3 MECHANICAL PROPERTIES

6.3.1 Elastic Moduli

Single crystal values

Five independent elastic constants are required to define the stress-strain relationships for the hexagonal graphite crystal. Kelly (1981) has reviewed the various measurements made on single crystals and highly oriented graphites (Moore, 1973) and gives the values in Table 1 as the most reliable current estimates. The Young's modulus parallel to the basal planes, E_a, to the hexagonal axis, E_c, and the shear modulus parallel to the basal planes, G, are related to the compliances, s_{ij}, by:

$$E_a = s_{11}^{-1} = 1020 \, GPa; \quad E_c = s_{33}^{-1} = 36.4 \, GPa; \quad G = s_{44}^{-1} = c_{44} = 4.5 \, GPa \quad (2.2.1)$$

The value of E_a, 1020 GPa, reflects the magnitude of the 'in-plane', covalent sp^2 C-C bonds, which are the strongest of any known material. The very weak interlayer, van der Waals type, bonding results in exceedingly low values of E_c. G. Baker and Kelly (1964) found a value for c_{44} significantly lower than that in Table 1 for natural graphite crystals, which was attributed to the presence of glissile basal plane dislocations. Irradiation increased the value by more than an order of magnitude and similar increases in this parameter have been reported by other workers (see Kelly, 1981). Using the data of Table 1, Kelly calculated the reciprocal Young's modulus as a function of the angle, ϕ, with respect to the hexagonal axis, from the equation

$$E^{-1} = s_{11}(1 - \gamma^2)^2 + s_{33}\gamma^4 + (2s_{13} + s_{44})\gamma^2(1 - \gamma^2) \quad (2.2.2)$$

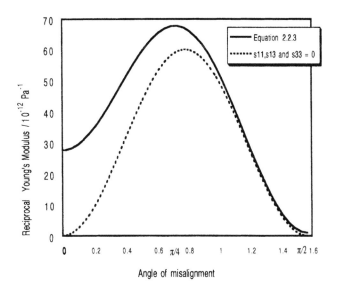

FIGURE 5. Variation of reciprocal Young's modulus with the angle of misalignment between the hexagonal and tensile axes.

where ($\gamma = 1 - \cos \phi$). The result is displayed in Figure 5, compared with an approximation in which all compliances in the equation other than s_{44} are taken as zero. Over a wide range of angles agreement is good, demonstrating that the s_{44} compliance is the dominant influence on the elastic and other mechanical properties of polycrystalline materials with crystallites misaligned with respect to the tensile axis.

The elastic constants decrease with increasing temperature. The nature of the bonding in the principal directions and the temperature dependence of the lattice parameters suggests that the temperature dependence of the c_{33} and c_{44} constants should be larger than the c_{11} and this is reported to be the case. A reduction of 50% in c_{33} is reported over the temperature range to 2000°C. Measurements of the shear strength of graphite crystals and pyrolytic graphite give very low values of 0.03 and 0.57 MPa respectively, consistent with the modulus values in Table 1.

Polycrystalline materials-general

In considering polycrystalline synthetic materials, we should first look at relatively pore free materials such as the pyrolytic graphites and high temperature fibres and filaments from graphitisable precursors. Fibres and pyrolytic deposits are more or less anisotropic as are most pitch-coke type graphites, which have been extruded. Thus, a similar set of elastic moduli and compliances, now defined relative to the axis of symmetry, are appropriate to characterise the elastic properties. Price (1965) and Goggin and Reynolds (1967) modelled the elastic moduli of polycrystalline materials of zero porosity on the basis of constant stress

(Reuss) and constant strain (Voigt) models modified to include orientation parameters. The constant stress model has found favour in interpreting the elastic properties of highly oriented carbon fibres. The model is also applicable to the range of pyrolytic graphites. It has been pointed out (Bokros and Price, 1966; Jenkins, 1973), however, that one problem in applying this model quite widely (especially to carbon materials) is in the variation in the crystallite elastic parameters, especially c_{44} and c_{33}, due to imperfections in structure. In the defective carbons and graphites, the c_{33} and c_{44} constants are expected to be lower than the single crystal values because of the larger interlayer spacing (part 6.3.1) and therefore the lower interlayer bond strength. However, The shear moduli should be considerably higher than for perfect graphite. Thus, the application to synthetic graphites is limited. Whilst the average lattice spacings and preferred orientation parameters are relatively easy to measure, the changes in the intrinsic crystallite properties are difficult to predict. Jenkins also considers that the mechanical properties of carbons and graphites are strongly influenced by the boundary restraints between crystallites, which involve C-C bonds and therefore can be very strong. Thus, where boundary restraints dominate, the effects of the intrinsic crystallite properties are not so important. This view is undoubtedly of importance for materials with very small crystallite dimensions and particularly for pregraphitic carbons in which the crystallites are very highly defective.

Another difficulty is that the elastic constants of polycrystalline graphites are actually strain dependent. This is discussed in some detail by Jenkins (1973). Stress-strain curves in compression and tension are non-linear on loading, showing a decrease in modulus with increasing stress, hysteresis on unloading and a 'permanent set' at zero stress (Figure 6). The behaviour is attributed to the ease of dislocation movement at room temperature, which allows a dramatic increase in the dislocation density when stressed. Subsequent loading-unloading cycles show a continuation of this process and an increase in the magnitude of the 'permanent set', as illustrated in Figure 6. The effect can be thermally annealed and the original properties recovered. Similarly, prestressing in compression also lowers the subsequently measured elastic modulus in tension. These effects are not so pronounced for carbon materials, but some hysteresis can still be observed.

Application of the constant stress model to an isotropic graphite with zero porosity comprising perfect crystallites predicts a Young's modulus of about 30 GPa. This low value is primarily due to the dominance of the s_{33} and s_{44} compliances. The constant stress model is probably the most appropriate of the two extreme mixture rule models, since it leads to the dominant influence of the shear modulus and low values of the constants, as is observed experimentally (see below). Jenkins (1973), on the other hand, disagrees with this view and calculates much lower modulus values for pore-free material from the constant stress model. However, he uses much lower values for c_{44} than are given in Table 1. Poisson's ratio is found to be different in the directions parallel and perpendicular to the extrusion axis for anisotropic graphites. The values are very low (0.14 in isotropic graphite) and also change (increase) with strain.

Highly Anisotropic Materials

The most perfect filamentary type of graphitic structure is probably that of a single wall nanotube. The Young's modulus of such a structure should be close to the theoretical

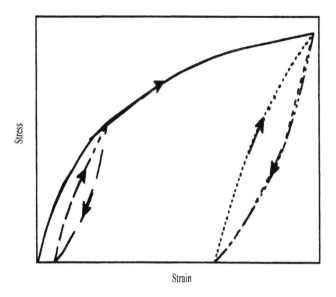

FIGURE 6. Schematic representation of a typical stress-strain curve for a polygranular graphite loaded in compression, showing hysteresis effects on unloading/loading and permanent 'set'.

maximum. Currently, attempts are being made to obtain reliable experimental values for such structures. The next most perfect structures arise from vapour grown filaments after high temperature graphitisation. Dresselhaus *et al.* (1988) discuss the elastic parameters of carbon fibres of different defect structure. Bacon (1960) was the first to produce and characterise such materials, obtaining values of 800 GPa. Graphitised catalytically grown VGCFs have also exhibited values of 760 GPa. The range of mesophase fibres gives rise to materials of systematically varying crystallinity and orientation when progressively graphitised. The most highly oriented can show Young's moduli of 900 GPa. The alignment of layer planes along the fibre axis is high in the high performance fibres from PAN and from mesophase pitch. Hence, in these materials the elastic modulus increases with heat treatment temperature as L_a increases and the average degree of orientation of planes with respect to the fibre axis correspondingly increases. The constant stress model is widely used to account for the elastic properties of carbon fibres, but Ruland (1968) proposed an 'elastic unwrinkling' model to account for the non-linear stress-strain behaviour of most fibres, where the modulus increases with strain. The fibres were considered to comprise an array of fibrils oriented along the fibre axis, but exhibiting some undulation. This undulation was considered to be gradually straightened by the applied stress, leading to the observed behaviour. The relative merits of this and the 'uniform stress' model have been reviewed by Dresselhaus *et al.* (1988). In isotropic carbon fibres the elastic modulus is very low, around 40 GPa. This value is close to, but a little higher than, the predicted value for a pore-free isotropic graphite. Boundary restraint effects may account for the higher value. The author is not aware of the existence of any isotropic graphitic fibres.

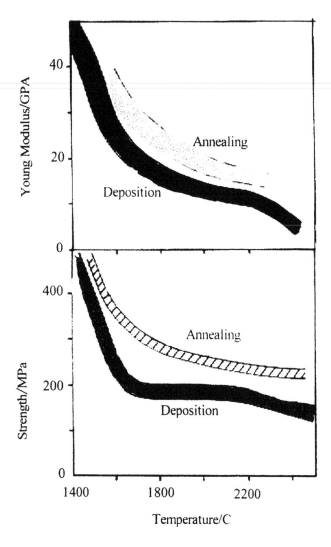

FIGURE 7. Schematic representation of the effect of deposition and annealing temperature on the flexural strength and Young modulus of pyrolytically deposited carbon.

The Young's moduli of pyrolytic graphites vary quite widely from the near single crystal values quoted above, obtained on HOPG, to values as low as 15 GPa, the actual values being dependent upon crystallite orientation and perfection (see Bokros, 1969 and Kotlensky, 1973 for reviews). The modulus increases with decreasing deposition temperature, as shown in Figure 7. The higher moduli from the low heat treatment temperature materials reflect the non-graphitic character and possibly show influences from strong boundary restraint effects. When materials deposited at relatively low temperatures are subsequently annealed at higher temperature the elastic moduli decrease in line with the effect of deposition temperature indicating the dependence on graphitic character. The low values in the more graphitic materials arise from the increasing influence of the shear modulus in misaligned lamellar regions.

Polygranular materials

In addition to the effect of preferred orientation of crystallites of varying size and perfection, the presence of porosity of wide size and shape distribution poses another more complex constraint on the interpretation of the elastic properties of bulk polygranular carbons and graphites. As discussed above, the anisotropy varies with the character of the coke grain employed and the processing conditions, which control the extent of alignment of the elongated grains during extrusion. However, the anisotropy ratio for Young's modulus rarely exceeds a value of three. The measured elastic moduli of this class vary from about 5 to 20 GPa, but most commercial materials have moduli lower than about 15 GPa. The experimental values are lowered by the influence of porosity, as discussed below. Generally, it is observed that the larger the size of the coke grain, the lower is the modulus. This is partly due to the influence of the shape of cracks that are associated with the coke grains.

Kelly (1981) considers that there are two main effects controlling the stress-strain behaviour of the range of polygranular graphites,

- the magnitude of the constant c_{44}, which controls the reaction of the crystallites to imposed stresses, and
- the crack morphology and distribution which changes the stress distribution in the body and hence the stresses experienced by the individual crystallites.

To this we should add the effect of boundary restraints, introduced by Jenkins, which are of major significance in the non-graphitic carbons. The porosity in polygranular pitch-coke carbons and graphites has a significant effect on properties. The larger (more equiaxed) pores act by generally lowering the Young's modulus in all directions, but the microfissures that influence the thermal expansion characteristics (eg the Mrozowski cracks) have a more dominant influence. Usually, empirical models of the type

$$E = E_0(1 - aP) \tag{2.2.3}$$

or

$$E = E_0 \exp(-bP) \tag{2.2.4}$$

have been used to describe the effect, where E_0 is the modulus at zero porosity, P is the porosity and a and b are empirical constants. The value of b in equation (2.2.4) has been shown to increase with the aspect ratio of the fissures and pores (Buch, 1983). For example, Cost et al. (1968) found that a value of b of 3.5 applied to a fine-grained (25 μm) isotropic graphite. Pickup et al. (1986) found that a value of $b \geq 7$ described the data for an oxidised graphite, which displayed slit shaped pores of high aspect ratio. An attempt has been made to

B. RAND

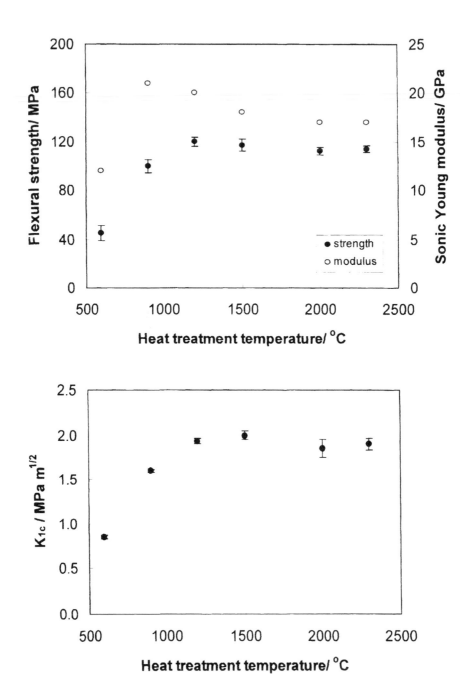

FIGURE 8. Variation of Sonic Young modulus, flexural strength and K_{1c} with heat treatment for carbons prepared from a mesophase microbeads precursor (Ting, 1995).

characterise the contributions of pores of different shape to the porosity effect by a modified form of Eq. (2.2.3)

$$E = E_0[1 - a(P - p)][1 - cp]/[1 - mcp] \qquad (2.2.5)$$

where p is the porosity in microfissures and c and m are constants. c is greater than a to account for the greater influence of the microfissures due to their higher aspect ratio.

The elastic properties of polygranular carbons have not been studied so extensively investigated as those of graphites. A number of studies (Andrew et al., 1960; Honda et al., 1966; Andrew and Distante, 1961) have shown that Young's modulus of carbons increases with heat treatment temperature up to about 1200°C after which there is a gradual decrease. Figure 8 shows some data for a polygranular 'sintered' carbon. The rise in E up to 1200°C has been attributed to the change from an essentially molecular structure to a more cross-linked para-crystalline carbon. This is associated with an increase in true density and the loss of hydrogen, which resides at the molecular boundaries. The fall from the maximum as the heat treatment temperature increases is almost certainly associated with the gradual elimination of defects which act as restraints to shear deformation, i.e. reduction in boundary restraint (Jenkins, 1973).

Temperature dependence

As a result of the porosity, particularly the microfissures, the Young's moduli of graphites also show a more complicated dependence on measurement temperature than expected from the single crystal behaviour. At temperatures below about 300°C, the normal decrease with increasing temperature is observed. But, above this value E increases to an extent that is dependent upon the microstructure of the graphite material, but which can reach values of 40% or so. Above about 2000°C the modulus tends to decrease again. However, the macroporosity does not affect the magnitude of the coefficient of thermal expansion, nor the temperature dependence of E. The increase in E with temperature is due very largely to the role of the microcracks which are aligned with the lamellar structure (the 'Mrozowski' cracks). These can accommodate the large c-axis expansion of the crystallites and thereby reduce the value of p, which changes the stress distribution within the polycrystalline aggregate. Thus, both E_0, c and p in equation 2.2.5 should be dependent upon temperature. Since the microcrack structure, which dictates c and p is determined by the processing conditions, the effect is strongly microstructurally dependent. The large increase in E with temperature has led Jenkins to suggest that the microfissures contribute 50% to the room temperature compliance of graphites of this type.

6.3.2 Strength and Fracture Behaviour

Highly anisotropic materials

Carbon and graphite materials fall into the general class of brittle materials, despite the evidence for dislocation movement cited in the previous section. The fracture stress, σ, is therefore interpreted in terms of the Griffith equation (2.2.6) in which the strength is controlled by the size of the critical stress raising flaw in the material, c, as well as the effective fracture energy, γ,

$$\sigma = (2E\gamma/\pi c)^{1/2} \qquad\qquad (2.2.6)$$

The most perfect graphitic material is probably a single-walled nanotube. Strength measurements of such materials are experimentally very difficult to obtain, but might be expected to achieve values approaching 100 GPa for a modulus of 1000 GPa, on the basis of the theoretical predictions of ideal strength of crystals arising from the Orowan-Polanyi equation

$$\sigma = (E\gamma/a)^{1/2} \qquad\qquad (2.2.7)$$

where a is the interatomic distance between the two crystallite planes which are to be separated by the applied stress. In fact, the strongest graphitic forms measured so far are the graphitic filaments produced by Bacon (1960), that displayed a strength of 20 GPa and a strain to failure of 2.5%. Continuous carbon and graphite fibres from mesophase pitch show significantly lower failure strains indicating highly defective structures. The Young's modulus increases dramatically with heat treatment temperature as the graphitic structure develops. The strength also increases but to a much smaller extent because the failure strains also fall from just under 2 to around 0.5%. The strength of fibres is undoubtedly controlled by flaws, identified as impurity particles and well graphitised, misaligned crystallites at the fibre surfaces (perhaps produced by catalytic graphitisation via the impurity particles, Johnson, 1987). However, the behaviour is not fully consistent with the predictions of the Griffith equation (see review by Dresselhaus *et al.*, 1988) which led Reynolds and Sharp (1974) and later Reynolds and Moreton (1980) to propose a mechanism in which misoriented crystallites have a strong influence on the fracture stress. Basal plane shear in the misoriented crystallite is restricted by neighbouring crystallites, which impose a compressive stress. The result is that rupture across the crystallites occurs at a stress, which lies in between that for tensile rupture of perfectly aligned crystallites and that for basal plane shear. Pitch-based fibres show distinct transverse textures, which are thought to influence the fracture behaviour. Some evidence for this was provided by Guigon *et al.* (1984) for fibres displaying a folded ribbon structure. It was considered that cracks were arrested at the folds and the strength increased as the ratio of radius of curvature of folds to crystallite width decreased. Control of transverse texture is an important area of technological development of carbon fibres currently. Some VGCFs display a microstructure comprising a 'nest' of graphitic cylinders. These tend to exhibit a characteristic 'sword in sheath' fracture mechanism. This arises from the propagation of a crack initiated in the outer layers being arrested and deflected at a basal plane as it moves into the fibre. The load transferred to the remaining core initiates fracture at a weak point in a different plane and so on leading to the progressive 'pull-out' of the core from the outer sheath and consequent energy absorption.

The strengths of pyrocarbon deposits vary markedly with deposition conditions, especially the temperature, Figure 7. The dependence is very similar to that of the Young's modulus, parallel to the deposition plane. The magnitude of tensile or flexural strength is lower than that for carbon fibres, but at relatively low deposition temperatures it may be quite high, in the region of 400 MPa. Annealing such low temperature deposits at higher temperatures reduces the strength, again in accord with the change in modulus. The increasing influence of shear deformation is the usual explanation. At high deposition/anneal temperatures, when the materials are more graphitic, the strengths fall to 100–200 MPa. Although much lower than the theoretical values and significantly lower than those of fibrous forms, these strengths are considerably greater than those displayed by the highly porous polygranular materials discussed below, largely because of the much lower porosity.

Polygranular materials

The fracture of polycrystalline bulk graphite is much more complicated as a result of its more complex microstructure. The strengths are low varying from about 5 MPa to a maximum of about 120 MPa for ultra fine-grained isotropic graphite. However, the majority of graphites used for engineering purposes in the metallurgical and nuclear industries have strengths at the lower end of this spectrum (5–30 MPa), mainly because they have large grain size and significant porosity. The strength and fracture mechanisms of grades used for nuclear purposes have been extensively studied, but even so there are still many uncertainties. Very generally, the strength tends to increase when the elastic modulus increases but the exact relationship varies as the crack networks within different graphites vary. There is strong dependence on porosity of the form of equation (2.2.3), with the constant a taking values of 4–7. The strength is anisotropic for materials with anisotropic structure. Impregnation increases the density, modulus and strength, whilst radiolytic and thermal oxidation reduce these parameters. The consensus from many studies is that fracture is initiated in the grains. An empirical relationship seems to exist in which strength varies as $d^{-1/2}$, where d is the average grain size. This expression resembles the Griffith equation and in some studies the application of this equation to fracture results has given values for the critical flaw size that are similar to the size of the coke grains. There is also some evidence (Brocklehurst, 1977) that the strength changes that arise from changes in the intergranular regions, i.e. from impregnation or oxidation, occur such that the strain energy density per unit volume, σ^2/E, in the grains, is constant. This implies that the ratio γ/c is constant.

It is well established that basal plane shear and slip lead to microcracking at subcritical stresses, as identified by acoustic emission studies and the hysteresis effects discussed earlier. The microcracks, ahead of the main crack, form a kind of process zone. They link and connect existing pores together eventually forming a main crack which propagates through the sample to promote catastrophic failure. Direct microscopical evidence has confirmed these processes. The shrinkage cracks running parallel to the layer planes within the coke grains are identified as the flaws in the structure which initiate fracture, with many cracks propagating outwards into the surrounding structure. These cracks may be deflected initially at the grain boundaries but then propagate via the binder phase and binder-coke interfaces to other intragranular, lamellar cracks and pores which are linked until a crack of critical dimensions is formed leading to the catastrophic failure. Rose and Tucker (1982) have modelled this behaviour, dividing the structure into blocks of uniform size, characteristic of the grains, which contain randomly oriented cleavage planes. These fail when the resolved stress on the plane exceeds a critical value. The model has the advantage of incorporating microstructural features, which can be adjusted to conform to measured structural features.

Although the fracture behaviour can be described as brittle, the sub-critical plastic deformation and microcracking events are energy absorbing processes that increase the apparent work of fracture. The surface energy of non-basal planes is around 4–5 Jm^{-2} and that of the basal plane is almost an order of magnitude less than this. However, as a result of the above processes, the measured values of work of fracture at room temperature are of the order of 100 Jm^{-2} for coarse-grained electrode and nuclear graphites. The fracture toughness for mode 1 fracture of brittle materials is commonly assessed via the measurement of the parameters K_{1c}, the critical stress intensity factor, and G_{1c}, the critical energy release rate. ($G_{1c} = 2\gamma$). These values are usually determined from fracture measurements using specimens into which well defined cracks of known size, c, have been introduced, utilising the Griffith equation in the form

$$\sigma = (K_{1c}/Yc^{1/2})$$
(2.2.8)

where Y is a geometrical factor, and

$$K_{1c} = (EG_{1c})^{1/2}$$
(2.2.9)

K_{1c} values are typically of the order of 0.5–1.5 MPa m$^{-1/2}$. These values are low when compared to other brittle materials, even ceramics. They reflect the low value of Young's modulus for these porous materials. The critical energy release rate values, G_{1c}, however, are of the same order as the work of fracture, i.e.100 Jm^{-2}, which is very much larger than the values obtained with un-reinforced ceramic materials (typically 10 Jm^{-2}). They also reflect the sub-critical fracture events and contributions from plastic deformation at crack tips. It is significant that at temperatures above 2000 C, when plastic flow at crack tips becomes very much easier, the work of fracture shows a very sharp increase.

Related to the toughness is the sensitivity of the material to notches, which is also important for engineering design purposes. This relates to the internal structure of the material and depends on whether the imposed notch creates a significant defect compared to the inherent defects already within the test material. Large grain graphites are found to be relatively insensitive to notches, which can be understood from the pre-existing crack structure within the coke grains. This will especially be the case when small specimens of coarse-grained graphites are tested. Notch sensitivity increases with decreasing average grain size. Slow crack growth influenced by plasticity at the crack tips takes place when polygranular graphites are subjected to cyclic stresses well below the static failure stress resulting in typical fatigue failure.

The flaw dominated failure of graphites results in the values of strength being a function of the volume of material tested, since the larger the volume the greater the probability of finding a critical flaw. The statistical variation in the strength of brittle materials is usually described by the Weibull equation, the basis of which is discussed in standard texts on fracture mechanics. The equation can be expressed in the form

$$F(\sigma, V) = 1 - \exp[-B(\sigma, V)]$$
(2.2.10)

where

$$B = \int [\sigma - \sigma_u)/\sigma_0]^m dV$$
(2.2.11)

V is the volume, σ the strength of the volume element, dV, σ_u is the minimum strength, σ_0 is a normalising factor. The parameter m describes the spread of the distribution, large values reflecting a narrow distribution typical of material with a narrow distribution of flaw sizes. The equation is in standard usage to describe the dependence of fibre strength on fibre test length. A dependence on diameter is well established for vapour grown fibres, which can have varying diameters, but not so for carbon fibres which tend to have fairly uniform diameters. In coarse polygranular graphites, the largest grain size may be of the order of millimetres and to obtain strength results characteristic of the microstructure, specimens with dimensions of the order of centimetres should be tested. However, in very many studies this has not been the case (Kelly, 1981), with the result that an anomalous behaviour is observed in which the strength increases with sample volume. Nevertheless with samples of similar volume or of appropriate size the Weibull model is a reasonable statistical approach to adopt. It has been reviewed by Tucker *et al.* (1986) and compared with other treatments of graphite fracture. Recently, Burchell (1996) developed a probabilistic approach to fracture, which also incorporates a microstructural approach similar to the Rose-Tucker model. Its general applicability remains to be established.

In accord with their higher elastic constants, non-graphitic carbons tend to have higher strengths than graphites. The variation of strength, fracture toughness and critical energy release with heat treatment temperature is broadly similar to that of the Young's modulus, as described above, Figure 8 illustrates the strength and K_{1c} variation for typical directly bonded ('sintered') carbon system derived from mesophase microbeads.

Temperature dependence of strength

The effect of increasing temperature is generally to increase the strength of the graphite to an extent much greater than that shown by the elastic modulus. This is mainly due to the closure of microcracks, which in addition to changing the stress distribution also reduces the concentration of defects, which can link to form the critical flaw. Diefendorf (1959) identified a stress corrosion effect at room temperature from adsorbed gases (particularly water) acting at crack tips to lower the surface energy. Desorption of adsorbed species at elevated temperature should reduce the effect on strength. The effect is consistent with a dependence of strength on the strain rate at low temperature, which was also confirmed by Diefendorf. At high strain rates, the strength is increased as the effect stress corrosion, which is time dependent, is reduced. Interestingly, a reverse strain rate dependence is exhibited when measurements are made at temperatures above 2000°C. This is due to the ductility displayed in this temperature regime which reduces the stress concentration at crack tips. At low strain rates only a few cracks are evident, extending to form the main fracture-controlling feature. In contrast, at the fast strain rates the relaxation time for plastic deformation is too long and the fracture processes are then similar to those at low temperatures, with much microcracking prior to fracture.

Creep

In addition to the low temperature transient creep associated with the permanent set, poly-granular graphites also show thermally activated creep both at temperatures approaching the previous heat treatment temperature and at temperatures above about 2200°C. In this latter region the creep rate increases continually with temperature. The strain shows a power law time dependence as does the rate at constant strain, i.e.

$$\varepsilon - \varepsilon_0 = A\sigma^P t^n \tag{2.2.12}$$

The creep appears to show two different activated processes. It is partially recoverable on removal of the stress (by about 20–35%), the recovery showing a logarithmic time dependence. A theory due to Green *et al.* (1970) explains these phenomena in terms of the opening a many small cracks which nucleate dislocations at their tips. These dislocations then move away from the tips by a climb process.

The process of stress annealing of pyrolytic graphite to produce HOPG involves plastic deformation. The characteristic growth cones are smoothed out and dewrinkling of the lamellar structure ultimately takes place. Elongational strains of 10–15% are reported.

Creep strains can also be induced at low temperatures by irradiation of stressed graphites. The strain is recoverable by irradiating with zero stress and there is no pronounced temperature dependence of the strain rate. The mechanism has not been fully worked out, but is thought to be due to the movement of interstitials, rather than the vacancies, which participate in high temperature creep.

Effect of irradiation

Irradiation has profound influences on the elastic moduli. The elastic constants c_{44} and c_{33} are changed in pyrolytic and single crystal graphite, when irradiated at low temperatures. The magnitude of the increase in c_{44} varies from one study to another but it can be as large as an order of magnitude. The effect is attributed to dislocation pinning by the interstitial defects and thus can be annealed. The relative magnitude of the pinning mechanism has been called into question by Jenkins who contends that the major effect is the linking of adjacent planes by the interstitial groups which increases the shear modulus. The majority of researchers appear to favour the dislocation mechanism, at least for highly graphitic materials, but the issue has not been satisfactorily resolved. The effects of irradiation on the polygranular graphites are more complex. In the single crystal, only c_{44} is altered to any significant degree, but since this parameter dominates the elastic constants in all principal directions in the polygranular block, they are all changed. This is further evidence in favour of the constant stress model being the best current approximation, since it predicts c_{44} dominance of properties. As a result of the change in c_{44}, the stress strain curves tend to become more linear and, in polygranular graphites, there is evidence of an increase in the work of fracture. The pinning of dislocations at low doses rapidly increases the elastic modulus of polygranular graphite at a rate that is dependent on the concentration of interstitials and hence on temperature. At low temperatures increases in E by factors of 2–2.5 are possible. At low temperatures the modulus reaches a peak value. The subsequent decline has been attributed to the lowering in the c_{44} value as the crystallites expand in the c direction. At higher temperatures, however, this peak is not present and the modulus tends to level out. In this region, the interstitial concentration is lower due to the nucleation of loops, so the modulus increase is also lower. The c-axis expansion does not reach levels where a significant reduction in c_{44} takes place so there is no fall in the modulus. As the irradiation dose increases, the modulus begins to rise again at all temperatures due to the closing of pores by the crystallite expansion. Eventually, however, the heterogeneous nucleation processes intervene, creating new pores and the modulus falls drastically. The strength changes tend to follow those in the modulus, but are generally smaller. There is a tendency for the changes in strength and modulus to follow those in volume, as expected from the above model. Figure 9 shows schematically the variations expected. The effects of irradiation on mechanical properties are discussed in detail by Kelly (1981) and Brocklehurst (1977). Often σ/σ_0 is proportional to $(E/E_0)^{1/2}$ in the swelling region, which would indicate no significant change in the fracture initiating flaws. Beyond the maximum, however, the creation of porosity tends to lead to a linear relationship between σ and E.

6.4 CARBON/CARBON COMPOSITES

The low strengths and moduli of bulk carbons and graphites are limiting factors in their industrial applications. The relatively poor properties have led to the development of a wide range of carbon fibre reinforced carbon matrix composites which utilise the high stiffness and strength of fibres. Table 2 lists some mechanical properties typical of this class of materials. Clearly the improvement over the porous polygranular materials is substantial, although this achieved at a dramatic increase in production costs. It is evident that extremely high moduli and strengths are possible, but even for unidirectional materials a very wide variation in these values exists. The major factors influencing these properties are:

- the fibre structure
- the matrix structure
- the strength of the fibre-matrix bond
- the fibre architecture (volume fraction and arrangement of the fibres)
- the porosity, ie the extent to which the fibre preform is filled with the carbon matrix.

It is beyond the scope of this chapter to discuss the mechanical properties of this class of materials in detail. Some key features only will be outlined. Accounts of the fabrication technology of C/C composites and their structures and general properties are given in a number of reviews (Thomas, 1993; Buckley and Edie, 1992; Savage, 1993). The reader is referred to the above references for reviews of mechanical properties, by McEnaney and Mays (1993), McCartney (1993) and Kibler (1992).

In contrast to the situation with polymer-matrix composites, there has been very little modelling of the properties of C/C composites. One reason is that the matrix structure is different from that in bulk polygranular materials. Hence, there are usually no independent measures of the matrix modulus and strength available. This arises because the matrix microstructure is strongly influenced by the fibres, there being a tendency for the layer planes to align with the fibre surfaces to an extent that varies with the particular fibre-matrix couple and in particular with the nature of the fibre-matrix bond. Even normally non-graphitising precursors like phenolic resins can give rise to quite highly aligned interphases near the fibre surface, but revert to their typical isotropic structure further out into the bulk matrix

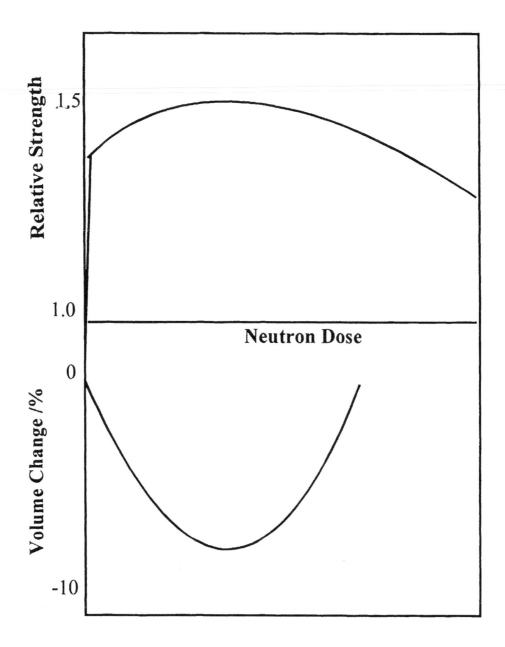

FIGURE 9. Typical variation in relative strength and volume with radiation dose at high temperature for polygranular nuclear graphite.

TABLE 2. Range of mechanical properties for carbon-carbon composites.

All properties are strongly dependent on:-

densification (i.e. no. of infiltration cycles), graphitisability and extent of graphitisation of the matrix, fibre properties, fibre volume fraction, fibre architecture, processing conditions including sequence of infiltrations where different precursors are employed

Bulk density	$1.3–2.0 \text{ gcm}^{-3}$
Porosity	5–35%
Young's modulus	
unidirectional	up to 300 GPa
2D and 3D	40–150 GPa
Flexural strength	
unidirectional	up to about 1.5 GPa
2D and 3D	up to about 350 MPa
felt & random short fibre	50–100 MPa
Tensile strength	approx. 1.4 times flexural
Fatigue	75–80% retention of strength after 10^7 flexural cycles reported
Interlaminar shear strength (2D)	8–20 MPa
Transverse strength (2D)	"
Torsional strength (2D)	up to about 30 MPa
3D & multi-D "off-axis" strengths	up to about 70 MPa
Work of fracture	up to about 10 kJm^{-2}
Thermal expansion coefficient (K^{-1})	$5 - 10 \times 10^{-6}$ perpendicular
(2D laminates)	$-2 - 2 \times 10^{-6}$ parallel
Thermal shock resistance	extremely high

(eg. Hishiyama *et al.* 1974; White and Zimmer, 1983). Mixture rule calculations of elastic properties may be significantly in error. For example, if the composite is not well densified there may be voids at the fibre-matrix interface resulting in poor load transfer and the modulus is then lower than predicted. On the other hand, in well-densified composites, if the matrix structure is highly aligned with the fibre surfaces and the composite is graphitised, the matrix may be intrinsically stiffer than the fibres and the elastic modulus may be significantly higher than expected. For example, Rhee *et al.* (1987) demonstrated values of the longitudinal Young's modulus of unidirectional composites as high as 500 GPa at a fibre volume fraction of about 60%. This is approximately 50% of the theoretical value of the graphite basal plane.

C/C composites are examples of brittle matrix composites. Therefore, it is to be expected that in line with other brittle systems, the strength will be strongly dependent on the fibre-matrix debond stress and the sliding frictional stress if fibres are extracted from the matrix (Evans, Zok and Davis, 1991). However, in contrast to the wealth of evidence that exists in the field of ceramic matrix composites there are relatively few good model studies of C/C composites on which to base a sound understanding of the important controlling factors. In order to attain full utilisation of the fibre properties it essential that fibre-matrix debonding should occur. The matrix failure strain is significantly less than that of the fibre. If there is strong fibre-matrix bonding, the composite will fail catastrophically at this strain. However, debonding of the fibre and matrix allows stress transfer from the matrix to the fibres accompanied by multiple-matrix cracking. Eventually the fibres begin to fracture, but, if sliding of the fibre within its matrix pocket is allowed, then fibre pull-out can lead to non-catastrophic failure and an extremely damage tolerant material. Unfortunately, this type of behaviour has not been well characterised for C/C composites. Only a few examples of fibre-matrix debonding, leading to near classical multiple-matrix fracture, have been observed. These have shown that with high modulus, non-surface treated fibres, when the surface has a strong basal character, and is thus relatively inactive, the fibres and matrix are not well bonded and high tensile strength can be achieved. In these cases the percentage utilisation of the fibre properties is high. Thomas and Walker (1978) showed that the strength is progressively reduced as the level of oxygen groupings at the fibre surface increases, i.e. as the strength of the fibre-matrix bond increases. Many of the commercially available C/C composites are manufactured from fibres of low heat treatment temperature. They are poorly graphitised and have reactive surfaces. Thus, the strengths of their carbon composites, whilst being significatly greater than the bulk polygranular materials and sufficient for the specific application, do not usually achieve the full potential of this type of material. Fortunately, carbon and graphite matrices generally have higher failure strains than brittle ceramics, typically 0.2–0.3%. The failure strains of carbon fibres are typically about 1%, so even so even when the fibre-matrix interaction is strong, and fibre strength utilisation of 20–30% is attained, the strengths are significantly greater than for unreinforced polygranular materials. For example, a unidirectional composite with 50 volume percentage of fibres of strength 2.5G Pa would display longitudinal tensile strengths of about 250–375 M Pa at the above typical failure strains. These strengths would be lower in laminate or cloth-based composites or in multidirectional structures.

A disadvantage of many carbon-carbon composites is that the interlaminar shear strength is often quite low (typically 10–30 M Pa), especially if the fibre-matrix bonding is weak or if laminates are not well bonded. The degree of densification, ie porosity, is significant here (Rhee et al., 1987). This has led to another uncertainty in evaluating the mechanical behaviour in some studies. Most workers employ flexural testing techniques, for which it is essential to ensure that the correct span to depth ratio is used to ensure tensile failure. The standard method is to use a value of about 25–40, but this can be too low if the shear strength is low when much higher ratios, ie 80–100, are required to ensure true tensile failure. Thus, true tensile measurements are more difficult to obtain. Delamination is a common feature in such tests. However, in very many studies the mode of fracture is not clearly stated, rendering the results of doubtful scientific merit. Figure 10 shows tensile measurements on two carbon-carbon composites with pitch-based carbon matrices. The HMU-PP composite contains high modulus PAN-based fibres with an unreactive surface. It displays evidence of multiple-matrix fracture at a strain of about 1.5% and good utilisation of fibre properties, failing ultimately at about a stress of 1.5 Gpa and a strain nearly 0.6%. On the other hand, the SMS-PP composite is made from a surface treated fibre, which bonds quite strongly to the matrix. This composite fails almost catastrophically at a lower strain, 0.2%, and consequently it displays an ultimate strength of 400–500 MPa.

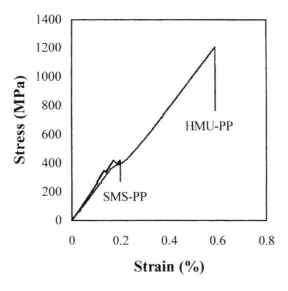

FIGURE 10. Stress-strain curves in tension of unidirectional panels of carbon/carbon composites fabricated at 1000°C from Pan-based fibres and a pitch matrix. The HMU-PP composite contains high modulus fibres with no surface treatment, whilst the SMS-PP composite contains intermediate modulus, surface treated fibres.

6.5 CONCLUSION

Carbon and graphite materials take many different forms and exhibit a very wide range of mechanical properties. This is the inevitable result of the extreme anisotropy in bonding, characteristic of the graphite crystal. Microstructural and crystallographic characteristics collectively control the response of the materials to imposed stresses. The key features are the degree of perfection of the crystallites, their orientation with respect to the applied stresses and the porosity of the material. Highly oriented materials of low porosity exhibit high elastic moduli, which can approach the theoretical maximum in certain types of fibre and in single walled nanotubes. On the other hand, a random orientation of crystallites, such as exists in isotropic carbons and graphites, results in materials of very low modulus due to the dominant influence of the constants, c_{33} and c_{44}. Most engineering carbons and graphites are porous and this lowers the elastic properties even further. In non-graphitic materials, strong intercrystallite bonding can modify the behaviour, but the influence of the intrinsic crystallography is still dominant. The strengths of the various types of material also vary widely. Highly oriented fibres can show extremely high strengths, which are strongly dependent on the flaws present. The combination of pores and lamellar cracks is critical to the strengths of the polygranular materials, which are low, exceeding 100 MPa only in very fine-grained materials. For most forms of carbon/graphite the strength is found to increase with measurement temperature, largely due to the high c-axis expansion coefficient which allows many of the micro-cracks to close on heating. Hence, graphite is an exremely important refractory material, especially in aerospace applications where its low density is also attractive. The generally low values of elastic modulus and strength displayed by typical engineering grades can be overcome in carbon fibre reinforced composites, in which the properties of the fibres may prevail. This imparts to bulk materials strengths in the region of hundreds of MPa, and with careful engineering of the microstructure, elastic properties approaching those of the fibres themselves. Carbon/carbon composites may also be designed to show extremely high damage tolerance through control of the fibre-matrix interaction. The variety of carbon and graphite materials available is probably unique in the vastness of the range of mechanical properties that can be displayed.

References

Amelinckx, S., Delavignette, P. and Heerschap, M. (1966) Dislocation and Stacking Faults in Graphite, in *Chemistry and Physics of Carbon*, Vol. 3, ed P.L. Walker Jr., Marcel Dekker, New York, p. 1.

Andrew, J.F. and Distante, J.M. (1961) *Proc. of 5th Carbon Conference*, Buffalo, Pergamon Press, New York, Vol. 2, 585.

Andrew, J.F., Okada, J. and Wobschall, D.C. (1960) *Proc. of 4th Carbon Conference*, Buffalo, Pergamon Press, New York, 559.

Bacon, R. (1960) *J.Appl. Phys.*, **31**, 283.

Baker, C. and Kelly, A. (1964) *Phil. Mag.*, **9**, 927.

Bokros, J.C., (1969) Deposition, Structure and Properties of Pyrolytic Carbon, in *Chemistry and Physics of Carbon*, Vol. 5, ed P.L. Walker Jr., M. Dekker, New York, pp. 1–118.

Bokros, J.C. and Price, R.J. (1966) *Carbon*, **4**, 441.

Brocklehurst, J.E. (1977) Fracture in Polycrystalline Graphite, in *Chemistry and Physics of Carbon*, Vol.13, ed P.L. Walker Jr. and P.A. Thrower, Marcel Dekker, New York, pp. 146–279.

Buckley, J.D. and Edie, D.D. (1992) *Carbon-Carbon Materials and Composites*, NASA Reference Publication 1254,1.

Buch, J.D. (1983) Extended Abstracts of 16th Biennial Conference on Carbon, California, American Carbon Society, 400.

Burchell, T.D. (1996) *Carbon*, **34**, 279.

Cost, J.R., Janowski, K.R. and Rossi , R.C. (1968) *Phil. Mag.*, 8th Series, **17**, 851.

Diefendorf, J.D. (1960) *Proc. of 4th Carbon Conference*, Buffalo, Pergamon Press, New York.

Dresselhaus, M.S., Dresselhaus, G., Sugihara, K., Spain, I.L. and Goldberg, H.A. (1988) *Graphite Fibers and Filaments*, Springer-Verlag, Berlin, Heidelberg, New York.

Evans, A.G., Zok, F.W. and Davis, J. (1991) *Composites Science and Technology*, **42**, 3.

Goggin, P.R. and Reynolds (1967) *Phil. Mag.*, **16**, 317.

Green, W.V., Weertman, J. and Zukas, E.G. (1970) *Mater. Sci. Eng.*, **6**, 199.

Guigon, M., Oberlin, A. and Desarmot, G. (1984) *Fibre Sci. Technol.*, **20**, 55 and 177.

Hishiyama, Y., Inagaki, M., Kimura, S. and Yamada, S. (1974) *Carbon*, **12**, 249.

Honda, H., Sanada, Y. and Furuta, T. (1966) *Carbon*, **3**, 421.

Hutcheon, J.M. (1970) Manufacturing Technology of Baked and Graphitised Carbon Bodies, in *Modern Aspects of Graphite Technology*, ed L.C.F. Blackman, Academic Press, London, 1.

Jenkins, G.M. (1973) Deformation Mechanisms in Carbon, in *Chemistry and Physics of Carbon*, **11**, eds. P.A. Thrower and P.L.Walker Jr., Marcel Dekker, New York, 189.

Johnson, D.J. (1987) in *Chemistry and Physics of Carbon*, **20**, ed P.A. Thrower, Marcel dekker, New York, 1.

Kelly, B.T. (1981) *Physics of Graphite*, Applied Science Publishers, London and New Jersey.

Kelly, B.T. and Burchell, T.D. (1994) Structure-related property changes in polycrystalline graphite under neutron irradiation, *Carbon*, **32**, 499–505.

Kibler, J.J. (1992) Mechanics of Multi-directional Carbon-Carbon Composites, in *Carbon-Carbon Materials and Composites*, NASA Reference Publication, 169.

Kotlensky, W.V. (1973) in *Chemistry and Physics of Carbon*, **9**, eds P.L. Walker and P.A. Thrower, Marcel Dekker, New York, 173.

McCartney, L.N. (1993) Mechanics of Cracking in Brittle Matrix Composites, in Thomas, C.R., ed., *Essentials of Carbon-Carbon Composites*, The Royal Society of Chemistry, Cambridge, 174.

McEnaney, B. and Mays, T.J. (1993) Relationships between Microstructure and Mechanical Properties in Carbon-Carbon Composites, in Thomas, C.R., ed., *Essentials of Carbon-Carbon Composites*, The Royal Society of Chemistry, Cambridge, 143.

Moore, A.W., (1973) Highly Oriented Pyrolytic Graphite, in *Chemistry and Physics of Carbon*, Vol. 11, ed P.L. Walker Jr. and P.A. Thrower, M. Dekker, New York, pp. 69–188.

Mrozowski, S. (1956) *Proc. First and Second Conferences on Carbon*, 1953/1955, Waverley Press, Baltimore, 31.

Price, R.J. (1965) *Phil. Mag.*, **12**, 561.

Pickup, I.M., McEnaney, B and Cooke, R.G. (1986) *Carbon*, **24**, 535.

Reynolds, W.N. (1966) Radiation Damage in Graphite, in *Chemistry and Physics of Carbon*, Vol. 2, ed. P.L. Walker Jr., M. Dekker, New York, pp. 121–196.

Reynolds, W.N. and Sharp, J.V. (1974) *Carbon*, **12**, 103.

Reynolds, W.N. and Moreton, R. (1980) *Philos. Trans. Roy. Soc.*, **A294**, 451.

Reynolds, W.N. and Thrower, P.A. (1965) *Phil. Mag.*, **12**, 573.

Rhee, B., Ryu, S., Fitzer, E. and Fritz, W. (1987) *High Temperatures-High Pressures*, **19**, 677.

Rose, A.P.G. and Tucker, M.O. (1982) *J. Nucl. Mater.*, **110**, 186.

Ruland, W. (1968) X-ray Diffraction Studies on Carbon and Graphite, in *Chemistry and Physics of Carbon*, Vol. 2, ed. P.L. Walker Jr., M. Dekker, New York, pp. 1–84.

Savage, G. (1993) *Carbon-Carbon Composites*, Chapman and Hall, London.

Simmons, J.H.W. (1965) *Radiation Damage in Graphite*, Pergamon Press, New York.

Thomas, C.R., ed. (1993) *Essentials of Carbon-Carbon Composites*, The Royal Society of Chemistry, Cambridge.

Thomas, C.R. and Walker, E.J. (1978) *High Temperatures and Pressures*, **10**, 79.

Thomas, J.M. and Roscoe, C. (1967) Non Basal dislocations in Graphite, in *Chemistry and Physics of Carbon*, Vol. 3, ed P.L. Walker Jr., M. Dekker, New York, pp. 1–??.

Thrower, P.A. (1969) The Study of Defects in Graphite by Transmission Electron Microscopy, in *Chemistry and Physics of Carbon*, Vol. 5, ed P.L. Walker Jr., M. Dekker, New York, pp. 217–320.

Ting, D.N. (1994) PhD. Thesis, University of Leeds, UK.

Tucker, M.O., Rose, A.P.G. and Burchell, T.D. (1986) *Carbon*, **24**, 581–602.

Walker, E.J. (1993) The Importance of Fibre Type and Fibre Surface in Controlling Compposite Properties, in Thomas, C.R., ed, *Essentials of Carbon-Carbon Composites*, The Royal Society of Chemistry, Cambridge, 37.

White, J.L. and Zimmer, J.E. (1983) *Carbon*, **21**, 323.

7. Carbon Surface Chemistry

H.P. BOEHM

Institut für Anorganische Chemie der Universität München, Butœnandt strasse 5–13 (Haus D), 81377 München, Germany

7.1 INTRODUCTION

The state of the surface of a carbon is decisive for its interfacial properties. In consequence, it is important for many applications of carbon materials. Whereas the surface of diamond is more or less uniform, due to its isotropic structure, there is an inherent inhomogeneity in the surface of graphite crystals. It consists of basal planes and of prismatic faces. The honeycomb network of the graphene layers is interrupted in the prismatic faces, and the carbon atoms have free valences, so-called "dangling bonds" which can be used for the binding of foreign atoms. In the case of diamond, each crystal face cuts the regular network of carbon-carbon bonds.

In an atomically clean diamond surface, the surface free energy is minimized by a distortion of the lattice in the topmost atom layers. The atoms are shifted from their regular, "ideal" positions, creating a new coordinative environment that allows saturation or partial saturation of the dangling bonds (Lander and Morrison, 1966). This phenomenon, observed with many crystal surfaces, is called "surface reconstruction". Reconstruction at the edges of the graphite layer planes, i.e. in the prism faces, has not been described, although there may be a small change of the interatomic distances.

The sp^2-bonded carbon atoms in the basal faces of graphite are saturated, and there is only a relatively weak chemical interaction possible of the π electron system with adsorbed atoms or molecules, in contrast to the formation of covalent σ bonds on the prismatic faces. However, defects, such as vacancies in the hexagon network, are sites of higher energy which are easily attacked, e.g. in the oxidation with molecular oxygen.

Most forms of carbon are derived from hexagonal graphite, the thermodynamically stable form of carbon. In the turbostratic carbons, the graphene layers are stacked in parallel, but without three-dimensional order. They are translated and rotated in the layer planes in an irregular way. Often, they show deviations from planarity, either by bending of the layers (as near edge dislocations in graphite) or by inclusion of pentagons or heptagons in the hexagon network (this leads to a curvature of the layers analogous to that in fullerene structures). Obviously, such deviations from planarity result in smaller sizes of the domains of parallel-stacked layers, and the "crystallite dimensions" determined from line broadening of the 002 and the two-dimensional hk X-ray diffraction lines become smaller than the actual size of the carbon layers is. Such carbons are also called disordered carbons.

The surface of carbons with relatively small graphene layers has a much higher proportion of edge sites than the surface of larger graphite crystals or highly-ordered pyrolytic graphite (HOPG). Typical examples are cokes, carbon blacks or activated carbons. The layers are often cross-linked by disordered carbon. i.e. by sp^3-hybridized carbon atoms or groups of atoms which may contain also sp^2 or sp carbon atoms. In many cases, the layer planes are terminated by "dangling" carbon atoms or groups of carbon atoms.

One distinguishes according to the forces of interaction between mostly irreversible chemisorption and and reversible physical adsorption (or physisorption). Carbons in the "as delivered" state contain almost invariably chemisorbed foreign elements, mostly hydrogen and oxygen. It appears useful to describe the surface compounds, also called surface complexes, separately for hydrogen, oxygen, nitrogen and halogens.

7.2 SURFACE-BOUND HYDROGEN

Hydrogen is the most common foreign element bound on the surface of carbons since most carbons are prepared from hydrocarbons or hydrogen-containing precursors. Obviously, the broken bonds at the edges of the graphene layers will be saturated by hydrogen to a large extent. Also methyl or methylene groups or other hydrocarbon chains may be present in chars that have not undergone heat treatment at temperatures in excess of $800°C$. Hydrogen chemisorbed on graphite is desorbed as H_2 on heating to quite high temperatures ($> 1000°C$) under conditions of high vacuum or under noble gases (He, Ar). The last remnants are lost at $\sim 1300°C$. With disordered carbons, hydrogen will be split off at much lower temperatures, mainly as methane. The resulting carbon surface is quite reactive, e.g. towards oxygen, and ESR studies showed the presence of free radicals, although their quantity was much lower than the number of "edge" carbon atoms (Lewis and Singer, 1981; Singer and Lewis, 1982; Leon y Leon and Radovic, 1994). In contrast, the hydrogenated carbon surface exhibits a much lower reactivity towards oxygen (Menéndez et al., 1996).

Also, vinyl-like groups, |–CH=CH_2, may exist on carbon surfaces. They would be π-bases, and it has been discussed that such groups can act as anchoring sites for metal complexes, e.g. of palladium (Ryndin et al., 1989).

Carbon surfaces can be hydrogenated by treatment with hydrogen at elevated temperatures. Temperatures and pressures as low as $300°C$ and 14 Pa, respectively, have been used for carbons outgassed at high temperatures (Bansal et al., 1974), although temperatures of $\sim 950°C$ and atmospheric pressure are normally preferred, depending on the nature of the carbon. Chemisorption of hydrogen at lower temperatures is considerably reduced when the carbon surface carries preadsorbed oxygen (Bansal *et al.*, 1974). Gasification of graphite as methane (and other small hydrocarbon molecules) becomes significant only above 1200°C. However, very disordered carbons react at much lower temperatures. With activated carbons, methane was already formed with hydrogen at 600°C, but the rate of its formation decreased after an initial surge (Egawa, 1996). On raising the temperature, a new rise in CH_4 evolution is observed that again decreases slowly; traces of ethylene have also been detected at 750°C. Obviously, reactive carbon atoms are preferentially removed from the surface. At higher temperatures, hydrogenolysis of the graphene layers to methane occurs. The mechanism of the uncatalyzed reaction and the intermediate steps have been studied by Pan and Yang (1990). The attack occurs by dissociated H_2 not only at the edges of the graphene layers, but also at vacancies, resulting in etch pits on the basal faces of graphite single crystals. These etch pits are only one atomic layer deep, and they form hexagons with an $\langle 101l \rangle$ orientation, i.e. a zig-zag conformation of the edge. Clearly, the "arm chair" edges, $\langle 112l \rangle$, have the higher reaction rate. Initially, single H atoms are added to the edge atoms, and $/CH_2CH_2\backslash$ bridges are formed by addition of two further H atoms. Finally, bond breakage occurs with hydrogenation to two CH_3 groups before CH_4 is released (Pan and Yang, 1990). Figure 1 shows a model of the zig-zag and armchair edges of the graphene layers.

Also the surface of diamond is hydrogenated on treatment with hydrogen at 800°C (Boehm and Sappok, 1968a). Signals for CH_2 groups have been detected by infrared spectroscopy. Hydrogenated diamond surfaces play an important role in the low-pressure synthesis of diamond by chemical vapor deposition (CVD) (Spear and Frenklach, 1994 a,b).

The surface of hydrogenated carbons, diamond as well as graphite or turbostratically disordered carbon materials, is hydrophobic as described in Section 7.3.4.

7.3 SURFACE OXIDES

7.3.1 Creation of Surface Oxides

The most studied surface compounds are those with oxygen. Carbons with an atomically "clean" surface are highly reactive towards oxygen at room temperature. Such surfaces are usually produced by outgassing in a high vacuum at temperatures near 1000°C, usually 950°C for practical reasons (the upper limit of the use of silica glass in high-vacuum apparatus), although traces of oxygen remain on the surface at this temperature. Highly reactive, high-surface area carbons can also be obtained by milling of graphite under inert conditions, e.g. under pure argon. The coefficient of friction of graphite increases drastically when oxygen and water vapor are excluded, and graphite can be converted to a very disordered "graphite wear dust" by sliding graphite against graphite. This graphite wear dust can be pyrophoric under ambient conditions. It has been reported that no oxygen is chemisorbed at 77 K, but there was a rapid exothermic reaction at −40°C (Fedorov *et al.*, 1963). The disordered structure of ball-milled graphite has been recently studied by Hermann *et al.* (1997) who found a dependence of the ignition temperature on the milling time and the material of the balls. Pyrophoric carbon materials can be stabilized by controlled oxidation at lower oxygen partial pressures.

zig-zag edge, (10 1̄*l*)

armchair edge, (11 2̄*l*)

FIGURE 1. Model of zig-zag edge and armchair edge of graphene layers.

As is the case with hydrogen, also molecular oxygen reacts only with the edges of the graphene layers and at vacancy sites in the basal faces of graphite. Both, zig-zag and armchair structures of the oxidized edges can prevail, depending on the oxidation conditions (Hennig, 1966; Thomas, 1965; Yang, 1983). However, atomic oxygen or ozone attack also the basal faces of graphite with creation of carbon vacancies.

Usually, the carbons are oxidized with dioxygen at elevated temperatures. The quantity of chemisorbed oxygen increases with increasing reaction temperature up to a temperature at which thermal decomposition of the surface oxides becomes a limiting factor. The optimal oxidation temperature depends on the nature of the carbon, usually temperatures between 300°C and 420°C are chosen for microcrystalline carbons. At higher temperatures, the surface oxides become increasingly unstable, and there is already slow gasification at these temperatures. However, often oxygen is chemisorbed again during the cooling under O_2. Generation of surface oxides occurs also when carbons are exposed to an oxygen plasma (Jones, 1993).

Surface oxides can also be created at or near room temperature by oxidation with dissolved oxidants, e.g. aqueous solutions of sodium hypochlorite, or ammonium peroxodisulfate. Quite popular is the use of concentrated nitric acid at 60°C to 80°C. The oxidizing power can be easily controlled by variation of the HNO_3 concentration and the reaction temperature.

It is often overlooked in the preparation of carbon-supported noble-metal catalysts that solutions of the metal salts or complexes have oxidizing properties. On treatment of activated carbons with solutions of, e.g. H_2PtCl_6 or H_2PdCl_4, the carbon surface is oxidized and metal is precipitated on it (Puri *et al.*, 1965; Suh *et al.*, 1992; Fu *et al.*, 1994). With inceasing surface oxidation, the carbon surface loses its reducing properties. Electrochemical oxidation is very convenient with carbon fibers, because of easy control and continuous operation in the production (Donnet and Bansal, 1990; Jones, 1993). The nature of the electrolytes used in industrial production is not divulged, but in the open literature mostly ammonium bicarbonate or acidic solutions are used (Kozlowski and Sherwood, 1987).

7.3.2 Aging of Carbons

Disordered carbons of high surface area are slowly oxidized by moist air at room temperature. This phenomenon is called "aging" of the carbons. It has been first described by Puri (1962). Aging plays a significant role in the uses of activated carbons and carbon blacks. The formation of surface oxides renders the surface hydrophilic (see below), and it has an adverse effect on the adsorption of noxious gases or radioactive methyl iodide on activated carbons (Billinge *et al.*, 1984; Billinge and Evans, 1984).

It has been shown that the presence of water vapor enhances considerably the surface oxidation of the carbons at low temperatures (Petit and Bahaddi, 1993). Hydrophilic adsorption centers are created in this reaction such as carboxylic or phenolic surface groups (see below). Figure 2 shows how the sodium hydroxide uptake due to such acidic functions increases with time on an activated carbon (Kuretzky and Boehm, 1994). Temperatures of $60°C$ and $110°C$ were used in these experiments in order to shorten the reaction times to a practical level. As shown in the figure, surface oxidation is faster at $60°C$ and a relative humidity of 80% than at $110°C$ under ambient air, that is at a very low relative pressure of water vapor. The oxygen functions on a surface-oxidized carbon (see Section 7.3.5) are decomposed on heating to $1000°C$ *in vacuo* or in a helium stream. As Carrasco-Marín *et al.* (1996) found, the functional groups were regenerated nearly identically after aging for two years under ambient conditions.

Aging is inhibited to a large extent when the reactive sites on the carbon surface are saturated by treatment at elevated temperatures with hydrogen (Verma and Walker, 1992; Menéndez *et al.*, 1996) or chlorine (Verma and Walker, 1992; Hall and Holmes, 1993).

7.3.3 Active Surface Area

Since it is known that oxygen is bound only at edge sites of the graphene layers, attempts have been made to determine this *Active Surface Area* (ASA) from the chemisorption of oxygen. Laine *et al.* (1966) used Graphon, a graphitized carbon black that had been activated with 66 Pa oxygen at $625°C$ to several levels of burn-off. The surface of graphitized carbon blacks consists almost exlusively of basal planes, and the activation treatment creates additional exposed edge sites. After outgassing at $950°C$, the samples were exposed to 66 Pa O_2 for 24 hours at $300°C$. The carbons were then outgassed and pyrolyzed at $950°C$. From the quantities of CO_2 and CO evolved, the ASA was calculated under the assumption that each oxygen atom had been bound to one carbon atom in the zig-zag edge of the layers.

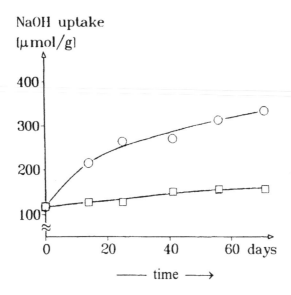

FIGURE 2. Formation of acidic groups upon aging of an activated carbon (Norit SGM). Acidic groups were determined by neutralization adsorption of 0.05 N NaOH (see Section 7.4.5). (○) storage at 60°C and 80% rel. humidity, (□) storage at 110°C under ambient air.

From the distance between the carbon atoms, 0.246 nm, and the interlayer distance of 0.335 nm, follows an area of 0.083 nm^2 per edge site (see also Walker, 1990). As expected, the ASA was much smaller than the total surface area, from 0.24 to 2.12%, depending on the burn-off.

This method of ASA determination works well with well-ordered, more or less graphitized carbons, but problems arise with microcrystalline carbons such as carbon fibers (Ismail, 1987). For instance, the time of exposure to oxygen becomes important, since gasification occurs concurrently with the relatively slow chemisorption process. However, Ehrburger *et al.* (1992) succeeded in studying the development of ASA during the activation of carbons.

7.3.4 Hydrophilic Properties

Whereas clean carbon surfaces are, in essence, hydrophobic, oxidized carbon surfaces become increasingly hydrophilic with increasing oxygen content. Adsorption of water vapor (and ammonia) is enhanced and the heat of immersion in water inceases. Highly oxidized carbons of colloidal dimensions, such as carbon blacks, are dispersed in the aqueous phase on agitation with mixtures of water and an unpolar solvent. This is also true for oxidized diamond powder (Sappok and Boehm, 1968a).

Water vapor adsorption isotherms on hydrophilic, non-porous carbons are of Type III of the BDDT classification (Brunauer *et al.*, 1940). There is little adsorption at low relative pressures p/p_0 (p_0 is the saturation pressure of water at the adsorption temperature), and the adsorbed quantities increase only at relatively high p/p_0 values. The reason is that very little hydrogen bonding will occur as long as only few water molecules are adsorbed, with a correspondingly low heat of adsorption. The water molecules are adsorbed by the weaker *van der Waals* forces on hydrophobic surfaces. The tabulated p/p_0 values are too high because p_0 is based on the properties of liquid, hydrogen-bonded water.

With increasing surface oxidation, increasingly more water is adsorbed at low relative pressures. Walker and Janov (1968) applied the BET equation for the determination of monolayer capacities in physical multilayer adsorption (Brunauer *et al.*, 1938) to the lower part of the adsorption isotherms of water on activated Graphon and found a good correlation with the ASA values when it was assumed that each water molecule adsorbed in the "monolayer" covered an area of $0.106\ nm^2$. Obviously, approximately one water molecule is adsorbed to one surface oxygen atom by hydrogen bonding. This was confirmed by Sappok and Boehm (1968b) with oxidized diamond powder. There was an excellent agreement with the content of active hydrogen (OH groups) on the surface as determined by the Zerewitinov method (see Section 7.3.5.2). The heat of adsorption of the first water molecules bound by hydrogen bonding to hydrophilic adsorption centers (surface oxygen atoms or OH groups) is significantly higher than for the following adsorption by van der Waals forces, and therefore application of the BET adsorption equation results in a "monolayer volume" corresponding to the quantity of hydrogen-bonding adsorption sites. The adsorption on the hydrophobic parts of the surface is almost negligible at low relative pressures. Clearly, the BET evaluation can give meaningful results only when the hydrophobic and the hydrophilic parts of the surfaces are uniform and quite different in their affinity for water. Similar water vapor adsorption isotherms have been observed for oxidized graphites (Miura and Morimoto, 1991). With activated carbons, the isotherms have a more complex shape, due to the effects of micropore filling (Gregg and Sing, 1982). The isosteric heat of adsorption of water on hydrophobic carbons drops rapidly, after adsorption on a few high-energy sites, below the heat of condensation, and increases gradually to this limiting value on further adsorption (Young *et al.*, 1954; Sappok and Boehm, 1968b; Miura and Morimoto, 1991).

Less water is adsorbed on a graphite surface after treatment in hydrogen at $1000°C$ than after vacuum outgassing at this temperature (Miura and Morimoto, 1991). After outgassing at $500°C$ or $1000°C$ water was irreversibly chemisorbed in the first adsorption, and a certain, relatively small amount of hydroxyl groups was created on the surface. The following adsorption – desorption cycles were fully reversible.

The formation of hydrophilic surface oxides can have a strong, often detrimental influence on the selectivity of adsorption of unpolar compounds in the presence of polar species. On the other hand, a hydrophilic character of the surface of carbon fibers is beneficial for an adhesive interaction between the fibers and the polymer matrix, e.g. epoxy resin, in composites. A good measure of the hydrophilicity is the contact angle between the carbon surface and water. Its measurement with carbon fibers has been described by Hüttinger *et al.* (1991a,b).

7.3.5 *Functional Groups in the Surface Oxides*

7.3.5.1   Acidic and Basic Properties of the Carbon Surface

The chemical nature of the chemisorbed oxygen has been the subject of many studies. The similarity of the graphene layers to very extended polycondensed aromatic systems

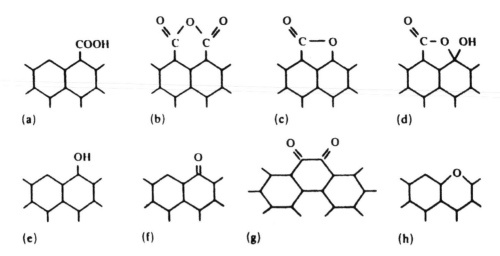

FIGURE 3. Possible structures of surface oxygen groups on the edges of a graphene layer. (a) Carboxyl group, (b) carboxylic anhydride, (c) lactone, (d) lactol, (e) phenolic hydroxyl group, (f) carbonyl group, (g) o-quinone-like structure, (h) ether-type oxygen. (From Boehm (1994), with permission)

suggests that the chemistry of the edge-bound oxygen will be similar to that in smaller aromatic molecules with oxygen functionality. The functional groups that are most likely to occur are shown in Figure 3. The groups (a) to (e) have acidic properties. A model with a few more possible groups has been published by Fanning and Vannice (1993).

However, carbons which have been treated at high temperatures (900–1200°C), either *in vacuo* or under inert gases or hydrogen, exhibit basic surface properties. The pH values of their aqueous suspensions are larger than seven, especially when small concentrations of neutral salts are present. Such carbons can bind acids by neutralization adsorption, and they have anion exchange properties. As explained in the beginning of Section 7.2, most surface compounds are destroyed by this heat treatment, and such carbons will chemisorb oxygen when they are exposed to air or oxygen after cooling to room temperature. The quantity of bound oxygen is relatively small, however, but the chemisorbed oxygen species seem to be involved in the basic surface properties ("basic surface oxides", see Section 7.3.5.3). On additional oxidation, with O_2 (or air) at elevated temperatures (or much more slowly under ambient conditions, "aging", see Section 7.3.2), the surface acquires increasingly acidic properties and cation exchange capacity. The same effect is observed after treatment with liquid oxidants at room temperature or moderately raised temperatures. It was shown by Kruyt and de Kadt (1929) and by Kolthoff (1932) that the *same* sample of carbon could acquire acidic or basic properties depending on the temperature of its reaction with oxygen after pretreatment at ca. 1000°C. When the carbon was contacted with oxygen at room temperature, basic properties and a positive surface charge in an aqueous dispersion were observed, whilst the surface was acidic and carried negative charges after oxidation at 400°C. The carbons had anion or cation exchange properties, respectively. Studebaker

(1957) showed a linear correlation of the pH of carbon black suspensions and their oxygen content; the pH was the lower, the higher the oxygen content was. Analogous observations have been made with an activated carbon (Barton *et al.*, 1997).

It seems practical to describe first the chemical nature of the acidic surface oxides and that of the basic surface oxides thereafter. Several reviews have appeared in the literature (Boehm *et al.*, 1964; Boehm, 1966; Boehm, 1994; Leon y Leon and Radovic, 1994).

7.3.5.2   Acidic Surface Oxides

Chemical methods were used in the earliest attempts to elucidate the nature of the acidic surface sites. A very simple, but still practical and frequently employed method is the neutralization adsorption of bases of different base strengths (Boehm *et al.*, 1964). This method is based on the age-old experience of chemists that carboxylic acids are neutralized by sodium bicarbonate, whereas phenols require sodium hydroxide for neutralization. The series of bases has been enlarged by sodium carbonate which opens lactone and lactol rings, and sodium ethoxide (in ethanolic solution).

Usually, an appropriate quantity of carbon is agitated with 50 or 100 ml of 0.05 N solutions of the bases (0.1 N in the case of sodium ethoxide). Equilibration times of several hours (over night) are often needed; the reason is thought to be a slow diffusion in narrow micropores and slow hydrolysis of carboxylic anhydrides and lactones. An aliquot of the supernatant is taken for titration after sedimentation of the carbons. An excess of 0.05 N HCl solution is added, and the evolved CO_2 is driven out by heating to near boiling temperature (this is also useful in the case of NaOH). The excess acid is then back-titrated with 0.05 N NaOH using a suitable indicator or potentiometric indication. Oxidized carbon blacks form frequently stable colloidal dispersions, and sedimentation has to be forced by centrifuging.

Of course, the acidity constant of a given type of functional groups, e.g. carboxyl groups, will be influenced by ist chemical environment, the size and form of the polyaromatic layers and the presence and position of other substituents. However, the differences in acidity of the different groups seemed to be sufficiently large to differentiate between them by this simple titration method. It has been shown that changes of slope can be observed in the conductometric titration curves when suspensions of the carbons are titrated with sodium methoxide in methanol. These breaks agree very well with the "standard" titration values (Boehm, 1994). Contescu *et al.* (1997a) performed a careful potentiometric titration with incremental addition of the alkali solution. A new increment was added only when the drift in pH was smaller than 0.01 pH units per minute. The pK_a distribution curves were calculated from the titration data by use of the method introduced by Bandosz *et al.* (1993). In this earlier paper, the titration was continuous, and it was shown in the later work (Contescu *et al.*, 1997) that a quite good agreement with the neutralization values with $NaHCO_3$, Na_2CO_3, and NaOH (the so-called "Boehm titration") was obtained when the actual ionization was taken into account of the groups that had pK_a values near the pH values of the above-mentioned solutions. Up to seven peaks in the acidity distribution were observed with some oxidized carbons. In one case, three groups were practically fully neutralized at the pH of the $NaHCO_3$ solution (pH 8.2), one to an extent of 80%, and one to 28%. With an other carbon, only three peaks could be observed below pH 11, the experimental limit of the method. Of course, there is a distribution of acidity for each species, and a Gauss distribution was assumed for the calculations. A drawback of this method is the long time needed for one measurement, up to 12 hours. The carbon particles should be small (< 180 nm) to avoid errors due to very slow equilibration.

Two closely spaced carboxyl groups can split off water, e.g. on heating, with formation of carboxylic anhydrides (Figure 3b). Such anhydrides may be directly formed in the reaction of carbons with oxygen at elevated temperatures, especially when water vapor is excluded. However, since most carbons contain hydrogen, some water will be present in most oxidation reactions. Otake and Jenkins (1993) presented evidence that carboxylic anhydrides were formed in the oxidation of a porous carbon with air at 375°C, whereas free carboxyl groups resulted from the oxidation with nitric acid at 67°C. The conclusions were based on TPD experiments (see Section 7.4.1) in combination with base titration data, including $Ba(OH)_2$ as a base.

Also, chemical methods used in the analysis of organic compounds have been applied to the elucidation of the structures of oxygen-containing surface groups. It is assumed that the most acidic groups which react with $NaHCO_3$ (pH 8.2) are carboxyl groups or their anhydrides, while carboxylic groups condensed to lactones or lactols are hydrolyzed only with the stronger base Na_2CO_3 (pH 10). More weakly phenolic hydroxyl groups are ionized by NaOH. The lactone rings can be opened in the reaction with boiling methanol in the prresence of HCl gas with formation of methyl esters of the carboxyl groups. The methoxy groups, $-OCH_3$, have been quantitatively determined by the well-established Zeisel method (reaction with concentrated HI solution to CH_3I that can be separated by distillation and determined titrimetrically). The methoxyl contents of the carbons agreed well with the difference of the Na_2CO_3 and $NaHCO_3$ titration values, i.e. with the lactone or lactol groups (Boehm *et al.*, 1964).

Reaction with diazomethane, CH_2N_2, in etheric solution results in the methylation of free carboxyl groups, lactones (lactols), and phenolic groups. Whereas the methyl esters of carboxylic acids can be easily hydrolyzed by boiling aqueous acids (e.g. HCl), the methyl ethers of phenols are resistant to such treatment. Consequently, the neutralization uptakes of NaOH become equal to those with Na_2CO_3 (Table 1) while the methoxyl content after hydrolysis is equaal to the difference in NaOH and Na_2CO_3 uptake, indicating that this difference corresponds to groups similar to phenolic hydroxyls (Boehm *et al.*, 1964). This assignment was confirmed by reactions of phenolic groups with 2,4-dinitrofluorobenzene or 4-nitrobenzoyl chloride which eliminate HF or HCl, respectively (Boehm *et al.*, 1966, Boehm, 1994).

The nature of lactone or lactol groups should be pointed out briefly: lactones are condensation products of carboxyl groups and neighboring hydroxyl groups (Figure 3c), either 5-membered or 6-membered rings can be formed. Such rings are opened by hydrolysis with stronger bases such as Na_2CO_3. However, the presence of lactols seems more plausible at the edges of oxidized graphene layers (Figure 3d). Newman and Muth (1951) found that free 2-benzoyl-benzoic acid is predominantly in the lactol form (see Reaction Scheme 1), whereas the opened ionized form has a carbonyl group. The situation is similar to that with the indicator phenolphthaleine, and every chemist knows from experience that phenolphthaleine shows the purple color of the opened form with Na_2CO_3 solutions but not with $NaHCO_3$.

TABLE 1. Reactions of oxidized carbons with diazomethane. Methoxyl group formation and changes in the neutralization behavior after hydrolysis (see text).

	Sample			
	CK 3	sugar char H.T. 450°C	sugar char H.T. 1100°C	sugar char activated
Methoxyl contents in μmol/g				
hydrolyzable	560	500	810	420
not hydrolyzable	330	320	430	300
Neutralization in μeq/g				
before reaction with CH_2N_2				
$NaHCO_3$	310	270	420	210
Na_2CO_3	600	500	820	430
NaOH	890	810	1240	710
after reaction with CH_2N_2 and hydrolysis				
$NaHCO_3$	320	270	400	210
Na_2CO_3	570	490	840	410
NaOH	550	540	810	420

Samples: CK 3: carbon black (Degussa, similar to channel blacks); H.T.: Heat treatment under N_2; activated: activated with CO_2 at 950°C, ca. 50% burn-off. All samples were oxidized with O_2 at 400–450°C.

(Reaction Scheme 1)

Free carboxyl groups, lactols and phenols have hydrogen atoms bonded to oxygen, and such groups are weak Bronsted acids. The acidity is sufficient to allow isotope exchange with deuteriium oxide or formation of methane with the Grignard reagent methylmagnesium iodide (Zerevitinov reagent). This hydrogen is called "active hydrogen", in contrast to non-reactive carbon-bonded hydrogen. The liberated methane is measured volumetrically, but the use of this method is limited to non-porous carbons, because microporous carbons will keep the methane in adsorbed form. In such cases, the isotope exchange method with heavy water (or tritiated water) is the method of choice. The surface is deuterated by reaction with excess D_2O, the sample is evacuated at e.g. 100°C, and the deuterium is back-exchanged with a known quantity of H_2O. If an appropriate excess of H_2O is used, virtually all of the deuterium will be in the water, and it can be determined in various ways. One method is reduction to hydrogen with metal hydrides, e.g. $LiAlH_4$, followed by determination of the HD formed by passing it in H_2 flow over a heat conductivity cell (Boehm and Knözinger, 1983).

As Table 2 shows, there was a fairly good agreement of the active hydrogen content with the NaOH neutralization values. This supports also the assumption that the groups neutralized by Na_2CO_3 are lactols, and not lactones. A similar agreement of active hydrogen with NaOH uptake has been described by Rivin (1962).

The presence of carbonyl groups was shown by the reaction with 2,4-dinitrophenylhydrazine to the 2,4-dinitrophenylhydrazone (Boehm *et al.*, 1966; Fryling *et al.*, 1995; Chen *et al.*, 1995). The quantity of carbonyl groups can be estimated from the nitrogen content or from the intensity of the Raman emission at 1140 cm^{-1} (Fryling *et al.*, 1995).

Chen *et al.* (1995) and Chen and McCreery (1996) presented convincing evidence that carbonyl groups on the surface of glassy-carbon electrodes enhance the electron transfer rate in redox reactions with hexaquo complexes, e.g. $Fe(H_2O)_6^{2+/3+}$. The transfer rate wass much reduced when chemisorbed oxygen was removed by Ar^+ sputtering or when the carbonyl groups were derivatized with 2,4-dinitrophenylhydrazine. Apparently, the carbonyl groups can replace a water ligand in the hydrated ions. There was only little change in the transfer rate when the surface OH groups were blocked by reaction with 3,5-dinitrobenzoylchloride.

The reaction of sodium ethoxide in ethanol is not a simple neutralization reaction but has been shown to be an addition of ethoxide ions to reactive carbonyl groups with formation of the salts of hemiacetals (Boehm, 1964; Boehm *et al.*, 1966):

$$\text{>C=O} \;+\; \text{OEt}^- \;\rightarrow\; \text{>C}^+\begin{smallmatrix}\diagup\text{O}^-\\[2pt]\diagdown\text{OEt}\end{smallmatrix}$$

(Reaction Scheme 2)

Closely spaced carboxyl groups can form carboxylic anhydrides. Such carboxylic anhydrides react with only one equivalent of sodium ethoxide, one of the carboxylic functions is esterified in the reaction according to reaction scheme 3. This must be observed in the evaluation of sodium ethoxide adsorption data (Boehm *et al.*, 1964).

(Reaction Scheme 3)

TABLE 2. Comparison of active hydrogen content and NaOH neutralization.

Sample	Surface area m^2/g	Determination of active H	Active H $\mu eq/g$	NaOH uptake $\mu eq/g$
Corax 3^a – oxid. w. $(NH_4)_2S_2O_8$	n.d.[b]	Zerewitinov	130	200
Corax 3, H.T. 1100°C – oxid. w. $(NH_4)_2S_2O_8$	n.d.	Zerewitinov	228	206
Corax 3, H.T. 1400°C – oxid. w. $(NH_4)_2S_2O_8$	n.d.	Zerewitinov	94	123
Graphite wear dust – oxid. w. air at 420°C	275	isotope exchange	670	580
oxid. w. NaOCl	345	isotope exchange	1040	1025
oxid. w. $(NH_4)_2S_2O_8$	330	isotope exchange	1440	1650

[a]Corax 3 is a furnace black (Degussa) with a surface area of 84 m^2/g. It was assumed that its surface area did not change much by wet oxidation with an $(NH_4)_2S_2O_8$ solution.
H.T.: heat treatment.
[b]n.d.: not determined.

Many more reactions taken from the behavior of organic compounds have been applied to carbons. The reader is refererred to the original literature (Studebaker, 1957; Boehm *et al.*, 1966; Donnet, 1968; Puri, 1970; Rivin, 1971; Boehm, 1994). In particular, the effects of chemical reduction and of reactions with reagents for functional-group analysis have been studied.

Carbonyl groups can be situated pairwise to form quinone-like structures, e.g. similar to o-quinones as in Figure 3(g). Reversible peaks in cyclovoltammetric experiments have been taken as evidence for quinone functions (Drushel and Hallum, 1958; Epstein *et al.*, 1971; Kinoshita and Bett, 1973). Of course, the redox potential of the corresponding quinone/hydroquinone couples depends on the geometrical configuration of the edges of the carbon layers, analogous to the acidity constants of acidic groups. The presence of quinone-like functions has also been inferred from chemical reactions (Donnet *et al.*, 1966; Matsumura and Takahashi, 1979).

It must be pointed out that the oxygen content accounted for by the sum of the functional groups estimated by the titration procedures and related methods is almost always considerably smaller than the analytically determined oxygen content or that found as $CO + CO_2$ in thermal decomposition experiments. One reason may be that ether-type oxygen (Figure 3h) is not determined by these methods. However, the difference unaccounted for seems usually rather large to be explained by ether-type oxygen only. The discrepancy has also been shown in thermal desorption experiments (Carrot *et al.*, 1988; Otake and Jenkins, 1993). It has been atributed to limited access of the reagents to very narrow pores (Molina-Sabio *et al.*, 1991).

7.3.5.3   Basic surface sites

It is known for a long time that carbon surfaces low in oxygen content have basic properties. Acids are adsorbed from aqueous solutions. Burstein and Frumkin (1929) observed that acid was adsorbed on carbons outgassed at 1000°C in a high vacuum only when oxygen gas was present. Oxygen was adsorbed irreversibly in the reaction.

Whereas the identification of the various acidic surface sites made good progress (see Section 7.3.5.2), the nature of the basic surface sites is still not as well understood. There are four theories that seem plausible at first sight, but they are contradictory to some extent.

In the Frumkin theory, as modified by Matskevich (1974), the carbon surface acquires a positive charge by oxidation (instead of O_2, also $FeCl_3$ may be used), and the anions of the acid are adsorbed as counter ions in the electric double layer near the surface. Oxygen is reduced to water in this reaction schme, and it is not necessarily adsorbed on the carbon surface (the surface is symbolized by a vertical line):

$$C| + X^- \rightarrow C^+|X^- + e^-$$

$$\frac{1}{2}O_2 + 2H^+ + 2e^- \rightarrow H_2O$$

or

$$Fe^{3+} + e^- \rightarrow Fe^{2+}$$

Garten and Weiss (1957 a,b) concluded that oxygen is essential for the basic properties, and suggested that it was bound in chromene-like structures. The positive charge can be delocalized in several resonance structures.

In a similar model, presented by Voll and Boehm (1971), the basic properties are due to pyrone-type structures as shown in the model of Reaction Scheme 4. The carbonyl group

(Reaction Scheme 4)

and the furan-type ether oxygen need not be located in the same hexagon ring, as indicated by the wavy line; possible other structures with a similar resonance system are shown in Figure 4. Such a system seems to be more plausible at the edge of graphene layers than chromene structures. However, it has been argued by Leon y Leon and Radovic (1994) that the basicity of pyrones is very low, corresponding to a pK_b of 12–15, but recent studies indicate that the basicity might be considered higher with pyrone-like structures on polycyclic aromatic molecules (see below).

When a carbon is freed from surface complexes by heating to 950°C in a high vacuum or under an inert gas, it chemisorbs oxygen when it is exposed to O_2 at room temperature. Voll and Boehm (1970) reported that the same quantity of oxygen, again, was taken up when this carbon was immersed in dilute aqueous hydrochloric acid. As Figure 5 shows, there was almost a 1:1 relationship between the chemisorbed oxygen atoms and adsorbed acid.

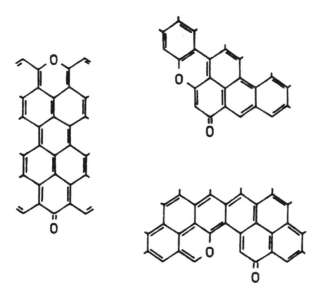

FIGURE 4. Various possible structures of basic surface sites of the pyrone type. (From Boehm (1994), with permission)

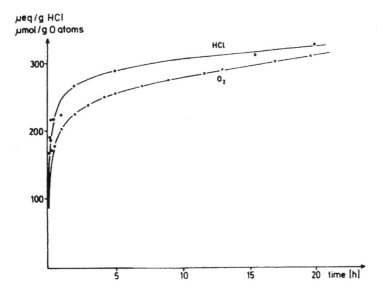

FIGURE 5. The course of uptake of hydrochloric acid and chemisorption of oxygen in the formation of basic surface oxides. (Sugar char, heat-treated at 1100°C and activated to ca. 50% burn-off. It was outgassed at 950°C before immersion in the acid. (From Boehm (1994), with permission)

The anions in the diffuse double layer can be exchanged for other anions. This chemisorption reaction occurs also in water since water is a sufficiently strong acid. The exchangeable anions are OH^- ions, in this case.

Clearly, one positive charge is located on the carbon surface for each oxygen atom taken up, or for two oxygen atoms if the oxygen chemisorbed in the first, "dry" reaction is taken into consideration in addition to that bound in the following, "wet" reaction in the presence of aqueous acid. This disagrees with the theory of Matskevich et al. (1974) in which two positive charges are created for each oxygen taken up in the presence of acid. Hydrogen peroxide is formed in the "wet" chemisorption of oxygen (Rivin, 1963), but its concentration decreases rapidly bacause carbon is an efficient catalyst for H_2O_2 disproportionation (Voll and Boehm, 1970).

Most of the HCl adsorbed in the reaction can be washed out on prolonged extraction of the carbons with water, but this hydrolysis is very slow, and complete removal of the adsorbed HCl is not possible in reasonable times, even at boiling temperature. However, it can be removed by outgassing *in vacuo*.

Rivin (1963) and Boehm and Voll (1970) discussed also the possibility that protons are adsorbed on systems which act as Lewis bases. According to the latter authors this might occur at higher acid concentrations due to the low basicity of π systems. Leon y Leon et al. (1992) concluded, however, that the π basicity of the basal planes is responsible to a significant degree for H^+ (or H_3O^+) adsorption from acidic solutions. The ratio of adsorbed HCl to oxygen content, as measured by temperature-programmed desorption, was determined for a series of pretreated carbon blacks and of microporous carbons. The ratio HCl/O was ca. 3 for the carbon blacks with a very low oxygen content (0.3 wt.%), indicating clearly that HCl must have been adsorbed on a non-oxygenated surface. With increasing oxygen content, the ratio dropped steeply to values below 0.25. The authors concluded that there are two kinds of basic sites, π systems of the aromatic graphene layers and oxygen at edge sites such as pyrone-type structures. Simonov et al. (1997) inferred the presence of isolated C=C double bonds of higher Lewis basicity from Pd^{2+} ion adsorption experiments.

Contescu et al. (1997b) applied their potentiometric titration method to the characterization of basic carbon surfaces. This technique had been very successful in differentiating the acidic surface functions (see Section 7.3.5.2). The authors could distinguish three peaks in the basicity distribution (within the "experimental window" of pH 3–11).The most basic and most abundant species had a pK_b of 5.4–5.6 (in this paper, the pK_a values of the conjugate acids are given). This peak is tentatively assigned to the bi-anions of hydroquinone-type structures. However, hydroquinone has a pK_{a1} of 10.35, the conjugate anion would be significantly more strongly basic ($pK_b = 3.65$).

However, there was indication of an other species at the edge of the "experimental window" with $pK_b < 4.5$. A further, broad peak in the pH range 4–7 consisted apparently of two components. It was assigned to chromene-type structures. Pyrone-type functions would be outside the experimental window because of their low basicity. However, recent ab-initio calculations predict that pyrone-like structures on polycyclic aromatic systems such as, e.g., anthracene or phenanthrene exhibit strong basicities, stronger than that of pyridine ($pK_b = 8.75$) (Menéndez et al., 1999; Suárez et al., 1999, 2000). For one tricyclic system a pK_a value of 12.7 has been estimated, corresponding to $pK_b = 1.3$ (Menéndez et al., 1999).

From all this follows that the nature of the basic surface sites on carbons is presently not sufficiently well understood. It should be remembered that all carbons have an acid adsorption capacity, regardless of whether they carry also acidic functions or not. Usually, the concentration of basic surface sites decreases with increasing concentration of acidic surface groups.

7.4 SELECTED TECHNIQUES FOR THE STUDY OF SURFACE GROUPS

7.4.1 Thermal Desorption

Many attempts are described in the literature to distinguish between the various types of functional groups by their different thermal stabilities. On pyrolysis, oxygen complexes on the carbon surface will decompose to CO_2 and/or CO and, usually, H_2O. The carbon samples are heated with a linear temperature ramp (10 K/min in many cases) in a vacuum or in a helium stream. The evolved gases are analyzed by use of (quadrupole) mass spectrometry or gas chromaography, and recorded as function of decomposition temperature. The method, called Temperature-Programmed Desorption (TPD) or Thermal Desorption Spectroscopy (TDS), has found wide application in studies of adsorption and catalysis (Cvetanovic and Amenomiya, 1967). At high temperatures, above 930°C, also small quantities of hydrogen are evolved from many carbons (Otake and Jenkins, 1993).

It is generally assumed that a given group will decompose to a definite product, for instance, carboxyl groups will give CO_2, and phenols or ethers will produce CO. However, neighboring carboxyl groups can dehydrate to cyclic anhydrides which will decompose to CO_2 plus CO. Furthermore, secondary reactions can not be excluded, in particular with microporous carbons because of the relatively slow diffusion in narrow pores. Primarily formed CO_2 molecules may react at higher temperatures with the carbon walls to form two CO molecules. On the other hand, at lower temperatures CO may disproportionate to CO_2 plus C, and, at higher temperatures, CO may react with undecomposed surface oxides (Hall and Calo, 1989):

$$2\,CO \rightarrow CO_2 + C$$

and

$$CO + C(O) \rightarrow CO_2 + C_f$$

(C(O) symbolizes chemisorbed oxygen on an active site, C_f is a free active site).

It is not surprising, therefore, that the results will depend strongly on the experimental conditions, e.g. sample size and heating rate, and on the nature of the carbon studied. Despite these difficulties, reproducible and plausible TPD spectra have been published in the recent years. The results can give a good overview of the state of the surface, but they do not provide an unambiguous analysis of the surface groups.

An increase of the heating rate results in a shift of the desorption maxima to higher temperatures. It is possible to estimate the heats of activation for the desorption or decomposition reactions from these shifts (Cvetanovic and Amenomiya, 1967).

Generally, more CO_2 is evolved from carbons that were wet-oxidized at relatively low temperatures (e.g. with nitric acid) than from carbons that were oxidized with air or oxygen at 300–400°C. This is shown in Figure 6 as an example. The reason is that a part of the carboxyl groups created at low temperature decompose already below 300°C (Boehm, 1966;

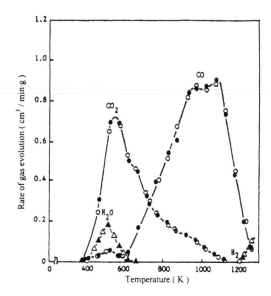

FIGURE 6. Example for the evolution pattern of CO_2, CO and H_2O during TPD from an oxidized activated carbon (oxidized with concentrated HNO_3 for 5 hrs. at $67°C$). N_2 stream, heating rate: $5°C$ / min. Open and filled circles represent duplicate measurements. (From Otake and Jenkins (1993), with permission)

Otake and Jenkins, 1993; Moreno-Castilla *et al.*, 1995). Water is also released at low temperatures. Such thermally labile carboxyl groups can obviously not be found with carbons oxidized at $300°C$ or higher. Carasco-Marín *et al.* (1996) reported that the ability to react with $NaHCO_3$ was lost after heating to $400°C$, whereas some Na_2CO_3-active lactone groups survived even at $1000°C$. Other CO_2 peaks are observed in the temperature range $400–650°C$, with a tailing up to $>850°C$. The peak positions vary in the various reports. This CO_2 may derive from the decomposition of lactones and carboxylic anhydrides which are thermally more stable than free carboxyl groups. As Calo *et al.* (1997) reported, the temperature of decomposition of carboxylic anhydrides is lowered by $\sim40°C$ in the presence of hydrogen. Hydrogen is bound on the carbon surface during the decomposition reaction. Menéndez *et al.* (1997) suggested that some CO_2 may also be formed by disproportionation of CO at high temperatures.

Evolution of CO occurs mainly at higher temperatures with the main peaks at $530–600°C$ and at ca. $800°C$, and it is not finished at $950°C$. The CO is produced not only in the decomposition of carboxylic anhydrides. Calo *et al.* (1997) analyzed the CO evolution curves and resolved two additional peaks at $\sim680°C$ and $\sim840°C$ ($70–90°C$ higher at a heating rate of $50°C/min$). Smaller quantities of CO are usually also found at the same temperatures at which CO_2 is evolved.

Otake and Jenkins (1993) measured the neutralization behavior of an activated carbon with different bases after decomposing the surface oxides at constant temperatures in an N_2 stream. There was a linear correlation of the NaOH consumption with the quantity of the CO_2-forming complexes on the surface, whereas no proportionality was found with the CO-forming complexes. Also the changes in $NaHCO_3$ and Na_2CO_3 uptake after thermal degradation of the surface complexes were measured in this study.

A related method, the observation of *transient kinetics* (TK), has been applied to studies of the role of the various oxygen groups in the continuous combustion of carbons (Zhuang *et al.*, 1994; 1995; 1996). The carbon is oxidized in an $^{18}O_2$ (5%)/He stream at constant temperature, e.g. 500°C, and the gas atmosphere is abruptly changed to $^{16}O_2$(in He). The resulting changes of $C^{18}O_2$, $C^{16}O^{18}O$ and $C^{18}O$ evolution are followed over time. The concentrations of $C^{18}O$ and $C^{18}O_2$ decreased in two steps, a fast one and a considerably slower one; $C^{18}O$ decayed even more slowly than $C^{18}O_2$ and $C^{16}O^{18}O$ and was still observable after 30 minutes. Carboxylic anhydrides and lactones desorb as $CO_{2'}$ (and CO) without direct interaction with $O_{2'}$, whereas carbonyl and ether groups produce CO_2 and CO with participation of $O_{2'}$. The latter decomposition reactions are the main route for carbon gasification (Zhuang *et al.*, 1995). In an other experiment, the $^{18}O_2$/He stream was switched to pure helium; the same authors conclude from the quantity of evolved carbon oxides that a part of the oxygen complexes, also lactones and anhydrides, are stable at 500°C, but they are destabilized when the carbon is gasified with $^{16}O_2$, that is when the surrounding carbon is removed.

7.4.2 Infrared Spectroscopy

Although infrared spectroscopy is a standard technique for the identification of functional groups in organic compounds, many problems arise when this method is applied to surface-oxidized carbons. Transmission methods are limited by the strong absorption of infrared radiation by carbons. A few weak signals could be observed with high-surface area carbon blacks that were finely dispersed in KBr pellets (Donnet, 1968). Good spectra could be obtained in transmission with films of charred cellulose (Zawadzki, 1980, 1989), but the absorption became too strong with increasing heat treatment temperatures of the carbons. Although the sensitivity could be improved by the photothermal beam deflection technique, useful spectra could be obtained only when the chars had not been heated to higher than 880°C (Morterra *et al.*, 1984). Also Internal Reflection Spectroscopy (IRS) has been used with some success (Mattson *et al.*, 1969; Sellitti *et al.*, 1990). A break-through came with the advent of commercial Fourier Transform infrared spectrometers, in particular of Diffuse Reflectance Infrared Fourier Transform Spectroscopy (DRIFTS). Many spectra can be accumulated, the sensitivity is much higher, and difference spectra can be plotted. It is also possible to perform *in situ* studies.

A further problem is the assignment of the infrared absorption bands to specific surface groups. Usually, it is done by comparison with the spectra of organic molecules of known structure. However, the assignments are not always unambiguous. For instance, the nature of a relatively strong band at 1590–1600 cm^{-1} has been discussed for a long time. It has been ascribed by many authors to stretching vibrations of C=C bonds of the aromatic structure that are polarized by electronegative oxygen atoms bound near one end and thus become infrared-active. An alternative explanation by hydrogen-bonded, highly conjugated carbonyl groups was only recently contradicted by experiments with oxygen-18 (Fanning and Vannice, 1993). Also, different authors reported often slightly different wave numbers

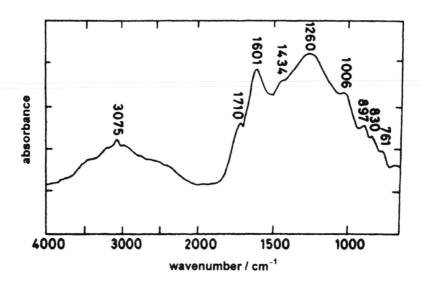

FIGURE 7. Typical DRIFT spectrum of an activated carbon (outgassed at 110°C). (From Meldrum and Rochester (1990b), with permission)

for apparently the same bands. As Meldrum and Rochester (1990a) found, the position of the bands changes a little when the carbons are imbedded in a KBr diluent. The reader is referred to the original literature (Starsinic *et al.*, 1983; O'Reilly and Mosher, 1983; Zawadzki, 1989; Sellitti *et al.*, 1990; Meldrum and Rochester, 1990 a-d; Fanning and Vannice, 1993).The ranges of observed wave numbers for the various surface functional groups were listed by Fanning and Vannice (1993). A typical DRIFT spectrum of an oxidized carbon is shown in Figure 7.

The main result of infrared spectroscopy is the confirmation of the existence of carboxyl groups and their anhydrides, of carbonyl groups on polyaromatic systems and of various groups with C–O single bonds. Cyclic anhydrides of carboxylic acids with absorption bands at 1840 and 1770 cm^{-1} were observed after oxidation of an activated carbon with dry oxygen (Meldrum and Rochester, 1990b,c). The anhydrides are hydrolyzed by water to carboxyl groups; this reaction is in part reversible on heating (Meldrum and Rochester, 1990c). With ammonia, the rings are also opened with formation of carboylic amides and ammonium salts of carboxylate ions. The evidence for the existence of cyclic lactones was not as conclusive. They give a single peak near 1760 cm^{-1} (Morterra and Low, 1985), and a clear distinction from cyclic anhydrides is not possible (Zhuang *et al.*, 1994). Strong alkali forms carboxylates (Starsinic *et al.*, 1983) and phenolates (Meldrum and Rochester, 1990a,d). Particular difficulties arise in the assignment of various bands in the range of C–O single-bond vibrations (ca. 1000–1500 cm^{-1}) to specific groups, such as C–OH bonds in carboxylic acids, phenols, alcohols (in disordered constituents of the carbons), or C–O of ethers. Meldrum and Rochester (1990a) thought the existence of epoxide groups to be very likely, but other authors did not discuss this possibility. In several publications, the existence of nitro groups, –NO_2, was claimed after oxidation with nitric acid, but other authors could not confirm this observation. As to be expected, there is evidence for C–H bonds in carbons prepared from organic precursors. In addition to the C–H stretching vibrations in the 3000 cm^{-1} region and a broad band near 1635 cm^{-1}, three peaks in the region of the wagging vibrations of aromatic C–H bonds increase in intensity after hydrogenation treatment at ca. 450°C in the presence of a dispersed nickel catalyst (Meldrum and Rochester, 1990c). They have been assigned to isolated aromatic C–H groups, pairs of C–H groups, and three neighboring C–H groups. The hydrogenation treatment lead to decreased intensity of the bands of oxygen-containing groups, as expected. Zawadzki (1989) reported that he found IR-spectroscopic evidence for the formation of superoxide ions, O_2^-, on adsorption of oxygen.

7.4.3 Photoelectron Spectroscopy

X-ray Photoelectron Spectroscopy (XPS) is a surface-sensitive analytical method. The sample is irradiated with monochromatic soft X-ray radiation (usually Mg-Kα or Al-Kα), resulting in the emission of electrons of the inner shells of the atoms. The kinetic energy (E_{kin}) of the emitted electrons corresponds to the energy of the exciting radiation ($h\nu$) minus the binding energy (E_b) and a correction (Φ) for the work function of the instrumental setup:

$$E_{kin} = h\nu - E_b - \Phi$$

The binding energies can thus be derived from the measured kinetic energies. The binding energy of an inner-shell electron is affected by the over-all electron density around the nucleus and, therefore. by the chemical environment. Binding of the atom to electronegative elements reduces the electron density and, correspondingly, increases the binding energy of the core electrons. The differences for differently bonded forms of a given element are usually similar to or not much larger than the line widths (in the order of one or a few eV), and curve-fitting procedures for deconvolution are necessary for a good resolution.

The kinetic energy of the photoelectrons is usually around 1000 eV or less, and they can penetrate only relatively thin layers without inelastic scattering and energy loss. The escape depth and, therefore, the depth of analysis corresponds to only a few atomic layers, usually 1–5 nm.

XPS can be used to determine the elemental composition of the surface with a precision of about 1 at.% (hydrogen is not detectable by XPS). The exact binding energy for the peaks gives information on the chemical surrounding of the heteroatoms in a carbon surface. However, the differences in the O1s binding energies for the various oxygen species are relatively small, due to the high electronegativity of oxygen, and the signals for the different species are not very well resolved (Desimoni et al., 1992). Desimoni et al. (1994) succeeded recently in resoving a spectrum of oxidized carbon fibers into four signals (Figure 8), and Papirer et al. (1993; 1994) found even five components in the O1s peak of an oxidized carbon. In practice, it is more convenient to measure the C1s signal. Carbon atoms linked to two oxygen atoms (e.g. carboxyl groups) or to one oxygen atom by a double-bond (carbonyl groups) or a single bond (ethers and phenols) differ in their C1s binding energies. The signals appear as satellites on the high-binding energy side of the main peak for the carbon matrix (see Figure 9). The C1s and O1s binding energies for the various species are listed in Table 3.

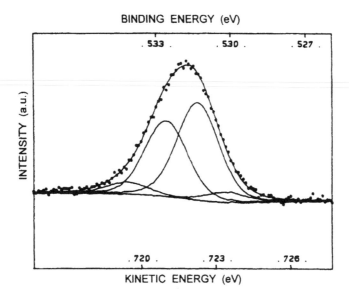

FIGURE 8. O1s photoelectron spectrum (XPS) of a PAN-based high-modulus carbon fiber after surface-oxidation in a corona discharge in an O_2/N_2 atmosphere. (From Desimoni *et al.* (1994), with permission)

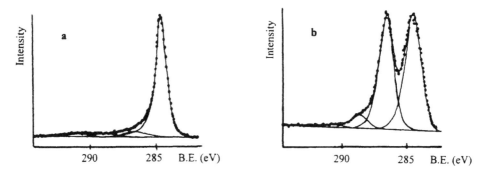

FIGURE 9. C1s photoelectron spectra (XPS) of a high-modulus carbon fiber, (a) untreated, (b) after electrochemical oxidation in HNO_3. (From Kozlowski and Sherwood (1987), with permission).

Despite these shortcomings and the relatively expensive instruments, XPS is the method of choice for studies of carbons with a small surface area such as carbon fibers (Kozlowski and Sherwood, 1987; Donnet and Bansal, 1990) or glass-like carbon (Kelemen and Freund, 1988).

XPS is also very useful in elucidating the nature of the nitrogen species in carbons prepared from nitrogen-containing precursors or carbons treated with ammonia at high

TABLE 3. C1s and O1s binding energies for the various surface groups (From Desimoni *et al.*, 1992).

Species	C1s signal eV	O1s signal eV
Graphite (and aromatics)	284.6	–
Alcohols, phenols, ethers	286.1	532.3–533.3[a]
Carbonyl groups, quinones	287.6	531.1–531.8[b]
Carboxyl groups	289.1	see [a] and [b]
Carbonate esters	290.6	
Water		535.6–536.1
Surface plasmon	291.3	–

[a] also the hydroxyl oxygen of carboxyl groups.
[b] also the carbonyl oxygen of carboxyl groups.

temperatures, e.g. 900°C (Stöhr *et al.*, 1991). The N1s spctra could be deconvoluted to signals from pyridine- or acridine-type nitrogen (binding energy 398.7 eV) and pyrrole-type nitrogen (400.4 eV), i.e. >N–H groups, bound at the edges of the graphene layers (Pels *et al.*, 1995). So-called quaternary nitrogen (401.3 eV), that had been first observed in coals, seems in the case of carbons to be substitutional nitrogen atoms within the graphene layers. The extra electrons must be taken up by empty states, i.e. in the conduction band. A further signal at binding energies near 403 eV has been tentatively assigned to structures similar to pyridine-N-oxide (Pels *et al.*, 1995).

7.4.4 Electrokinetic Measurements

As has been mentioned in Section 7.4.5.1, the presence of acidic or basic surface sites can be deduced from the electrophoeretic behavior of the carbons. A surface with acidic groups has cation exchange properties, the protons of the Brønsted acids can dissociate in aqueous suspensions, and the carbon surface acquires a negative charge. Due to thermal effects, the counter ions, H^+ or other cations, may overcome the Coulombic attraction and form a diffuse double layer near the surface. Their concentration decreases with increasing distance from the surface (Shaw, 1980; Leon y Leon and Radovic, 1994).The extension of the diffuse double layer decreases with increasing ionic strength of the electrolyte (see text books of colloid chemistry, e.g. Shaw, 1980). Analogously, a basic surface has a positive surface charge with anions in the diffuse double layer. In addition to pyrone-type structures (see Section 7.3.5.3) and π bonds, also carbonyl groups, ethers etc. are weak bases that are protonated in strongly acidic suspensions.

Since the acidity or basicity of the surface groups is relatively weak, their dissociation and the surface charge are dependent on the pH of the electrolyte. There is a pH value at which the positive and negative surface charges are equal and the net surface charge is zero. This pH is called the *point of zero charge* (PZC). Hence, H^+ ions in the electrolyte determine the potential of the surface (versus the electrolyte at infinite distance), they are potential-determining ions.

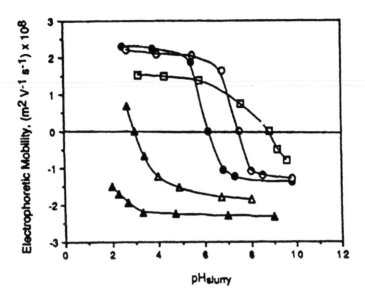

FIGURE 10. Electrophoretic mobility in 10^{-3} M KNO_3 of a furnace black after various pretreat-
ments. Filled circles: as received, filled triangles: oxidized with boiling 6 M HNO_3 (1 hr.), open
circles: heat-treated under argon at 2500°C (after 1 year storage in air), open triangles: heat-treated at
2500°C and oxidized with 6 M HNO_3, open squares: heat-treated at 2500°C, stored in air for 1 year
and then heat-treated in N_2 at 1200°C. (From Solar *et al.* (1990), with permission)

Charged particles in a suspension will migrate in an electric field in the direction to
the oppositely charged electrode. This phenomen is called *electrophoresis*. Due to friction
forces, the particles move at constant velocity.

However, the electrophoretic mobility, i.e. the ratio of velocity to field strength, is not
determined by the charge and potential directly at the interface. Water (or other electrolyte)
molecules adjacent to the surface will move with the particle, and a gradient of relative
velocity is established between the interface and the bulk electrolyte. The effect is as if a thin
layer of electrolyte were fixed to the particle surface and some of the electrolyte, including a
part of the counter ions, will migrate with the particle. A hypothetical shear plane is assumed
between this fixed surface layer and the bulk electrolyte, and the (usually) smaller charge
and potential at this shear plane determine the electrokinetic behavior. The potential at the
hypothetical shear plane is called the *electrokinetic potential* or *zeta (ζ) potential*. The pH at
which the ζ-potential is zero, the *isoelectric point* (IEP), is not necessarily identical with the
the PZC. A further complication is that specific adsorption of multivalent ions or charged
molecules, in particular tenside ions, can affect the surface potential and the ζ-potential.

The ζ-potential can be calculated from the electrophoretic mobility by application of the
equations of Hückel or v.Smoluchowski (for details, see text books of colloid chemistry,
e.g. Shaw, 1980). The electrophoretic mobility and the ζ-potential depend strongly on the
pretreatment of a carbon as shown in Figure 10 for a carbon black.

The PZC of solids can be determined from plots of pH vs. the quantity of acid or base equivalents added to the suspension. The curves are flattened by higher ionic strengths (concentrations) of the supporting electrolyte (usually $NaNO_3$ solutions). The curves for different ionic strengths intersect ideally in one point, corresponding to the PZC. *De facto*, the curves (usually measured for three $NaNO_3$ concentrations) often do not intersect at the same point, and determination of the PZC becomes ambiguous.

This problem can be circumvented by the technique of "mass titration" introduced by Noh and Schwarz (1989, 1990). The pH of an electrolyte is changed in the direction towards the PZC by addition of the solid powder. It approaches asymptotically the PZC with an increasing ratio of solid to electrolyte. Noh and Schwarz (1990) have shown that nearly the same pH is established at about 10 wt.% addition of activated carbon to solutions of NaOH or HNO_3 with pH values of 3, 6, and 11. The PZC values determined by this method agreed within the experimental scatter with those measured by acid/base titration.

The PZC values reflect an average over the acidic and basic surface sites of a carbon. In various studies, the PZC and IEP values decreased with increasing surface oxidation (Noh and Schwarz, 1990: Radovic and Rodriguez-Reinoso, 1997). Moreno-Castilla *et al.* (1995) reported that both, PZC and IEP increased with increasing burn-off during steam activation of a carbon. The surface became more basic with increasing weight loss. However, the IEP values of activated carbons tend to be considerably smaller than the PZC values (see Table 4). Mostly, values in the acidic range, pH 2–4.5, are observed even when the PZC values of the carbon are in the basic range. The difference is usually smaller with non-porous carbon blacks. The reason for the large differences between IEP and PZC has been discussed by Corapcioglu and Huang (1987) and, in more detail, by Menéndez *et al.* (1995). Electrokinetic behavior and IEP are determined by the charge on the surface of the porous carbon particles, whereas the properties of the much larger surface area in the micropore system are included in the measurements of the PZC. Hence, false information on the surface properties of an activated carbon is obtained from electrophoretic measurements. The external surface of freshly activated or reduced acivated carbons is oxidized on exposure to the air (see Section 7.3.1) and is further changed by slow oxidation in ambient air (aging). Obviously, such changes are very much slower in the micropore system, due to the slow diffusion of oxygen and water vapour. The gases are consumed near the entrances of the micropores, and there is initially mostly nitrogen in the pores. As well basic as acidic PZC values have been reported for activated carbons taken "from the shelf" without further pretreatment (except low-temperature outgassing), see Table 4. Very likely, this is caused by different times of storage.

Menéndez *et al.* (1997) showed that the PZC of Norit C carbon increased with increasing temperature of heat treatment (under N_2 as well as under H_2 up to a maximum at $1880°C$. At $2600°C$ the PZC was lower (Table 4). The authors assume that double bonds (localized bonds) at the edge of the graphene layers are more basic than the system of the basal faces. At $2600°C$, a considerable ordering of the carbon layers and a significant increase of the average diameter of the layers (crystallites) occurred as evidenced by a decreased broadening of the X-ray diffraction lines. This results in a decreased ratio of Lewis base sites on the edges to those on the basal planes, leading to an overall decrease of basicity.

Knowledge of the PZC is of practical importance in the preparation of carbon-supported metal catalysts (Solar *et al.*, 1990). The catalysts are prepared by deposition of the precursor

TABLE 4. Points of zero charge and isoelectric points for selected activated carbons.

Carbon and pretreatment	PZC	IEP	Reference
Calgon BPL, outgassed at 110°C	9.16	–	Barton *et al.* (1997)
N$_2$ at 900°C	10.4	–	"
ox. with 6 N HNO$_3$ at 85°C, 2 hrs.	3.02	–	"
Norit C, as received	2.5	2.6	Menéndez *et al.* (1995)
N2 at 950°C	5.2	3.5	"
H$_2$ at 950°C	9.0	4.9	"
H$_2$ at 950°C, then 30 d ambient air	8.4	3.3	"
ox. with air at 250°C	2.2	2.2	"
ox. w. conc. HNO$_3$ at b.p., 30 min.	2.8	1.4	"
HT at 1800°C[a]	8.9	–	"
HT at 2600°C[a]	7.6	–	"
Activated carbon[b], as prepared	10.4	2.4	Moreno-Castilla *et al.* (1995)
ox. with H$_2$O$_2$	4.8	2.3	"

[a] Atmosphere not given, must have been an inert gas.
[b] Prepared from carbonized almond shells by steam activation, 25% burn-off.

metal compounds, such as H$_2$[PtCl$_6$], on the carbon, either by ion exchange adsorption from excess solution or by the so-called "incipient wetness impregnation", when just the right volume of solution is added to fill the pore volume of the activated carbon. A third method is impregnation with a larger amount of solution, followed by evaporation of the solvent. The metals are reduced, after drying, with hydrogen, usually at 200–400°C. It is desirable to have a high dispersion of the supported metal, defined as the fraction of metal atoms exposed in the surface of the metal particles, since only the surface atoms are catalytically active sites. Evidently a fine dispersion of the reduced metal is favored by a molecularly-disperse distribution of the precursor ions. Therefore, it would be advisable to use negatively charged, i.e. oxidized carbon surfaces for the anchoring of cationic metal complexes, such as hydrated metal ions or ammine complexes, e.g. [Pd(NH$_3$)$_4$]$^{2+}$. For anionic complexes, such as [PtCl$_6$]$^{2-}$, a basic carbon surface would be appropriate. However, H$_2$[PtCl$_6$] or H$_2$[PdCl$_4$] solutions are usually employed (often even with excess hydrochloric acid) which are strongly acidic and have pH values smaller than the PZC of the carbon. The surface is positively charged under these conditions, and a high content of oxygen-containing surface groups provides a high concentration of H$^+$-binding sites. High dispersions are obtained with carbons with acidic surface oxides, in this case. The extensive literature on this subject has been briefly reviewed by Roman-Martínez *et al.* (1995) and, in detail, by Radovic and Rodriguez-Reinoso (1997).

    Clearly, it is also useful to know the nature of the internal surface when activated carbons are used as adsorbents for acidic and basic gases.

7.5 SURFACE–BOUND NITROGEN

Surprisingly, there exists relatively little literature on nitrogen surface compounds on carbons. In principle, several nitrogen functions should be possible on carbon surfaces.

Treatment of oxidized carbons with ammonia results in the formation of ammonium salts of carboxylic acids which dehydrate to carboxylic amides on heating:

$$|-COOH + NH_3 \rightarrow |-COO^- NH_4^+ \xrightarrow{\Delta} |-CO-NH_2 + H_2O.$$

The amides could, in principle, give nitrides on further heating

$$|-CO-NH_2 \xrightarrow{\Delta} |-C \equiv N + H_2O,$$

but no conclusive evidence for this has been described.

When carbons are treated with ammonia at higher temperatures, e.g. 600– 900°C, nitrogen is bound by the carbons (Boehm et al., 1983). In this temperature range, ammonia decomposes with formation of reactive species such as NH_2 and NH radicals and atomic H which attack the carbon surface. At 900°C, there is considerable gasification of the carbon, resulting in an increase of the micropore volume and the apparent (BET) surface area. The reaction products CH_4, HCN and $(CN)_2$ have been detected in the gas phase. There was no significant gasification at 600°C. Investigations of such carbons by XPS (see Section 4.3) detected the presence of pyridine-type nitrogen and pyrrole-type nitrogen, i e. >NH groups as part of aromatic rings, in addition to substitutional (quaternary) nitrogen atoms (Figure 11). The extra electrons of substituted nitrogen atoms are more or less delocalized and can be transferred to adsorbed species (Stöhr et al., 1991). The same XPS signals are observed with carbons prepared by carbonization of nitrogen-containing polymers at ca. 1000°C. Pyridine- and pyrrole-type sites have basic character and such carbons had a larger uptake of hydrochloric acid than similar nitrogen-free carbons.

The binding of $-NH_2$ groups on the surface of carbon films has been described by Jones and Sammann (1990) and by Jones (1993) who treated the films in an ammonia plasma. The X-ray photoelectron spectrum was very similar to that observed after treatment of carbons with ammonia at high temperatures. An N1s signal with a binding energy of 398.9 eV was assigned to surface amines, whereas most authors ascribe this signal to pyridine-type nitrogen (see above and Section 7.4.3). Amine groups have been attached to the carbon surface by etching the carbon in an argon plasma and then reacting the free-radical sites with ammonia or with ethylene diamine. The existence of amine functions was proven by the condensation reaction with carboxylic acids, e.g. nicotinic acid, to form carboxylic amides (Oyama et al., 1978; Oyama and Anson, 1979). No amine groups could be detected, however, after reaction of chlorinated carbon blacks with ammonia (Boehm, 1966).

7.6 SURFACE-BOUND HALOGENS

7.6.1 Chemisorption of Chlorine and Bromine

Chlorine molecules, as dioxygen, can not react with the basal faces of graphite. Chemisorption is possible only at the edges of the graphene layers. The basal planes are attacked, however, by atomic chlorine, as by other free radicals. As described by Schlögl and Boehm (1988), chemisorbed chlorine can be detected by XPS (Cl2p and C1s spectra) after UV

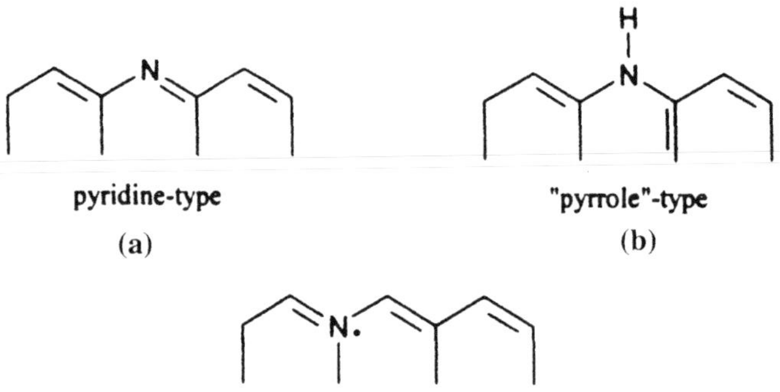

FIGURE 11. Model of (a) pyridine-type nitrogen, (b) pyrrole-type nitrogen, and (c) substitutional (quaternary) nitrogen in graphene layers.

irradiation of a graphite suspension in carbon tetrachloride. CCl_4 is photolyzed to Cl and CCl_3 radicals under these conditions. On prolonged irradiation, even graphite flakes of several mm diameter were converted to an evil-smelling sludge. This is only possible if the attack occurs also at the basal crystal faces.

Chlorine gas reacts, however, with the edges of the graphene layers. Nikitina *et al.* (1970) reacted graphite wear dust with Cl_2 at room temperature and pressures up to $7 \cdot 10^{-3}$ Pa. They reported that $1.1 \cdot 10^{15}$ Cl atoms were fixed per cm^2 of prismatic faces. This corresponds to 0.091 nm^2 per chemisorbed Cl atom which is very close to the value of 0.082 nm^2 calculated per C atom in the active surface area (ASA, see Section 7.3.3).

The edge sites of graphite have been also chlorinated by heating to 800–900°C in the presence of chlorine-containing organic vapors (McKee and Spiro, 1985). The reactive sites on the ASA are apparently blocked by the chemisorbed Cl atoms; the gasification rate in air at 650°C was lowered by more than 50%. Also for carbon fibers, a drastic decrease of the oxidation rate at 800°C by admixture of Cl_2 to a 10% O_2 (in Ar) stream has been described by Ehrburger *et al.* (1994).

Disordered, turbostratic carbons react with Cl_2 by several mechanisms (Tobias and Soffer, 1985a). In addition to fixation by free-radical sites on the surface, it can be added to olefinic bonds which may exist on the edges of the graphene layers. The main reaction, especially with carbon blacks, is by substitution of edge-bound hydrogen; one molecule of HCl is evolved for each Cl atom fixed. A reaction of minor extent on hydrogen-bearing edges is dehydrogenation without chemisorption of Cl.

Carbon blacks are heavily chlorinated by treatment with Cl_2 at elevated temperatures. The maximum uptake was found at 400–500°C (Puri and Bansal, 1967; Puri, 1970). HCl is evolved in the reaction, and its quantity was always a little higher than corresponded to the fixed chlorine. Papirer *et al.* (1995) showed in an XPS study that chlorine is bound not only on the surface of carbon blacks, but that some of it diffuses also into the depth of the particles. The $Cl 2p$ signal could be resolved into five components. The major one was from single Cl atoms bonded to carbon atoms, but there were also 11–16% of CCl_2 and CCl_3 groups. Small mounts of Cl_2 and HCl were entrapped in molecular form, and there was also some evidence for the existence of Cl_2^- ions.

Apparently, at elevated temperatures, e.g. 350–400°C, Cl_2 molecules can penetrate by activated diffusion into very narrow micropores which are inaccessible to nitrogen at 77 K (Tobias and Soffer, 1985a). They can not escape during outgassing at room temperature (Barton et al., 1987). However, this strongly physisorbed chlorine is removed by washing with water. Surface oxides are generated in this process due to the formation of strongly oxidizing hypochlorous acid by disproportionation of the Cl_2 with H_2O (Tobias and Soffer, 1985a; Barton et al., 1987).

The chemisorbed chlorine is thermally stable up to 700–800°C (Puri and Bansal, 1967; Puri, 1970; Tobias and Soffer, 1985b). With carbon blacks a substantial part of it was lost after heating in vacuo at 1000°C (Puri and Bansal, 1967), but with activated carbons, most of the chlorine remains even after outgassing at 1200°C (Puri, 1970). Complete dechlorination was achieved by heating in hydrogen to 800°C. Chlorine bound to the carbon surface is chemically quite inert. It is not hydrolyzed by washing with dilute alkali, and only a small fraction passed into solution on treatment with boiling 2.5 N NaOH for several hours (Puri, 1970). Carbon surfaces with chemisorbed chlorine are hydrophobic (Hall and Holmes, 1993).

The reaction of carbon surfaces with bromine is, in principle, analogous. However, considerably less Br than Cl is bound on the carbon surface under comparable conditions (Puri, 1970). An XPS study by Papirer et al. (1994) of brominated carbon blacks showed also five different Br components and sub-surface reaction.

7.6.2 Surface-Bound Fluorine

Graphite reacts with fluorine to carbon tetrafluoride at temperatures above ~ 600°C. Although disorderd carbons react at lower temperatures, and carbons with a high surface area may ignite spontaneously with F_2 at atmospheric pressure, surface fluorination is posible at lower (partial) pressures. Carbon black and activated carbon have also been fluorinated by means of a CF_4 plasma (Nakahara et al., 1994) with similar results as found in the reaction with F_2 gas.

The reaction with fluorine is complicated by the fact that graphite reacts with F_2 in a solid-state reaction near 600°C to the layer compound graphite fluoride, $(CF)_n$ ($x \approx 1$). One F atom is covalently bonded to each sp^3-hybridized C atom of the preserved, but buckled layers (Watanabe et al., 1988). Graphite fluoride is electrically insulating and white when x is approaching 1. Also, dark compounds of the composition $C_y F$ ($y \approx 2$) with carbon double layers are known. Furthermore, graphite forms true graphite intercalation compounds with hydrogen fluoride, $C_{24n}^+ HF_2^- \cdot m\,HF$ (n is the stage number, i.e. the number of graphene layers between the intercalated layers). The electrically conducting compounds are generated when graphite is oxidized under hydrogen fluoride; e.g. with F_2. Such compounds are also formed with well-ordered turbostratic carbons.

Hall and Holmes (1991) described the fluorination of activated carbon with F_2 diluted with N_2 (5% F_2) at temperatures at or below 200°C. The surface fluorination caused a decrease of the micropore volume and apparent (BET) surface area. Some elementary fluorine was strongly physisorbed; KI solution was oxidized by the fluorinated carbons.

A detailed XPS study of fluorinated carbon blacks has been described by Nansé et al. (1997a,b). The main species was F atoms bonded to sp^2 and sp^3 C atoms at the surface. There were also a few CF_2 groups and minor quantities of CF_3 groups. Interestingly, also some fluorine corresponding to semi-ionic fluorine was observed similar to that in C_yF graphite fluoride.

The fluorination of carbon fibers has become of interest recently. Chong and Ohara (1992) reported that the surface of pitch-based carbon fibers became more hydrophilic after a weak fluorination treatment. The contact angle with water (see end of Section 7.3.4) decreased from 95° to a minimum of 77° after three minutes exposure to F_2 at 1.33 10^3 Pa and ca. 25°C. This angle was lower than that observed after electrochemical oxidation of the fiber. On more intensive fluorination, the contact angle increased again. This observation is interesting because the wetting of the carbon fibers by polar phases can be improved by such treatment without significant weakening of the fibers that may occur by normal oxidation. This may be important for the mechanical behavior of carbon fiber composites. A similar decrease of the contact angle with water in the beginning of fluorination was observed by Bismarck et al. (1997) with one type of fiber but not with an other product. Possibly, this phenomenon is related to the occurrence of "semi-ionic" fluorine which was detected by XPS by Chong and Ohara (1992) and Bismarck et al. (1997) or to an activation of the surface by slight fluorination for rapid 'aging', see Section 7.3.2 (Chong and Ohara, 1992). In principle, one would expect fluorinated carbon surfaces to be hydrophobic in analogy to poly(tetrafluoroethylene), and, indeed, a water-repellent nature of the surface has been described by Li et al. (1995) for activated carbon fibers that had been fluorinated at 100°C.

7.7 GRAFTING OF SPECIFIC SURFACE GROUPS

The creation of simple surface groups has been described in the preceding sections. However, there is not much variability and control of the surface properties, apart from hydrophobic or hydrophilic and acidic or basic behavior. There has been great interest in bonding well-defined groups or molecules to the carbon surface that are amenable to specific interactions with other chemical entities. For instance, carbons with organo-metallic complexes or enzymes anchored to their surface might be useful as catalysts, or specific surface groups might be beneficial for strong fiber-matrix interactions in carbon fiber composites.

The oldest and easiest way of attaching larger molecules to the carbon surface is by condensation reactions with carboxyl or hydroxyl groups on the surface of oxidized carbons. As shown in Section 7.3.5.2, carboxyl groups can be esterified with alcohols. An acid catalyst such as HCl or BF_3 can facilitate the reaction (Papirer et al., 1978):

$$|-COOH + HO-R \rightarrow |-COOR + H_2O$$

A facile condensation reaction is also possible with acyl chloride groups. Such $-COCl$ groups can be easily created from carboxyl groups by reaction with suitable compounds such as thionyl chloride, $SOCl_2$, phosphorus pentachlride, PCl_5 (Boehm and Knözinger, 1983) or oxalyl chloride, $(COCl)_2$ (Bouwman et al., 1981):

$$|-COOH + SOCl_2 \rightarrow |-COCl + HCl + SO_2$$

followed by

$$|-COCl + ROH \rightarrow |-COOR + HCl$$

The advantage of using thionyl chloride is that the reaction products HCl and SO_2 are volatile. However, it is better to apply this reagent in solution in benzene or toluene since pure $SOCl_2$ begins to decompose at reflux temperature with production of SCl_2, S, and other reaction products (Boehm and Knözinger, 1983; Jester *et al.*, 1980; Koval and Anson, 1978; Watkins *et al.*, 1975). Of course, acyl chloride molecules can also be reacted with OH groups of the carbon surface. Surface amides are produced in an analogous reaction of surface acyl chlorides with amines. A review of such reactions has been published by Murray (1984).

Free carboxyl groups can be condensed with amines in a mild way by addition of carbodiimides, e.g. dicyclohexylcarbodiimide (DCC), $C_6H_{11}-N=C=N-C_6H_{11}$, which binds the water of condensation with formation of cyclohexyl-substituted urea:

$$|-COOH + H_2N-R \xrightarrow{DCC} |-CO-NHR + H_2O.$$

Among the many compounds fixed on the carbon surface by these reactions are benzidine (Evans and Kuwana, 1977), 4-aminomethylpyridine (Koval and Anson, 1978), and tetra(aminophenyl)porphyrine (Lennox and Murray, 1978). In the last case, it was shown by use of XPS that, in average, two of the amine functions were bound to the surface.

An other way to make use of the $-OH$ and $-COOH$ functions of oxidized carbons is to use cyanuric chloride as a linking agent:

(Reaction Scheme 5)

One of the chlorine atoms is used for a condensation reaction with the surface functions, and the remaining two chloride functions can be utilizd for the reaction with alcohols or amines (Yacynych and Kuwana, 1978; Harttig and Hüttinger, 1980).

Also chiral groups can be attached to a carbon surface in this way. This may be useful in the application of carbon adsorbents in liquid chromatography or in enenatioselective electrochemical reductions or oxidations at carbon electrodes (Firth *et al.*, 1976).

A handicap of surface anchoring by ester or amide bonds is their susceptibility to hydrolysis in acidic or alkaline media. However, carbons esterified with *n*-hexanol or *n*-decanol were hydrophobic and resistant to hydrolysis (Matsumura *et al.*, 1976).

Organic amines have been immobilized on carbon fiber surfaces by simple refluxing, in some cases even at room temperature (Buttry *et al.*, 1999). The reaction was assumed to be an addition of the nucleophilic amine molecules to reactive, electropositive C=C double bonds on the surface, possibly vinylic groups (see Section 7.2):

$$-CH=CH- + RNH_2 \rightarrow -CH(NHR)-CH_2-$$

When carbons are heated to ca. 1000°C *in vacuo*, existing surface compounds, e.g. surface oxides, are decomposed. Very reactive sites remain, and these sites can be used for the linking of orgnic molecules to the surface. Mazur *et al.* (1977) concluded that half-filled orbitals (free radicals) or empty orbitals should stick out from the edges of the graphene layers, and that olefins should add onto such sites. There was, indeed, irreversible adsorpion of olefins on carbon films that had been outgassed at 1020°C and 10^3 Pa, e.g. of *iso*-octene, allene, cyclopentadiene, acrylyl chloride or acrylic methyl ester. Ethylene and propylene were chemisorbed in much smaller quantities, however. The grafted acrylyl chloride could be used for the binding of 2-aminoanthraquinone:

(Reaction Scheme 6)

The chemisorption of alkenes and alkanes on graphitized carbon black, outgassed at 950°C, has also been described by Hoffman *et al.* (1984). The quantity of irreversibly adsorbed alkenes was about twice as high as that of alkanes. The chemisorbed species could not be thermally desorbed without decomposition.

As mentioned in the beginning of Section 7.6.1, not only the edges of the graphene layers, but also the basal planes are attacked by free radicals. Consequently, it should be possible to attach organic radicals to carbon surfaces. Donnet and Henrich (1975) reported that free radicals generated by the thermolysis of azo*iso*butyronitrile were bound on the surface of carbon blacks. The radical formation reaction is

$$Me_2(NC)C-N=N-C(CN)Me_2 \xrightarrow{\Delta} N_2 + 2\ Me_2(NC)C^\bullet$$

Lauryl radicals, $C_{11}H_{23}CO$ from the decomposition of lauroyl peroxide reacted in an analogous way. The derivatization of carbon electrodes has been reviewed by McCreery (1991). Barbier *et al.* (1990) used electrochemical oxidation of amines in acetonitrile solution to generate radical sites on the nitrogen atoms

$$R-NH_2 \xrightarrow{-e^-} R-NH_2^{+\bullet} \longrightarrow H^+ + R-NH^\bullet$$

The radicals were grafted onto glass-like carbon rods and high-tensile strength carbon fibers by formation of covalent C−N bonds. The surface-treated rods were used as electrodes in cyclovoltammetry. When the amines contained reducible functions, e.g. nitro groups in 4-nitrobenzylamine, the presence of these groups could be established by the corresponding reversible reduction peaks in cyclovoltammetry. The quantitty of grafted molecules was estimated from the integrals of the current peaks during voltammetric reduction. It corresponded well to the areas occupied by upright-standing molecules in a dense monolayer. It had been hoped that free amine functions could be grafted to the carbon surface by use of ω-diamines, e.g. ethylenediamine. However, subsequent reactions with a reagent for free NH$_2$ groups, 4-fluorobenzoyl chloride, showed that that 75% of the fixed ethylene diamine units were bound to the surface with both nitrogen atoms.

Delamar *et al.* (1992; 1997) created p-substituted phenyl radicals by electrochemical reduction of 4-nitrophenyldiazonium salts

$$O_2N-C_6H_4-N_2 + e^- \rightarrow N_2 + O_2N-C_6H_4^{\bullet}$$

The nitrophenyl radicals were grafted to carbon electrodes. Also in this case, the NO$_2$ groups in the reaction product could be electrochemically reduced to NH$_2$ groups. Many other derivatives of aniline with, e.g., –CN, –COOH, –COPh groups instead of the –NO$_2$ group could be converted in the usual way to the corresponding diazonium salts and thereafter electrochemically reduced to the radicals (Delamar *et al.*, 1992; Allongue *et al.*, 1997). After reaction with the nitrophenyl radicals, XPS showed the presence of nitro groups, but also of some amine nitrogen. The unintentional reduction of a part of the nitro groups was confirmed by Liu and McCreery (1995). These authors found also that the density of grafted molecules was higher on the edge faces of highly oriented pyrographite (HOPG) than on the basal faces. The radicals attack also the basal faces, but the prismatic faces of graphite are more reactive. The grafted molecules occupied an area of 0.256 nm^2. Recently, Delamar *et al.* (1997) developed continuous process. for the grafting of 4-nitrophenyl groups onto the surface of carbon fibers. Furthermore, the reaction can be controlled in such a way that the nitrophenyl groups were reduced to aminophenyl groups in the same step. Embedded in an epoxy resin, the amine functions reacted with the epoxy functions of the polymerizing resin, resulting in a considerably increased adherence between the fibers and the matrix.

Grafting of free radicals created by anodic oxidation of aryl acetates was recently described by Andrieux *et al.* (1997). Various p-substituted phenylacetates were oxidized by the Kolbe reaction

$$X-C_6H_4-CH_2-COO^- \xrightarrow{-e^-} CO_2 + X-C_6H_4-CH_2^{\bullet}$$

It was shown in cyclovoltammetric experiments that the electrode surface from glass-like carbon became increasingly unreactive on repeated cycling, due to coverage of the surface with the grafted species. Also naphthylmethyl and anthracenylmethyl radicals could be fixed on the surface of HOPG, and a regular, ordered packing of the anchored species on the basal planes could be shown by scanning tunneling microscopy (Andrieux *et al.*, 1997).

The grafting of *polymers* to carbon surfaces has been described in a review by Donnet *et al.* (1975). When vinyl monomers are polymerized by a radical mechanism in the presence of carbon black, polymer chains are grafted to the surface (Ohkita *et al.*, 1975). The polymerization is started by the addition of dilauroyl peroxide or azo*iso*butyronitrile which producee free radicals on warming. Polystyrene and poly(methylmethacrylate) chains have been anchored on the surface of furnace blacks. Channel blacks with an oxidized surface were less suitable. Tsubokawa and Koshiba (1997) produced free radical sites on the surface of an activated carbon by decomposition of azo groups that had been attached to the surface by condensation reactions (see above) with suitable azocompounds. These radical sites were used for graft polymerization of vinyl monomers. Alternatively, grafting of polymers carrying the radical sites onto the surface of carbon blacks has been achieved by thermolysis of polymers containing azo groups, peroxy groups or TEMPO groups (= 2, 2, 6, 6-tetramethyl-1-piperidinyloxy groups) (Hayashi *et al.*, 1998; Yoshikawa *et al.*, 1998).

Anionic polymerization of olefins can be induced by lithium organic compounds. The "living polymers" may graft onto carbonyl groups or carbxylic methylesters on a carbon surface (Donnet *et al.*, 1971). Also a cationic graft polymerization reaction has been described (Tsubokawa *et al.*, 1980).

Attachment of *biomolecules* on carbon surfaces has become of interest, too, e.g. for the preparation of specific electrodes. Glucose oxidase (Bourdillon *et al.*, 1980) and lactate dehydrogenase (Laval *et al.*, 1984) have been anchored on an oxidized carbon surface by condensation reactions with the aid of carbodiimide (see above). Anchoring by the radical route (by electrochemical reduction of surface diazonium salts, see above) has been used for the immobilization of glucose oxidase on the surface of glass-like carbon (Bourdillon *et al.*, 1992). However, the catalytic acivity was considerably lower than with free glucose oxidase. This was explained by the proximity of the enzyme to the carbon surface.

References

Allongue, P., Delamar, M. Desbat, B., Fagebaume, O., Hitmi, R., Pinson, J. and Savéant, J.-M. (1997) *J. Am. Chem. Soc.*, **119**, 201.

Andrieux, C.P., Gonzalez, F. and Savéant, J.-M. (1997) *J. Am. Chem. Soc.*, **119**, 4292.

Bandosz, T.J., Jagiello, J., Contescu, C. and Schwarz, J.A. (1993) *Carbon*, **31**, 1193.

Bansal, R.C., Vastola, P.J. and Walker, P.L., Jr. (1974) *Carbon*, **12**, 357.

Barbier, B., Pinson, J., Desarmot, G. and Sanchez, M. (1990) *J. Electrochem. Soc.*, **137**, 1757.

Barton, S.S., Evans, M.J.B., Koresh, J.E. and Tobias, H. (1987) *Carbon*, **25**, 663.

Barton, S.S., Evans, M.J.B., Halliop, E. and MacDonald, J.A.F. (1997) *Carbon*, **35**, 1361.

Billinge, B.H.M. and Evans, M.G. (1984) *J. Chim. Physique*, **81**, 779.

Billinge, B.H.M., Docherty, J.B. and Bevan, M.J. (1984) *Carbon*, **22**, 83.

Bismarck, A., Tahhan, R. Springer. J., Schulz, A., Klapötke, T.M.. Zell, H. and Michaeli, W: (1997) *J. Fluorine Chem.*, **84**, 127.

Boehm, H.P. (1966) *Angew. Chem., Internat. Ed. Engl.*, **5**, 533.

Boehm, H.P. (1966) in *Advances in Catalysis*, Vol. 16, Ed. D.D. Eley, H. Pines and P.B. Weisz. (New York, Academic Press) pp. 179–274.

Boehm, H.P. (1994) *Carbon*, **32**, 759.

Boehm, H.P. and Knözinger, H. (1983) in *Catalysis – Science and Technology*, Vol. 4, Ed. J.R.Anderson and M.Boudart. (Berlin, Heidelberg, New York, Springer-Verlag) pp. 39–207.

Boehm, H.P. and Voll, M. (1970) *Carbon*, **8**, 227.

Boehm, H.P., Diehl, E., Heck, W. and Sappok, R. (1964) *Angew. Chem., Internat. Ed. Engl.*, **3**, 669.

Boehm, H.P., Diehl, E. and Heck, W. (1966) *Proc. 2nd Internat. London Carbon and Graphite Conference.* (London, Soc. Chem. Ind.) p. 360.

Boehm, H.P., Mair, G., Stöhr, T., de Rincón, A.R. and Tereczki, B. (1983) *Fuel*, **63**, 1061.

Bourdillon, C. ; Bourgeois, J.P. and Thomas, J.P. (1980) *J. Am. Chem. Soc.*, **102**, 4231.

Bourdillon, C., Delamar, M., Demaille, C., Hitmi, R., Moiroux, J. and Pinson, J. (1992) *J. Electroanal. Chem.*, **336**, 113.

Bouwman, R., Freriks, I.L.C. and Wife, R.L. (1981) *J. Catal.*, **67**, 282.

Brunauer, S., Deming, L.S., Deming, W.E. and Teller, E. (1940) *J. Am. Chem. Soc.*, **62**, 1723.

Brunauer, S., Emmett, P.H. and Teller, E. (1938) *J. Am. Chem. Soc.*, **60**, 309.

Burstein, R. and Frumkin, A. (1929) *Z. Physik. Chem. (Leipzig)*, **A 141**, 219.

Buttry, D.A., Peng, J.C.M., Donnet, J.-B. and Rebouillat, S. (1999) *Carbon*, **37**, 1929.

Calo, J.M., Cazorla-Amorós, D., Linares-Solano, A., Román-Martínez, M.C. and Salinas-Martínez de Lecea, C. (1997) *Carbon*, **35**, 543.

Carasco-Marín, F., Rivera-Utrilla, J., Joly, J.P. and Moreno-Castilla, C. (1996) *J. Chem. Soc., Faraday Trans.*, **92**, 2779.

Carrott, P.J.M., Drummond, F.C., Roberts, R.A. and Sing, K.S.W. (1988) *Chem. and Ind.*, 371.

Chen, P. and McCreery, R.L. (1996) *Anal. Chem.*, **68**, 3958.

Chen, P., Fryling, M.A. and McCreery, R.L. (1995) *Anal. Chem.*, **67**, 3115.

Chong, Y.-B. and Ohara, H. (1992) *J. Fluorine Chem.*, **57**, 169.

Contescu, A., Contescu, C., Putyera, K. and Schwarz, J.A. (1997a) *Carbon*, **35**, 83.

Contescu, A. Vass, M. Contescu, C. Putyera, K. and Schwarz, J.A. (1997b) *Carbon*, **36**, 247.

Corapcioglu, M.O. and Huang, C.P. (1987) *Carbon*, **25**, 569.

Cronce, D.T., Mansour, A.N., Brown, R.P. and Beard, B.C. (1997) *Carbon*, **35**, 483.

Cvetanovic, R.J. and Amenomiya, Y. (1967) in *Advances in Catalysis*, Vol. 17, Ed. R.D.Eley, H.Pines and P.B.Weisz. (New York, Academic Press) pp. 103–149.

Delamar, M., Hitmi, R., Pinson, J. and Savéant, J.-M. (1992) *J. Am. Chem. Soc.*, **114**, 5883.

Delamar, M., Désarmot, G., Fagebaume, O., Hitmi, R., Pinson, J. and Savéant, J.-M. (1997) *Carbon*, **35**, 801.

Desimoni, E., Casella, G.I. and Salvi, A.M. (1992) *Carbon*, **30**, 521.

Desimoni, E, Salvi, A.M., Biader Ceipidor, U. and Casella, I.G. (1994) *J. Electron Spectr. Relat. Phenom.*, **70**, 1.

Donnet, J.-B. (1968) *Carbon*, **6**, 161.

Donnet, J.-B. and Bansal, R.C. (1990): *Carbon Fibers*, 2nd Ed. (New York, Marcel Dekker).

Donnet, J.-B. and Henrich, G. (1960) *Bull. Soc. Chim. France* , 1609.

Donnet, J.-B., Lahaye, J. and Schultz, J. (1966) *Bull. Soc. Chim. France*, **1966**, 1769.

Donnet, J.-B., Riess, G. and Majowski, G. (1971) *European Polym. J.*, **7**, 1065.

Donnet, J.-B., Papirer, E. and Vidal (1975) in *Chemistry and Physics of Carbon*, Vol. 12. Ed. P.L. Walker, Jr. and P.A. Thrower. (New York, Marcel Dekker) pp. 171–207.

Drushel, H.V. and Hallum, J.V. (1958) *J. Phys. Chem.*, **62**, 1502.

Egawa, M. (1996) *Ph.D. thesis*, University of München, Germany.

Ehrburger, P., Pusset, N. and Dziedzinl, P. (1992) *Carbon*, **30**, 1105.

Ehrbuerger, P., Rabillaud, F. and Dantzer, J. (1994) *Carbon*, **32**, 537.

Elliott, C.M. and Murray, R.W. (1976) *Anal. Chem.*, **48**, 1247.

Epstein, S.D., Dalle-Molle, E. and Mattson, J.S. (1971) *Carbon*, **9**, 609.

Evans, J.F. and Kuwana, T. (1977) *Anal. Chem.*, **49**, 1632.

Fanning, P.E. and Vannice, M.A. (1993) *Carbon*, **31**, 721.

Fedorov, G.G., Zarif'yants, Yu.A. and Kiselev, V.F. (1963) *Zh. Fiz. Khim.*, **37**, 2344.

Firth, B.E., Miller, L.L., Mitani, M., Rogers, T., Lennox, J. and Murray, R.W. (1976) *J. Am. Chem. Soc.*, **98**, 8271.

Fryling, M.A., Zhao, J.A. and McCreery, R.L. (1995) *Anal. Chem.*, **67**, 967.

Fu, R., Zeng, H. and Lu, Y. (1994) *Carbon*, **32**, 593.

Garten, V.A. and Weiss, D.E. (1957 a) *Austral. J. Chem.*, **10**, 309.

Garten, V.A. and Weiss, D.E. (1957 b) *Rev. Pure Appl. Chem.*, **7**, 69.

Gregg, S.J. and Sing, K.S.W. (1982) *Adsorption, Surface Area and Porosity*, 2nd ed., (London and New York, Academic Press).

Hall, P.J. and Calo, J.M. (1980) *Energy Fuels*, **3**, 370.

Hall, C.R. and Holmes, R.J. (1991) *Colloids Surf.*, **58**, 339.

Hall, C.R. and Holmes, R.J. (1993) *Carbon*, **31**, 881.

Harttig, H. and Hüttinger, K.J. (1980) *J. Colloid Interface Sci.*, **78**, 295.

Hayashi, S., Naito, A., Machida, S., Okazaki, M., Maruyama, K. and Tsubokawa, N. (1998) *Appl. Organometal. Chem.*, **12**, 743.

Hennig, G.R. (1966) in *Chemistry and Physics of Carbon*, Vol. 2, Ed. P.L. Walker, Jr. (Marcel Dekker,New York), pp. 1–49.

Hermann, H., Schubert, Th., Gruner, W.and Mattern, N. (1997) *Nanostructured Materials.*, **8**, 215.

Hoffman, W.P., Vastola, F.J. and Walker, P.L., Jr. (1984) *Carbon*, **22**, 585.

Hüttinger, K.J., Höhmann, S. and Seiferling, M. (1991a) *Carbon*, **29**, 449.

Hüttinger, K.J., Höhmann-Wien, S. and Krekel, G. (1991b) *Carbon*, **29**, 1281.

Ismail, I.M.K. (1987) *Carbon*, **25**, 653.

Jester, C.P., Rocklin, R.D. and Murray, R.W. (1980) *J. Electrochem. Soc.*, **127**, 1979.

Jones, C. (1993) *Surface Interface Anal.*, **20**, 357.

Jones, C. and Sammann, E. (1990) *Carbon*, **28**, 509.

Kelemen, S.R. and Freund, H. (1988) *Energy Fuels*, **2**, 111.

Kinoshita, K. and Bett, J.A.S. (1973) *Carbon*, **11**, 403.

Kolthoff, I.M. (1932) *J. Am. Chem. Soc.*, **54**, 4473.

Koval, C.A. and Anson, F.C. (1978) *Anal. Chem.*, **50**, 223.

Kozlowski, C. and Sherwood, P.M.A. (1987) *Carbon*, **25**, 751.

Kruyt, H.R. and de Kadt, G.S. (1929) *Kolloid-Z.*, **47**, 44.

Kuretzky, T. and Boehm, H.P. (1994) *Extended Abstracts, Carbon '94, International Carbon Conference*, Granada, Spain, p. 262.

Laine, N.R., Vastola, F.J. and Walker, P.L., Jr. (1963) *J. Phys. Chem.*, **67**, 2030.

Lander, J.J. and Morrison, J. (1966) *Surface Sci.*, **4**, 241.

Laval, J.M., Bourdillon, C. and Moiroux, J. (1984) *J. Am. Chem. Soc.*, **106**, 4701.

Lennox, J.C. and Murray, R.W. (1978) *J. Am. Chem. Soc.*, **100**, 3710.

Leon y Leon, C.A. and Radovic, L.R. (1994) in *Chemistry and Physics of Carbon*, Vol. 24, Ed. P.A.Thrower. (New York,. Marcel Dekker) pp. 213–310.

Leony Leon, C.A., Solar, J.M., Colemas, V. and Radovic, L.R. (1992) *Carbon*, **30**, 797.

Lewis, I.C. and Singer, L.S. (1981). in *Chemistry and Physics of Carbon*, Vol. 17. Ed. P.L.Walker, Jr. and P.A.Thrower. (New York, Marcel Dekker) pp. 1–88.

Li, G., Kaneko, K., Ozuki, S., Okino, F., Ishikawa, R., Kanda, M. and Touhara, H. (1995) *Langmuir*, **11**, 716.

Liu, Y.C. and McCreery, R.L. (1995) *J. Am. Chem. Soc.*, **117**, 11254.

Matskevich, E.S., Strazhesko, D.N. and Globa, V.E. (1974) *Adsorptsiya, Adsorbenty*, **2**, 36.

Matsumura, Y. and Takahashi, H. (1979) *Carbon*, **17**, 109.

Matsumura, Y., Hagiwara, S. and Takahashi, H. (1976) *Carbon*, **14**, 247.

Mattson, J.S. and Mark, H.B. (1969) *J. Colloid Interface Sci.*, **31**, 131.

Mazur, S., Matusinovic, T. and Camman, K. (1977) *J. Am. Chem. Soc.*, **99**, 3888.

McCreery, R.L. (1991) in *Electroanaytical. Chemistry*, Ed. A.J. Bard, (New York, Marcel Dekker) pp. 221–374.

McKee, D.W. and Spiro, C.L. (1985) *Carbon*, **23**, 437.

Meldrum, B.J. and Rochester, C.H. (1990 a) *J. Chem. Soc., Faraday Trans.*, **86**, 2997.

Meldrum, B.J. and Rochester, C.H. (1990 b) *J. Chem. Soc., Faraday Trans.*, **86**, 861.

Meldrum, B.J. and Rochester, C.H. (1990 c) *J. Chem. Soc., Faraday Trans.*, **86**, 1881.

Meldrum, B.J. and Rochester, C.H. (1990 d) *J. Chem. Soc., Faraday Trans.*, **86**, 3647.

Ménendez, J.A., Illán-Gómez, M., Leon y Leon, C.A. and Radovic, L.R. (1995) *Carbon*, **33**, 1655.

Menéndez, J.A., Phillips, J., Xia, B. and Radovic, L.R. (1996) *Langmuir*, **12**, 4404.

Menéndez, J.A., Xia, B. Phillips, J. and Radovic, L.R. (1997) *Langmuir*, **13**, 3414.

Menéndez, J.A., Suárez, D., Fuente, E. and Montes-Morán, M.A. (1999) *Carbon*, **37**, 1002.

Miura, K. and Morimoto, T. (1991) *Langmuir*, **7**, 374.

Molina-Sabio, M., Muñecas, M.A. and Rodriguez-Reinoso, F. (1991). in *Proc. Conf. on the Chracterization of Porous Solids (COPS) II*, Ed. F.R. Rodriguez-Reinoso, J. Rouquerol, K.S.W. Sing and K.K. Unger. (Amsterdam, Elsevier) p. 329.

Moreno-Castilla, C., Ferro-García, M.A., Joly, J.P., Bautista-Toledo, I., Carrasco-Marín, F. and Rivera-Utrilla, J. (1995) *Langmuir*, **11**, 4386.

Morterra, C. and Low, M.J.D. (1985) *Langmuir*, **1**, 320.

Morterra, C., Low, M.J.D. and Severdia, A.G. (1984) *Carbon*, **22**, 5.

Murray, R.W. (1984) *J. Electroanal. Chem.*, **13**, 191–368.

Nakahara, M., Ozawa, K. and Sanada, Y. (1994) *J. Mater. Sci.*, **29**, 1646.

Nansé, G., Papirer, E., Fioux, P., Moguet, F. and Tressaud, A. (1997a) *Carbon*, **35**, 371.

Nansé, G., Papirer, E., Fioux, P., Moguet, F. and Tressaud, A. (1997b) *Carbon*, **35**, 515.

Newman, M.S. and Muth, C.W. (1951) *J. Am. Chem. Soc.*, **73**, 4627.

Nikitina, O.V., Kiselev, V.F. and Lejnev, N.N. (1970) *Carbon*, **8**, 402.

Noh, J.S. and Schwarz, J.A. (1989) *J. Colloid Interface Sci.*, **130**, 157.

Noh, J.S. and Schwarz, J.A. (1990) *Carbon*, **28**, 675.

O'Reilly, J.M. and Mosher, R.A. (1983) *Carbon*, **21**, 47.

Ohkita, K., Tsubokawa, N., Saito, E., Noda, M. and Takashina, N. (1975) *Carbon*, **13**, 443.

Otake, Y. and Jenkins, R.G. (1993) *Carbon*, **31**, 109.

Oyama, N. and Anson, F.C. (1979) *J. Am. Chem. Soc.*, **101**, 1634.

Oyama, N., Brown, A.P. and Anson, F.C. (1978) *J. Electroanal. Chem.*, **87**, 435.

Pan, Z.J. and Yang, R.T. (1990) *J. Catal.*, **123**, 206.

Papirer, E., Guyon, E. and Perol, N. (1978) *Carbon*, **16**, 133.

Papirer, E., Challamel, X., Wu, D.Y. and Fouletier, M. (1993) *Carbon*, **31**, 129.

Papirer, E., Lacroix, R. and Donnet, J.-B. (1994) *Carbon*, **32**, 1341.

Papirer, E., Lacroix R., Donnet, J.-B., Nansé, G. and Fioux, P. (1995) *Carbon*, **33**, 63.

Pels, J.R., Kapteijn, F., Moulijn, J.A., Zhu, G. and Thomas, K.M. (1995) *Carbon*, **33**, 1661.

Petit, J.C. and Bahaddi, Y. (1993) *Carbon*, **31**, 821.

Puri, B.R. (1962) *Proc. 5th Biennial Conference on Carbon*, Pennsylvania State Univ., Vol. 1 (Oxford, Pergamon Press) p. 165.

Puri, B.R. (1970) in *Chemistry and Physics of Carbon*, Vol. 6, Ed. P.L.Walker, Jr. (New York, Marcel Dekker) pp. 191–282.

Puri, B.R. and Bansal, R.C. (1967) *Carbon*, **5**, 189.

Puri, B.B., Singh, S. and Mahadjan, O.R. (1965) *Indian J. Chem.*, **3**, 54 (1965).

Radovic, L.R. and Rodriguez-Reinoso, F. (1997) in *Chemistry and Physics of Carbon*, Vol. 25, Ed. P.L. Thrower. (New York, Marcel Dekker) pp. 243–358.

Rivin, D. (1962) *Proc. 4th Rubber Technology Conference*, London, p. 1.

Rivin, D. (1963) in *Proc. 5th Bienn. Conference on Carbon*, Pennsylvania State University, 1961, Vol. 2. (Oxford, Pergamon Press) p. 199.

Rivin, D. (1971) *Rubber Chem. Technol.*, **44**, 307.

Roman-Martínez, M.C., Cazorla-Amoros, D., Linares-Solano, A., Salinas-Martínez de Lecea, C., Yamashita, H. and Anpo, M. (1995) *Carbon*, **33**, 3.

Ryndin, Yu.A., Alekseev, G.C., Simonov, P.A. and Likholobov, V.A. (1989) *J. Mol. Catal.*, **55**, 109.

Sappok, R. and Boehm, H.P. (1968 a) *Carbon*, **6**, 283.

Sappok, R. and Boehm, H.P. (1968 b) *Carbon*, **6**, 573.

Schlögl, R. and Boehm, H.P. (1988) *Synth.Met.*, **23**, 407.

Sellitti, C., Koenig, J.L. and Ishida, H. (1990) *Carbon*, **28**, 221.

Shaw, D.J. (1981) *Introduction to Colloid and Surface Science*, 3rd Ed. (London, Butterworths) pp. 145–182.

Singer, L.S. and Lewis, I.C. (1982) *Appl. Spectroscopy*, **36**, 52.

Simonov, P.A., Romanenko, A.V., Prosvirin, I.P., Moroz, E.M., Boronin, A.I., Chuvilin, A.L. and Likholobov, V.A. (1997) *Carbon*, **35**, 73.

Solar, J.M., Leon y Leon, C.A., Osseo-Asare, K. and Radovic, L.R. (1990) *Carbon*, **28**, 369.

Spear, K.E. and Frenklach, M. (1994a) *Pure Appl. Chem.*, **66**, 1773.

Spear, K.E. and Frenklach, M. (1994b) in *Synthetic Diamond: Engineering, CVD Science and Technology*, Ed. K.E. Spear, M. Frenklach and J.P. Dismukas. (New York, J. Wiley) pp. 243–304.

Starsinic, M., Taylor, R.L., Walker, P.L., Jr. and Painter, P.C. (1983) *Carbon*, **21**, 69.

Stöhr, B., Boehm, H.P. and Schlögl, R. (1991) *Carbon*, **29**, 707.

Studebaker, M.L. (1957) *Rubber Chem. Technol.*, **30**, 1401.

Suárez, D., Menéndez, J.A., Fuente, E. and Montes-Morán, M.A. (1999) *Langmir*, **15**, 3897.

Suárez, D., Menéndez, J.A., Fuente, E. and Montes-Morán, M.A. (2000) *Angew. Chem. Internat. Ed. Engl.*, **39**, 1376.

Suh, D.J., Park, T.-J. and Ihm, S.-K. (1992) *Ind. Eng. Chem. Rev.*, **31**, 1849.

Thomas, J.M. (1965) in *Chemistry and Physics of Carbon*, Vol. 1, Ed. P.L. Walker, Jr. (New York, Marcel Dekker) pp. 121–202.

Tobias, H. and Soffer, A. (1985 a) *Carbon*, **23**, 281.

Tobias, H. and Soffer, A. (1985 b) *Carbon*, **23**, 291.

Tsubukawa, N. and Koshiba, M. (1997) *J. Macromol. Sci.*, **A34**, 2509.

Tsubokawa, N., Takeda, N. and Kanamaru, A. (1980) *Carbon*, **18**, 326.

Verma, S.K. and Walker, P.L., Jr. (1992) *Carbon*, **30**, 837.

Voll, M. and Boehm, H.P. (1970) *Carbon*, **8**, 741.

Voll, M. and Boehm, H.P. (1971) *Carbon*, **9**, 481.

Walker, P.L., Jr. (1990) *Carbon*, **28**, 261.

Walker, P.L., Jr. and Janov, J. (1968) *J. Colloid Interface Sci.*, **28**, 449.

Watanabe, N., Nakajima, T. and Touhara, H. (1988): *Graphite Fluorides*. (Amsterdam, Oxford etc., Elsevier).

Watkins, B.F., Behling, J.R., Kariv, E. and Miller, L.L. (1975) *J. Am. Chem. Soc.*, **97**, 3549.

Willman, K.W. Rocklin, R.D., Nowak, R., Kuo, K.-N., Schultz, F.A. and Murray, R.W. (1980) *J. Am. Chem. Soc.*, **102**, 7629.

Yacynych, A.M. and Kuwana, T. (1978) *Anal. Chem.*, **50**, 640.

Yang, R.T. (1983) in *Chemistry and Physics of Carbon*, Vol. 19, Ed. P.A.Thrower. (New York, Marcel Dekker) pp. 163–210.

Yoshikawa, S., Machida, S. and Tsubukawa, N. (1998) *J. Polymer Sci., Part A Polymer Chem.*, **36**, 3165.

Young, G.J., Chessick, J.J., Healey, F.H. and Zettlemoyer, A.C. (1954) *J. Phys. Chem.*, **58**, 313.

Zawadzki, J. (1980) *Carbon*, **18**, 281.

Zawadzki, J. (1989) in *Chemistry and Physics of Carbon*, Vol. 21, Ed. P.A. Thrower. (New York, Marcel Dekker) pp. 147–380.

Zhuang, Q.-L., Kyotani, T. and Tomita, A. (1994) *Energy Fuels*, **8**, 714.

Zhuang, Q.-L., Kyotani, T. and Tomita, A. (1995) *Energy Fules*, **9**, 630.

Zhuang, Q.-L., Kyotani, T. and Tomita, A. (1996) *Energy Fuels*, **10**, 169.

8. Applications of Polycrystalline Graphite

MICHIO INAGAKI

Aichi Institute of Technology, Toyota 470-0392, Japan

8.1 CHARACTERISTICS OF POLYCRYSTALLINE GRAPHITE MATERIALS

All graphite materials which have some applications in industry are polycrystalline, most of them consisting of grains which are also composed from the units of parallel stacking of carbon hexagonal layers similar to graphite, called crystallites. Therefore, the properties of polycrystalline graphite materials depend strongly on the statistical distribution of grains and also on that of crystallites inside the grains, because the crystallite is highly anisotropic, as graphite single crystal does, and as a consequence the grains are also more or less anisotropic. Polycrystalline graphite materials have the following fundamental characteristics, most of them being due to the fact that they are composed from one kind of specific carbon atoms;

(1) High thermal resistance in non-oxidizing atmosphere,
(2) High chemical stability,
(3) Non-toxic,
(4) High electrical and thermal conductivities,
(5) Small thermal expansion coefficient and, as a consequence, high thermal shock resistance,
(6) Very light weight, as bulk density of 1.5–2.2 g/cm^3,
(7) High mechanical strength at high temperatures,
(8) High lubricity,
(9) Highly reductive at high temperatures and easily dissolved into iron,
(10) Biocompatible,

179

(11) Radiation resistance,
(12) Low absorption cross-section and high moderating efficiency for neutron.

The properties of polycrystalline graphite materials in the bulk are strongly governed by the preferred orientation of crystallites, which are due to the alignment of grains and also to the textures inside the grains, and moreover by thermal history which they were experienced, primarily the maximum treatment temperature. The former depending on the precursor and the condition of their forming process, and the latter deciding the size and structural perfection of crystallites. In Figure 1, electrical conductivity, bulk density, thermal expansivity and tensile strength of polycrystalline graphite are compared with other carbon materials, natural graphite, various fibrous carbon materials and graphite intercalation compounds (GIC).

Polycrystalline graphite is a good electrical conductor, but its conductivity of roughly 2×10^5 S/m is inferior to classical metals. By intercalation of different species into the gallery between graphite layer planes, however, it is much improved, as high as copper. Thermal expansion coefficient of graphite single crystal is very anisotropic, high and positive along the c-axis (perpendicular to the graphite layer plane), but negative (i.e., shrinkage) from room temperature to about $500°C$ along the a-axis (parallel to the layer). In polycrystalline graphite materials, this anisotropic thermal expansion is spatially averaged, depending strongly on the arrangement and size of crystallites (i.e., texture). In fibrous carbons, it is mainly governed by expansion along the layer planes and so rather small values. In most of physical properties such as electrical and thermal properties, as shown on electrical conductivity and thermal expansion coefficient in Figure 1, their upper and lower limits are those parallel and perpendicular to graphite layers in a single crystal.

Mechanical properties, such as tensile strength, and bulk density are texture sensitive and so they exhibit a wide range of values. The practical values for various carbon materials including polycrystalline graphite materials (high-density isotropic graphite and graphite electrode) are much inferior than the theoretical values for graphite layer plane, because of their polycrystalline nature, i.e., the existence of various defects including grain boundaries.

8.2 FABRICATION

Generalized flow sheet for the fabrication of polycrystalline graphites is shown in Figure 2. Fillers and binders, particle size distribution of the former and their mixing ratio having to be controlled in accordance with the requirements from their applications, are mixed at a temperature higher than softening point of the binder. For fillers, petroleum and coal tar pitch cokes are mostly used, but natural graphite, carbon blacks and also recycled graphites are sometimes used. For binder, petroleum and coal tar pitches are used in most cases, because of their relatively high carbon yield as about 60 wt%, but some thermosetting resins, such as phenol and epoxy resins, are also employed. The mixtures thus prepared, which is usually called as carbon pasts, are formed after warming up at a temperature around $150°C$ either by extrusion, molding or cold isostatic pressing (CIP or rubber pressing). The formed blocks are sent to calcination process at a temperature of $700–1000°C$ (carbonization) and then to graphitization at high temperature of $2600–3000°C$.

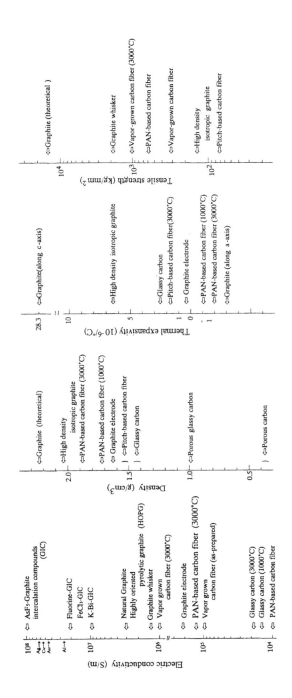

FIGURE 1. Variable ranges of electrical conductivity, bulk density, thermal expansivity and tensile strength of polycrystalline graphite materials.

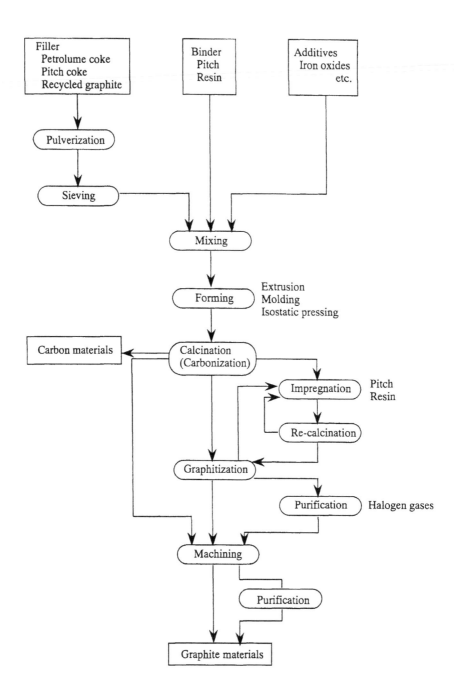

FIGURE 2. Flow chart for the fabrication of polycrystalline graphites.

Forming process is one important process for the fabrication of polycrystalline graphites, because it determines the preferred orientation of crystallites. One of three methods, extrusion, molding and isostatic pressing, is employed, their principle being schematically illustrated in Figure 3. The extrusion process is applied for the carbon pasts with thermoplastic binders, such as pitches, and gives the preferred orientation of flaky and needle-like filler particles along with the direction of extrusion, as shown in Figure 3a. Electrodes for metal processing with large diameter, various jigs with different sizes and also leads for automatic pencil with a diameter as thin as 0.3 mm are fabricated. In the molding process, crystallites are statistically aligned perpendicular to the compressing direction (Figure 3b). Carbon brushes for electric motors and electric contacts are fabricated by molding the carbon pasts. By isostatic pressing, compressive force is applied equally from all the directions and so isotropic nature is obtained in the resultant carbon blocks (Figure 3c). High-density isotropic graphites are produced by this process.

After the calcination (carbonization), three dimensional graphite structure is not developed yet, except when natural graphite or graphitized recycled materials are used, and the products obtained after calcination are called carbon materials and some of them are sent to machining for applications. But most of carbon materials are heat-treated at high temperatures as 2600–3000°C in so-called Acheson-type furnace, in which carbon materials are packed with a mixture of coke and sand, and get high temperature by the contact resistance of the packed mixture. Recently, heating of the calcined rods by direct Joule effect (so-called lengthwise graphitization furnace, LWG) is developed. For the process using conventional Acheson furnace, about one month is required to complete the graphitization treatment, but only a little more than one week for newly developed LWG process. Induction heating of the carbon rods is also used in order to avoid contamination from the packing mixtures.

In order to obtain high density, impregnation of molten pitches is applied and carbonization, impregnation and graphitization processes are repeated, as shown in Figure 2. If necessary, purification at high temperatures by using halogen gases is employed either before or after machining.

8.3 APPLICATIONS OF POLYCRYSTALLINE GRAPHITES

8.3.1 Applications for Metal Processing

Large amounts of graphite have been used as electrodes for metal processing, reaching to 2,250 million US dollars for steel production and 1,200 million dollars for aluminium production in 1991.

In Figures 4 and 5, some of graphite electrodes for electric-arc furnaces and the sketch of the furnaces are shown, respectively. Either three graphite electrodes or one are placed perpendicular to the metal melt in the furnace (Figure 5) and the electric arc between these graphite electrodes (poles) and ferrous scrap gives high temperature to melt this scrap. These electrodes must have good electrical conductivity and good refractory properties, such as low thermal expansion and high thermal shock resistance. In order to have these properties, highly developed graphite structure is required and therefore high temperature treatment (graphitization) is essential. They are gradually eroded by dissolving into molten steel and

a) Extrusion

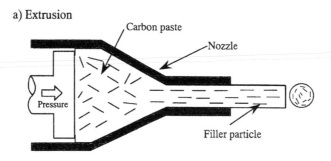

b) Molding

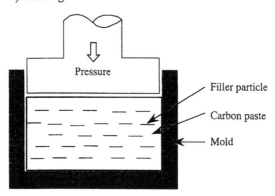

c) Isostatic pressing

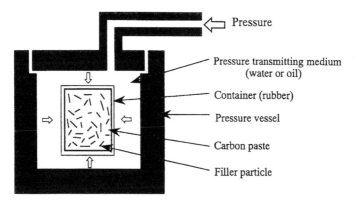

FIGURE 3. Scheme for forming processes of carbon pastes. In order to illustrate their preferred orientation, flaky filler particles are shown.

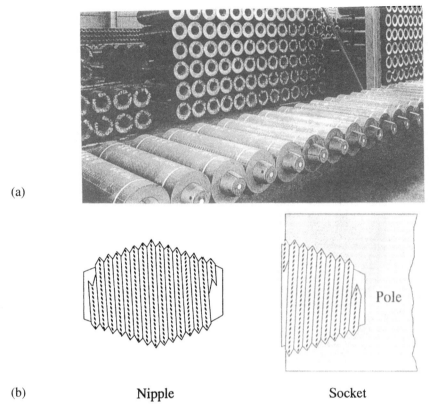

(a)

(b)   **Nipple**   **Socket**

FIGURE 4. Graphite electrodes for electric-arc furnaces. (a) General appearance (by the courtesy of SEC Co., Ltd.) (b) Connection of poles by nipple.

also by oxidation. Therefore, the graphite poles have to be lowered in regular periods and also new poles have to be connected by using nipples, (shown in Figure 4b), which are made from the same kind of polycrystalline graphite with a little higher mechanical strength.

The worldwide production of steel is slowly increasing, but its production using electric-arc furnace is increasing at more rapid rate. By the improvement of operation conditions of the furnace using high electrical power, however, the consumption of the graphite electrodes in order to produce one ton of steel is decreasing gradually, from about 7.5 kg in 1975 to about 5 kg in 1990. For the operation under high power, the graphite electrodes were required to be improved in their properties, particularly reduction of the thermal expansion and resulting improvement in thermal shock resistance. For this purpose, so-called needle-like cokes were necessary to be used.

Aluminium metal is produced by electrolysis of molten alumina (aluminium oxide) using some fluorides as flux (Figure 6). Polycrystalline graphites, therefore, are used as both anode and cathode materials. The anodes are similar to the electrodes described above, and the cathodes are usually made of calcined carbon blocks. For cathodes, carbons based anthracite were also used, but coke-based carbons were applied in recent years.

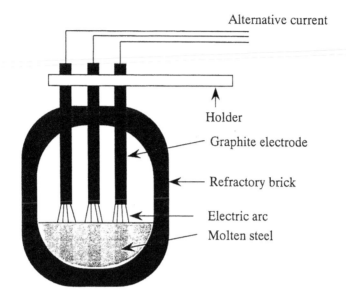

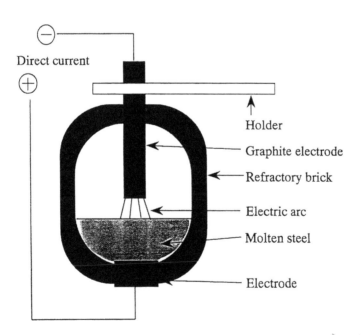

FIGURE 5.   Scheme of electric-arc furnaces for steel refining.

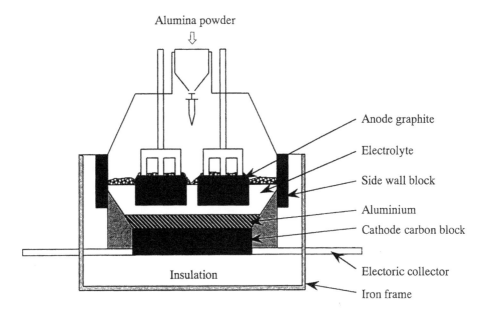

Alumina powder

Anode graphite

Electrolyte

Side wall block

Aluminium

Cathode carbon block

Insulation

Electoric collector

Iron frame

FIGURE 6. Electrolysis furnace for aluminium metal production using prebaked carbon electrodes.

Graphite materials are used as molds, dies, guides, run-out tables, furnace linings, boats and crucibles for the processing of not only ferrous metals and their alloys, such as grey and ductile irons, but also non-ferrous metals and alloys, such as copper, nickel, brass, bronze, zinc, noble metals, etc. For most of these purposes, high density isotropic graphites are selected, because of their high thermal conductivity, high mechanical strength and high density.

8.3.2 Semiconductor and Related Applications

Recent development in semiconductor technology and industries is remarkable and it leads to various microelectronics devices. Polycrystalline graphites are giving important back-up for semiconductor production in various respects.

For the synthesis of single crystals of silicon, germanium, and III–V and II–VI semiconductors by either ribbon or Czochralski crystal-pulling techniques, graphite heater and crucible (Figure 7) are essential. For these applications, high purity of graphite materials are strongly required and so purification at high temperatures by using halogen gases has to be applied to reduce the impurity content as low as 5 ppm for the most critical applications. High density isotropic graphites are mostly employed because of their accurate and easy machining, high strength and isotropy in electrical resistivity. In addition, there are so many applications, such as boats and assemblies for liquid epitaxy, susceptors and wafer trays for different CVD processes, and shields, electrodes and ionic sources for ion implantation, etc.

FIGURE 7. Graphite crucible and heater for single crystal growth of silicon (by the courtesy of Tokai Carbon Co., Ltd.).

8.3.3 Applications for Electrical and Electronic Devices

Electrical applications of polycrystalline graphites are well established and some of them are well specified in industrial standards. Brushes for electric motors have been used since 1890 and their usage is increasing now mostly due to automatization, for example, window of automobiles, and various daily necessaries, such as refrigerator, electric oven, video, etc. Some brushes are shown in Figure 8. The sizes and shapes of brushes are widely spreaded and even minute brushes are produced. To produce brushes, different raw materials, such as carbon blacks (soots), cokes and natural graphite, are employed as fillers to be formed by using either pitches or phenol resin as binder and calcined between 600 and 1000°C in inert atmosphere. Some of them heated up to 3000°C to be graphitized. Also some metals, such as copper and silver, are mixed, in order to control the electrical resistivity and lubricity.

Graphite plates are used as current-collecting shoes for electric trains, though their usage is decreasing because of high-speed driving. Graphite materials have high electrical conductivity and lubricity by themselves, do not form insulation oxides and also do not melt, and so they are used as electrical contacts.

Recent remarkable development in microelectronics opens new application fields for carbon materials, such as membrane switches, variable resistors and rechargeable batteries. In Figure 9, some membrane switches and a schematic illustration of their construction are shown. The membrane consists of fine graphite powders printed by using some binder onto polyester films. They are usually very thin, as thin as 10–15 μm, and very light. By pushing the key in keyboard of computer or switch in the control panel of an equipment, two membranes get contact and the signals are sent (Figure 9b). Their usage is going to increase, accompanied by the development of microelectronic devices. Variable resistors, which are produced by screen printing of carbon pasts of a mixture of natural graphite,

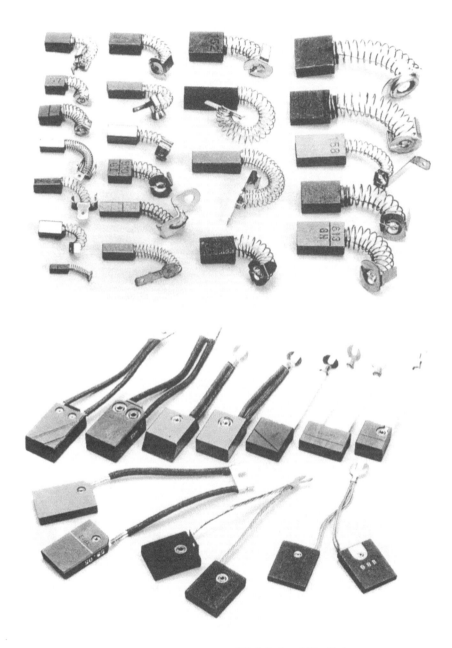

FIGURE 8. Electric brushes (by the courtesy of Fuji Carbon Mfg. Co.).

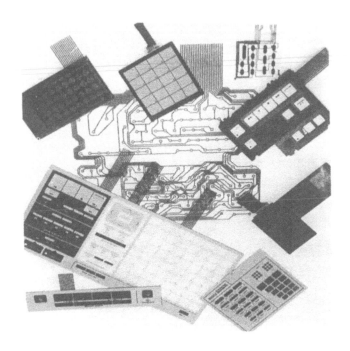

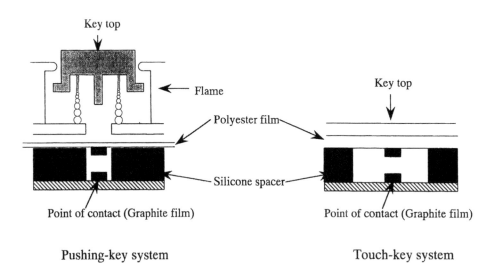

Key top

Flame

Polyester film

Silicone spacer

Point of contact (Graphite film)

Point of contact (Graphite film)

Key top

Pushing-key system **Touch-key system**

FIGURE 9. Membrane switches. (top) Some examples (by the courtesy of Nihon Kokuen Mfg. Co.) and (bottom) Schematic illustration of switch systems.

FIGURE 10. Various carbon resistors for audio equipments (by the courtesy of Tohoku Alps Co. Ltd.).

carbon black and other carbon materials using some thermosetting resin on substrate of phenol resin or alumina, are also widely used in electronic equipments. Some of them are shown in Figure 10.

Since the invention of Lclanche cell in 1864, various carbon materials, including polycrystalline graphite rods, have been used in various primary cells, manganese, maganese-alkali, air-zinc and lithium batteries. In a lithium primary cell, graphite fluoride $(CF)_n$, which presents covalent bonding between carbon atoms in graphite layers and fluorine, is successfully used as electrode materials. However, the strong demand of secondary rechargeable batteries in the relation to worldwide energy and resource problems has developed a new application of graphite materials in lithium ion batteries. The fundamental performance of these batteries are schematically illustrated in Figure 11. In rechargeable lithium ion batteries, lithium ions are extracted from cathode materials, such as ordered rock-salt-type $LiCoO_2$ and spinel-type $LiMn_2O_4$, and inserted into graphite anode during charging and *vice versa* during discharging. The anode reaction is the intercalation and deintercalation of lithium ions into graphite gallery. Therefore, the theoretical capacity of the anode is 374 Ah/kg, which corresponds to the formation of the intercalation compound with stage-1 structure LiC_6. However, a strong dependence of anode capacity on the crystallinity of graphite materials are reported. Some carbon materials with very disordered structure was reported to have much higher capacity than graphite, but also a significant irreversible capacity which is their major drawback.

Not only in lithium ion batteries, but also in various secondary batteries, different carbon materials are used; carbon fabrics for sodium-sulfur cells, electrodes of carbon fibers and

a) CF_x / Li (primary cell)

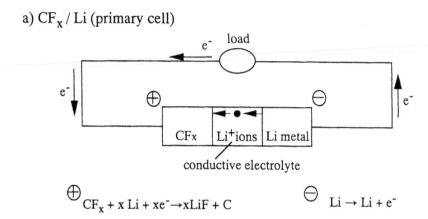

$$\oplus\quad CF_x + x\,Li + xe^- \rightarrow xLiF + C \qquad\qquad \ominus\quad Li \rightarrow Li + e^-$$

b) $LiMn_2O_4$ / C (secondary cell)

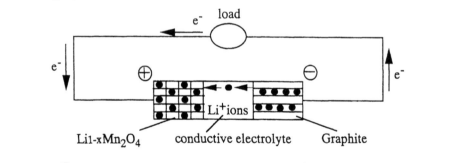

$$\oplus\quad LiMn_2O_4 \rightleftarrows LixMn_2O_4 + xLi + xe^- \qquad \ominus\quad C_6 + xLi+ + xe^- \rightleftarrows Li_xC_6$$

FIGURE 11. Scheme of operating systems of batteries. (a) Primary battery using graphite fluoride, (b) Rechargeable secondary battery using graphite as anode.

FIGURE 12. Graphite separator for the fuel cell of phosphoric acid type (by the courtesy of Showa Denko Co. Ltd.).

bipolar plates for zinc-bromine cells, carbon fiber cloths with high surface area for redox flow cells, etc.

If one of these batteries would be employed for automobile and electric power storage, a big demand of carbon materials would be expected.

Graphite materials are important structural components in the fuel cell of phosphoric acid type. A separator for fuel cell has to be gas impervious because of the separation of two fuel gases, such as oxygen and methane, and so is made of glass-like carbon. One of examples of this separator is shown in Figure 12. The electrodes for collecting electricity are also made of graphite materials and the porous carbon materials, such as felt of carbon fibers, are used as susceptor for minute particles of platinum catalyst.

8.3.4 Nuclear Applications

Polycrystalline graphite is one of the best materials for nuclear-fission applications because of its high moderating efficiency and low absorption cross-section for neutron, in addition to their high mechanical strength, chemical stability, high machinability and light weight. It has been used as the building block of many fission reactors, since the first one at Chicago, during the World War II. It is going to be used in high-temperature gas-cooled reactor. In Figure 13, the construction of a test reactor of this type in Japan is shown, where graphite materials are used as permanent reflectors at outer part, replaceable reflector at

TABLE 1. Specification on graphite materials for Japanese high-temperature gas-cooled test nuclear reactor (by the courtesy of Japan Atomic Energy Research Institute).

	Replaceable Reflectors Fuel-Element Blocks	Permanent Reflectors
Bulk density (g cm^{-3})	1.78	1.73
Tensile strength (kg cm^{-2})	258	83*1
Compressive strength (kg cm^{-2})	784	321*1
Thermal expansion coefficient (10^{-6} C)	4.06	2.34*1
		2.87*2
Thermal conductivity (cal cm^{-1} sec^{-1} C^{-1})	0.19	0.18*1
Ash content (ppm)	< 100	<7,000
Grain size (μm)	20	<800

*1 radial direction
*2 axial direction

inner part, fuel-element blocks at the center, and coating of small fuel particles. For the nuclear applications, graphite materials have to be of high purity, high strength and isotropic nature. In Table 1, specification on graphite materials used for Japanese test reactor is summarized.

Graphite materials, such as high-density isotropic graphite and carbon fiber/carbon composite, are also used in nuclear fusion reactors as interior liners, movable limiters and divertors. In Figure 14, the inner structure of Japanese test reactor JT-60 is shown. Low atomic number of graphite materials is the most important factor in reducing the interference with plasma, but many problems for practical use, such as physical and chemical sputtering resistance, thermal shock resistance, recycling of hydrogen, etc., have to be solved.

8.3.5 Mechanical Applications

Fine-grain high-density graphite materials are extensively used for bearing and seals, particularly in high-temperature circumstance. This application is mostly due to the following properties of graphite materials; lubricating and wear properties, high thermal conductivity, high compressive strength at high temperatures. However, their practical usage is somewhat limited, because they are easily oxidized and also reacted with some chemicals, such as *aqua regia*, perchloric acid, fumic sulfuric acid, chromic oxide, etc. Seal rings for gas-turbine engines, chemical pumps for corrosive fluid transfer, water pump of diesel engine, air compressor of air conditioning, and water pumps of home washing machine and dishwasher are also mentioned as commercial applications.

Flexible graphite sheets have large applications as sealing and gasket in power plants, chemical plants and engines. Gaskets and seals with flexible graphite tapes are shown in Figure 15. They are prepared from small flakes of natural graphite firstly by intercalation of mostly sulfuric acid, their deintercalation by quick heating to exfoliate (exfoliated graphite) and then rolling to make thin sheets. Their thermal and chemical stability developed a wide range of application in different industrial fields. A further expansion of its applications is also expected.

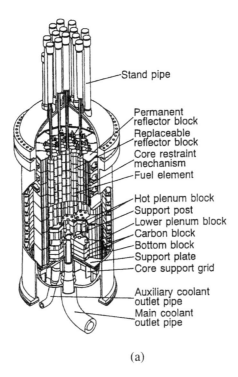

(a)

(b)

FIGURE 13. Test reactor of high-temperature gas-cooled type (by the courtesy of Japan Atomic Energy Research Institute). (a) Scheme of construction, (b) Fuel element blocks.

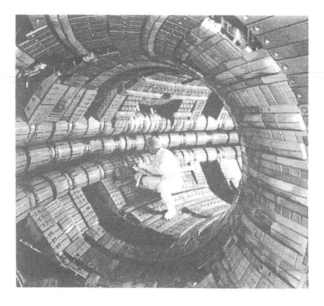

FIGURE 14. Inner structure of Japanese test fusion reactor JT-60 (by the courtesy of Japan Atomic Energy Research Institute).

8.3.6 Application for Aerospace

Polycrystalline graphite materials, specially carbon fiber/carbon composites (C/C), are one of important materials for aerospace systems, because of their high specific strength and Young's modulus (strength and Young's modulus per unit weight) and also of high thermal shock resistance. The different parts of space carriers, such as nose cone, tipfin, radar and frap are made in C/C. During re-entry into the atmosphere, these parts are expected to be exposed in very severe conditions, namely high temperatures up to 1700°C in oxidative atmosphere, very strong shock waves by hypersonic fluid, etc. In order to overcome these problems, different ideas, such as antioxidation coating of refractory oxides on the surface, etc., are necessary.

8.4 CONCLUSION

Polycrystalline graphites have been widely used in industries, because of their different characteristics and also of their wide range of properties. They have been important materials for the development of advanced technologies. This situation for polycrystalline graphites will not be changed and much severe demands for their performance are expected, such as higher functionality, higher reliability, etc. Oxidation resistance at high temperatures in oxidative atmosphere on carbon materials is strongly required in various industries, though it does not yet reach a satisfactory level.

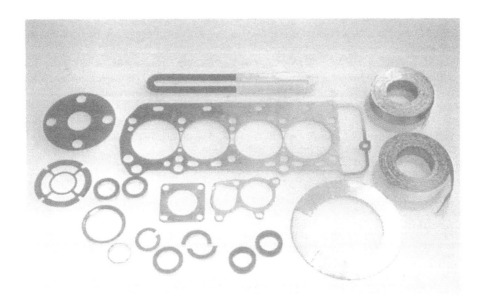

FIGURE 15. Flexible graphite sheets (by the courtesy of Toyo Tanso Mfg. Co.).

References

Buckley, J.D. and Edie, D.D. (1993) *Carbon-Carbon Materials and Composites*, Noyes Publications, Park Ridge.

Inagaki, M., and Hishiyama, Y. (1994) *New Carbon Materials* (in Japanese) Gihoudo Shuppan, Tokyo.

Legendre, L. (1991) *Le Materiau Carbone, des Ceramiques Noires aux. Fibres de Carbone*, Eyrolles.

Marsh, H. and Rodorigez-Reinoso, F. (1997) *Introduction to Carbon Materials*.

Pierson, H.O. (1993) *Handbook of Carbon, Graphite, Diamond and Fullerenes*, Noyes Publications, Park Ridge.

Thrower, P.A., A series of *Chemistry and Physics of Carbon*, Vols. 1–25, Marcel Dekker Inc., New York.

9. Carbonization and Graphitization

AGNÈS OBERLIN[1] and SYLVIE BONNAMY[2]

[1] *Mas Andrieu, 34380 Saint Martin de Londres, France*
[2] *C.R.M.D. – CNRS/Université d'Orléans, 1B rue de la Férollerie, 45071 Orléans Cedex 2, France*
e-mail: bonnamy@cnrs-orleans.fr

9.1 GENERAL REMARKS

Graphite (Bacon, 1950) is a single crystal produced by natural heating (geothermal gradient) applied to carbonaceous rocks already strained by tectonic stresses: stress graphitization (Bonijoly *et al.*, 1982). Less crystallized "graphite" is obtained by pyrolysis of suitable organic matters under an inert gas flow, up to about 3000°C. Natural heating without stresses (coalification) only produces almost pure carbon (Bonijoly *et al.*, 1982). The corresponding laboratory process (HTT $\leq$ 2000°C) is carbonization. In a first approximation both processes are equivalent. Any organic matter being made of C, H, O, N, S, it provides pure carbon by heteroatom release. Carbon precursors are thus organic sediments such as coals (van Krevelen, 1993) or kerogens (parent rocks of oil) (Durand, 1980), their derivatives such as oils, tars, heavy oil residues, pitches, or compounds like cellulose, anthracene, organic polymers, etc... Carbonization and coalification as well are thermally activated processes (Fitzer *et al.*, 1971) following kinetics laws. Pressure and residence time (also heating rate) must be fixed in addition to HTT.

During primary carbonization (van Krevelen, 1993) a part of heteroatoms is released as volatiles, among which mostly hydrocarbons evaporate with a violent outgassing (defined as oil window). Precursors only containing carbon and hydrogen such as oil derivatives have a low molecular mass. Increasing carbonization process develops heavier products up to the brittle solid state: end of primary carbonization (Ihnatowicz *et al.*, 1966; Brooks and

TABLE 1. Classes of increasing molecular mass of resins according to Dryden solvent classification.

Resins	γ	β	α
Solvent	Toluene soluble (TS)	Toluene insoluble (TI) Quinoline soluble (QS)	Quinoline insoluble (QI)
Molecular mass amu	<230–400	400–1100	—

Taylor, 1968). The various materials are classified (Table 1) according to their solubility into solvents of increasing strength (Dryden, 1952). Increasing molecular mass ranges are thus defined under the name of γ, β and α resins. During secondary carbonization (van Krevelen, 1993) only non condensable gases are produced (CH_4, H_2) by aromatic CH groups loss (defined as gas-window).

Except for some aromatics (70–80 kcal.mole^{-1}), the apparent activation energy of primary carbonization is not determined because of the complexity of the reactions involved. The activation energy values of secondary carbonization are 120–150 kcal.mole^{-1} (Fitzer et al., 1971). Theoretically, at the end of secondary carbonization pure carbon is obtained. However, the limit is difficult to set since it depends on the sensitivity of the analysis. A sudden change systematically occurring in the activation energy values near 2000°C will be chosen as the end of the process. Between 2000°C and 3000°C, activation energy of 250–280 kcal mole^{-1} is measured (Fischbach, 1971; Pacault, 1971).

Since carbon precursors are numerous, innumerable carbons are obtained. To increase again this wide variety, it is possible to repeatedly recarbonize the hydrocarbons produced as volatiles: coking of coals produces tars, distillation of tars provides pitches, carbonization of pitches produces again hydrocarbons etc... However some classification has to be introduced in this jungle, from chemical, structural and textural points of view.

A first successful attempt is due to van Krevelen, 1993, through elemental composition diagrams. Figure 1 represents the coalification paths of natural products. Coals and various kerogens are distinguished by their $(H/C)_{at.}$ ratio plotted versus $(O/C)_{at.}$. Depending on the slope of the paths, release of water can be distinguished from CO_2 and hydrocarbon release. The same patterns can be established for industrial products such as PAN-based carbon fibers as an example (Deurbergue and Oberlin, 1991).

From the structural point of view, no classification can be obtained, since among sp^2 carbons none are amorphous (they are devoid of sp^3 bonds) and all have a two dimensional crystalline structure. It is due to their stacking disorder either rotational (Warren, 1941) or produced by glides (Endo et al., 1995). Nevertheless, such structures are called turbostratic. Despite this structural uniformity, only a reduced number of carbons could graphitize, so that an other question without answer arises. Moreover what made the things more tricky, is the fact that all intermediates are known, between graphitizing and non graphitizing carbons, which keep a persistent two-dimensional structure (Oberlin, 1984; Iwashita and Inagaki, 1993). It is thus still quite justified to ask why carbons are so versatiles.

The first reason is their considerable anisotropy. The C-C bond length of an aromatic layer is 0.142 nm (versus 0.154 nm for diamond). The graphite spacing (0.335$_4$ nm) is due to van der Waals forces. The distances are even larger for disordered carbons ($\geq$ 0.4 nm (Oberlin et al., 1973; 1998)). The second reason lies in the smallness of the newly formed

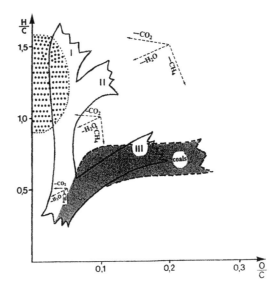

FIGURE 1. van Krevelen diagrams of various natural products. Kerogens are gathered inside bands in solid line, coals occupy the dashed area, industrial products are in the dotted area. The arrows indicate the slopes resulting from the loss of CO_2, H_2O, CH_4.

aromatic layer stacks. It can thus be expected that their three dimensional arrangements are infinitely variable. The final answer would thus be found in the determination of carbon textures (macrometric, micrometric and nanometric scales).

To characterize reliably carbonaceous materials, beyond all possible physicochemical techniques (elemental analyses, infra-red spectrometry, rheological measurements, thermogravimetry, differential thermal analysis, ...) all diffraction techniques should be added describing the elemental units and their arrangements. Among them statistical techniques describe average units, whereas imaging techniques restore the three-dimensional arrangement of individual components. The former are X-ray wide angle scattering or WAXS, electron diffractions i.e. selected area (SAD) or microarea (μD). X-ray small angle scattering or SAXS and neutron small angle scattering or SANS apply to large entities, inside plastic or liquid phases. Among imaging techniques, are optical microscopy in polarized light (OM) and scanning electron microscopy (SEM) for microtextures, joined to transmission electron microscopy (TEM) and tunneling or atomic force microscopies (STM, AFM) for nanotextures and structure. Useful references are found in Oberlin (1989) and Oberlin *et al.* (1998).

9.2 CARBON PRECURSORS

9.2.1 *Macromolecular Models*

Very low rank coals, kerogens, oil derivatives such as refinery residues, asphaltenes and pitches are actually described as macromolecules (Oberlin, 1984; van Krevelen, 1993)

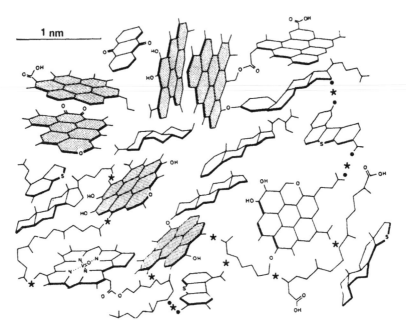

FIGURE 2. Molecular model of a kerogen. Stars suggest that covalent bonding between such molecules makes a macromolecule (Oberlin *et al.*, 1980).

made of polyaromatic molecules, whether or not stacked, connected into a continuum by various functional groups (Figure 2, Oberlin *et al.*, 1980). At that stage they appear as glass-like products viscoelastic or highly viscous.

At the very beginning of carbonization or coalification a more or less marked softening occurs corresponding to a breakage of this macromolecule. The materials contain random individual elemental units or BSU dispersed inside lighter suspension medium.

9.2.2 Elemental Units

Diffraction (Franklin, 1950), as well as imaging techniques, i.e. TEM (Oberlin, 1989) or STM (Saadaoui *et al.*, 1993; Donnet and Qin, 1993) evidence similar elemental units present in carbon precursors. These small bricks were defined as Basic Structural Units (BSU) in TEM (Figure 3a). In 002 dark-field image, each BSU seen edge-on appears as a bright dot (Oberlin *et al.*, 1973, 1974). BSU are made of two to four piled-up polyaromatic molecules less than 1 nm in diameter. Their size is near the resolution limit of the microscope, so that the lower limit is unknown. However, the polyaromatic molecules susceptible to pile up cannot be too small, since below coronene a face to face piling-up is forbidden. For small molecules, associations are possible only as edge to edge or edge to face. On the contrary, for coronene, up to larger molecules (Figure 3b, Vorpagel and Lavin, 1992), the energies calculated are −6.0 Kcal/mole for edge to face, 0.0 for edge to edge and −9.2 for face

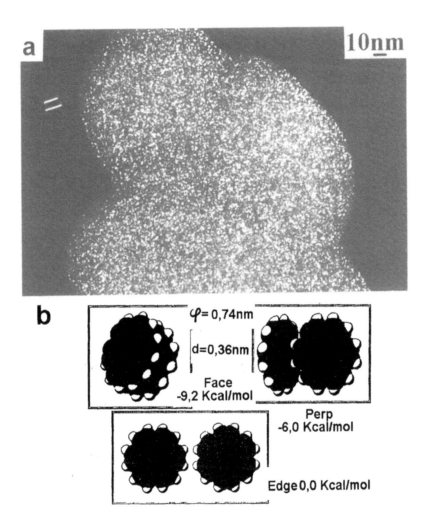

FIGURE 3. (a) TEM imaging of BSU (002 dark-field). Each bright dot is a BSU seen edge-on. BSU orientation is sketched by a double bar (Oberlin *et al.*, 1973). (b) Sketch of dicoronene (Vorpagel and Lavin, 1992).

to face. This configuration is thus the most favorable and dicoronene represents the smallest possible polyaromatic brick of aromatic layer stacks. On the other hand, the largest values of size given by TEM for BSU never trespass diekacoronene. In addition, chemical models based on the concept of colloid indicate that along the edges of polyaromatic molecules, i.e. BSU, are grafted various side-chains or groups, so that micelles or aggregates could be formed after macromolecule breakage.

The concept of colloid (Graham, 1861; van Olphen, 1963), already admitted for heavy oil products (Pfeiffer and Saal, 1940; Espinat and Ravey, 1993) coals and coal derivatives

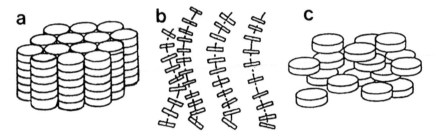

FIGURE 4. Sketches of various liquid-crystal textures made of disc-like units, from Nguyen *et al.* (1979) and Billard (1980). (a) Columnar order. (b) Distorted columnar order, (c) Nematic order.

(Chermin and van Krevelen, 1957; van Krevelen, 1993) was extended to other carbon precursors (Lafdi *et al.*, 1991). Various types of liquid crystals (Bonnamy, 1999) were recognized as well as other kinds of associations.

9.3 LIQUID CRYSTAL PHASES

Liquid crystals (mesophases) are intermediates between liquid and crystal (Friedel, 1922). Disc-like mesogenic molecules are usually made of a flat polyaromatic core, surrounded by grafted side chains (Billard, 1980). They associate into homogeneously oriented volumes limited by spherical or digitized contours (Nguyen *et al.*, 1979; Billard, 1980). Such associations were evidenced under the name of LMO, i.e. local molecular orientation (Boulmier, 1976; Villey, 1979; Oberlin *et al.*, 1980; 1989), inside oriented domains with digitized contours found in kerogens. Local molecular orientations were also found as various spherical domains (Bonnamy, 1987; 1999) among which are Brooks and Taylor mesophases (Brooks and Taylor, 1968). The highest ordering usually corresponding to spherical particles is a rectilinear arrangement (Figure 4a) of these disc-like molecules. A lower ordering is still columnar but with a total absence of lateral coherence and a more or less pronounced bending of the columns (Figure 4b). The lowest order (Figure 4c) corresponds to a nematic arrangement without columns but aligning almost in parallel the symmetry axis of the polyaromatic molecules (in Figure 4, side chains are omitted). Each disc could be there considered as a BSU. Bended columns can be realized only by a distorted piling up of BSU, like pile of plates, admitting very limited values of tilt and twist misorientations.

9.3.1 Primary Carbonization

During the hydrocarbon release of primary carbonization, i.e. outgassing of aliphatic CH groups (Rouxhet *et al.*, 1979), softening occurs first. Then a minimum of viscosity happens during which various types of liquid crystals demix and increase in size by coalescence before solidification. Minimum of hardness and solidification are precisely determined by Vickers microhardness measurements (van Krevelen, 1993; Bonnamy, 1999). Solidification is marked by breakage of the sample.

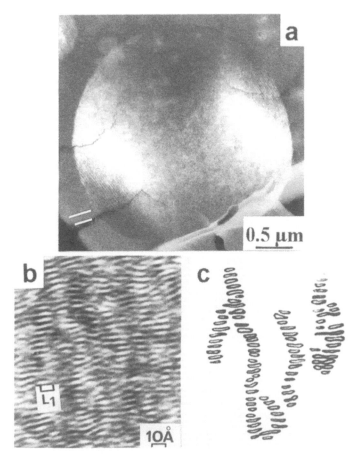

FIGURE 5. Brooks and Taylor single mesophase sphere (thin section). (a) TEM imaging (002 dark-field) from Bonnamy, 1987. BSU edge-on are oriented as indicated by double bar. (b) High resolution lattice imaging (002 lattice fringes) of a sphere extracted from its pitch by hot anthracene oil (some columns are sketched in Figure 5c), from Auguié *et al.* (1981).

9.3.1.1 Brooks and Taylor mesophases (mesophases A)

The first type of mesophases, here named A, is under the form of spherical anisotropic bodies, within which local molecular orientation occurs. Mesophase A was discovered by Taylor (Brooks and Taylor, 1968), then Ihnatowicz *et al.* (1966). It is already detailed in an other chapter of this book (chapter 10, Mochida *et al.*). Mesophase A is typical of isotropic pitches such as petroleum or coal-tar pitches. It is illustrated in TEM in Figure 5a. Each spherical body corresponds to a very large local molecular orientation (LMO) with sizes up to 50 μm and a characteristic PAN AM microtexture. At the nanoscale, inside a sphere extracted from its pitch, the BSU misorientation is reduced to $\pm 15°$. BSU are arranged in distorted columns (Figure 5b and c) (Auguié *et al.*, 1981). Columnar arrangement was

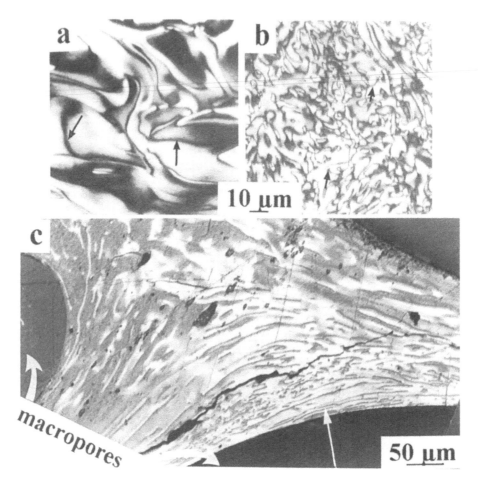

FIGURE 6. Optical microscopy of mosaics. (a) coarse mosaics. (b) medium mosaics. (c) oriented mosaics (bands) from Bonnamy (1999).

then systematically found by image analysis in TEM images of the mesophase spheres A observed in situ in their pitch (Oberlin *et al.*, 1998). Mesophase spheres are liquid crystals belonging to type b of Figure 4.

As carbonization progresses, the spherical bodies coalesce and loose their shape. At last, solidification occurs. At that point mesophase coalescence stops irreversibly under the form of mosaics or bands (Garza, 1982). Mosaics (LMO final size) are composed of isochromatic domains (Figure 6). In optical microscopy, an isochromatic domain is the area where aromatic layers are oriented in parallel and sectioned by the polished section plane more or less obliquely. It is the projection of a LMO. Its contours are delimited by grain boundaries where defects, i.e. disclinations, are gathered (White and Zimmer, 1978). These defects (arrows in Figure 6a and b) appear as isochromatic lines or extinction contours. Mosaics are classified according to the size of isochromatic domains into coarse (Figure 6a), medium (Figure 6b) and fine (not represented here). Bands, i.e. oriented mosaics (Figure 6c), are produced by the development of macropores preceding the solidification but concomitant to the establishment of total anisotropy. The overpresssure induced by the last hydrocarbon release elongates the plastic pore walls (arrow in Figure 6c) thus producing the oriented anisotropic bands (Bonnamy, 1999). There is no limitation of isochromatic domain contours since disclinations elongated by shear stresses are randomly distributed. Isochromatic domain sizes are thus not definite. Bands are the major component of solidified pitches (issued from the mesophases spheres A) in which final LMO sizes cannot be defined. Brooks and Taylor mesophases are typically produced by precursors devoid of cross-linkers, such as some kind of oxygen or sulfur bonds.

9.3.1.2 Other types of liquid crystals

As cross-linkers amount increases in the precursor (such as oil heavy products), softening temperature increases and different liquid crystals precipitate (Bonnamy, 1999). At first, to mesophase A are associated other spherical bodies here named B and C, respectively $< 1 \, \mu$m and > 200 nm (Figure 7a). Figure 7b is a larger magnification of B and C anisotropic bodies alone. Mesophase A can be entirely dark but never entirely bright in TEM 002 DF (Auguié et al., 1981). In opposition with mesophase A of Figure 7a, B and C spherical bodies are systematically entirely bright, then progressively darkened, then entirely dark (Figure 7b). They have thus not the PAN AM texture of mesophase A. The misorientation of their BSU is also much larger ($< 60°$ for B, $> 60°$ for C, versus ($30°$ for A). Then during further coalescence, they are never miscible neither with each other nor with A. They are partially quinoline soluble. They are thus of a different chemical composition. They are liquid crystals with columnar ordering (Oberlin et al., 1999), but they are quite distinct from mesophases A and belong to other categories.

As cross-linking increases again in the precursor (kerogens, coals), spheres A, B and C disappear successively and at LMO occurrence bodies not spherical but irregular (digitized contours), smaller than 200 nm, succeed (Figure 7c) (Bonnamy, 1999). As the maximum cross-linking increases, the size of the irregular domains decreases.

At the end of primary carbonization, the coalescence of the various liquid crystals provides final LMO. For spherical bodies, final LMO are coarse ($> 5 \, \mu$m), medium ($> 2 \, \mu$m) and fine ($< 2 \, \mu$m) mosaics. When mosaics are ground, they break along their disclination boundaries and they provide lamellae. Bodies with digitized contours break into micrometric fragments inside which LMO are measured. For this latter, final LMO are classified in 8 classes, from class 8 (≥ 200 nm) down to class 1 (< 5 nm) for the smallest (Oberlin, 1989). Figure 8 illustrates final LMO with digitized contours of class 5 (Figure 8a) and class 3 (Figure 8b). The log of final LMO size (Figure 9) is a linear function of a factor $F_{LMO} = ((O+S_r)/H)_{at.}$ measured at the LMO occurrence, i.e. at the liquid crystals demixtion (Bonnamy, 1987; 1999). O is the oxygen stable at LMO occurrence and supposed to be a cross-linker, S_r is the cross-linking sulfur known to be stable above $1700°$C (Bourrat, 1982; et al., 1987), H is the hydrogen remaining at LMO occurrence. Sulfur stable below $1700°$C is a promoter (Fitzer and Weisenburger, 1976). In the areas where it is released, it lets behind perfect graphite crystals accompanying swelling (puffing effect). The carbon-rich solid semi-cokes,

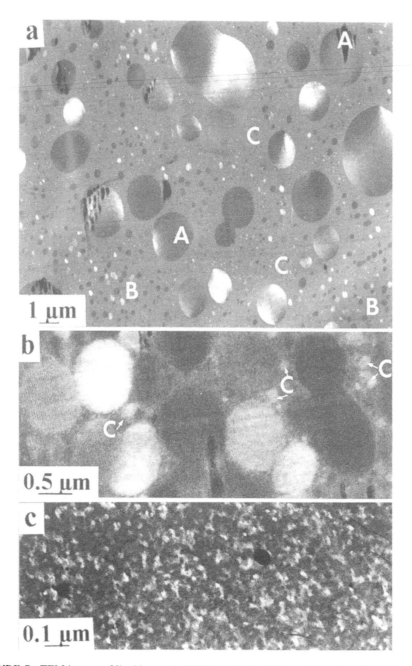

FIGURE 7. TEM images of liquid-crystals (002 dark-field of thin sections, from Bonnamy, 1999). (a) Association of liquid-crystal spheres A Brooks and Taylor mesophase, B and C: different smaller spheres (not miscible with A or between them). (b) Higher magnification of B and C liquid crystal phases. (c) Volumes with digitized contours at LMO occurrence.

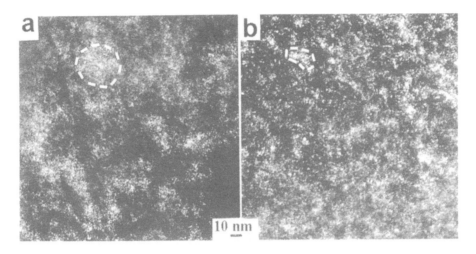

FIGURE 8. TEM imaging of LMO final sizes (002 dark-field), individual LMO domain with digitized contours is marked. (a) Class 5 (25–35 nm). (b) class 3 (10–15 nm); from Villey (1979).

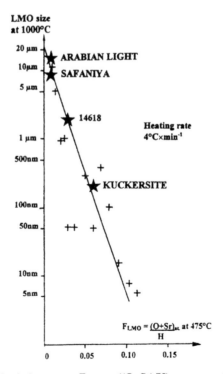

FIGURE 9. log LMO final size versus $F_{LMO} = ((O+Sr)/H)_{at.}$, oxygen O and residual sulfur Sr are cross-linkers. O is measured at 475°C, Sr at 1700°C then calculated at 475°C. H is the antagonistic hydrogen from CH groups grafted upon BSU at 475°C. Stars correspond to F_{LMO} measured at LMO occurrence; from Bonnamy (1999).

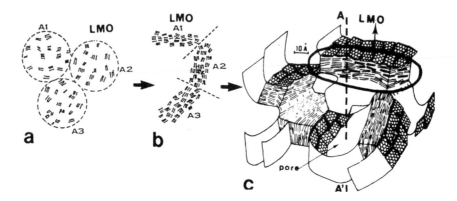

FIGURE 10. Sketch of evolution of local molecular preferred orientation (LMO). (a) at the end of primary carbonization (compact semi-coke). (b) and (c) during secondary carbonization (opening of smoothly curved pores); from Oberlin (1989).

obtained at the end of primary carbonization, still have a columnar texture inherited from the various mesophases (Oberlin *et al.*, 1999). At the micrometric scale, LMO such as A1, A2, A3,... entirely fill-up the space with continuous changes of orientation from one to the other (Figure 10a).

9.3.2 Secondary Carbonization

It produces gases by aromatic CH groups loss (Rouxhet *et al.*, 1979). Due to dangling bonds, BSU become heavy radicals so that spin concentration reaches a maximum then decreases by radical recombination (Mrozowski and Gutze, 1977; Durand, 1980). ESR (Mrozowsky, 1979) also shows that secondary carbonization is a two step process emphazized by a transition in the range 1500–1600°C.

9.3.2.1   First step (below 1500–1600°C)

During this step, the columnar arrangement detected in all the anisotropic bodies after their demixtion is maintained (Oberlin *et al.*, 1999). Only minor changes occur, due to BSU reordering. On the one hand, columnar ordering improves by elimination of defective BSU forming wedges between columns (Rouzaud and Oberlin, 1989). On the other hand, reordering of BSU decreases LMO thickness (Figure 10b), so that smoothly curved micro or even nanopores are formed. At that point, the shape of the pores is that of crumpled sheets of papers (de Fonton, 1980) (Figure 10c). The more cross-linked are the semi-cokes, the smaller are the final LMO, i.e. the smaller are the pore walls. This mechanism is common to all bodies with digitized contours. On the contrary, the early macropores typical of pitches are produced by a different mechanism since they occur before solidification. They were explained in Section 9.3.1.1.

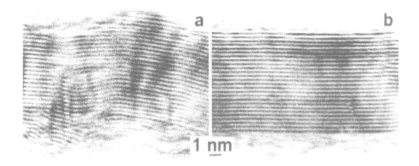

FIGURE 11. TEM imaging of secondary carbonization final steps (002 lattice fringes) from Rouzaud and Oberlin (1989). (a) Between 1600 and 2000°C. (b) End of the secondary carbonization, at the magic temperature (2000°C).

9.3.2.2   Second step (above 1600°C)

Columns suddenly disappear (Rouzaud and Oberlin, 1989). Lateral coherence newly produced between adjacent BSU, let them coalesce into continuous but distorted layers of about 5 to 20 nm in diameter (Figure 11a). N_1, the number of layers in a stack, jumps to over 40 (Oberlin and Terriere, 1973; Goma *et al.*, 1980). It is the very moment where BSU disappear, but where «crystallites» are not yet there. The carbon is still turbostratic.

The end of secondary carbonization ($\sim$2000°C) corresponds to a sudden wiping-out of all in-plane defects annealing all distorsions (Figure 11b). At this «magic» temperature carbon atoms become mobile. There is an increase in apparent activation energy from 120–150 kcal.mole^{-1} up to 250–280 kcal.mole^{-1} (Fischbach, 1971; Pacault, 1971). This self-induced plasticity is evidenced by mechanical properties measured at high temperature (up to 2600°C). The material changes suddenly from fragil to ductile (Figure 12, Souma *et al.*, 1995). Mobility of carbon atoms is responsible for the drastic change in electronic and crystallographic properties generally observed (Marchand *et al.*, 1965, 1997; Inagaki *et al.*, 1975). The resistivity decreases, the Hall effect is maximum, the magnetoresistance turns to be positive, etc... It is also at this «magic» temperature that the sudden perfection acquired by the aromatic layers transforms the macro and nanopores from smoothly curved into polyhedrons (Figure 13, Villey, 1979; Oberlin, 1989). The newly formed flat faces replacing final LMO are turbostratic grains inside which three dimensional ordering could be produced. But again, «crystallites» are not yet there.

9.3.3   Graphitization

Between 2000°C and 3000°C, graphitization eventually occurs through single carbon atoms or vacancies displacements. The lamellae issued from coarse mosaics (precursors almost devoid of cross-linkers) are equivalent to grains which could progressively graphitize, because their very large size allows carbon atoms to move freely (Warren, 1941; Franklin, 1951; Maire and Mering, 1970; Fischbach, 1971). $\bar{d}_{002}$ noticeably decreases whereas

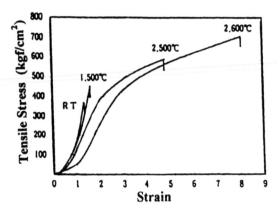

FIGURE 12. Tensile stress-strain of an isotropic graphite at various temperatures; from Souma *et al.*
(1995).

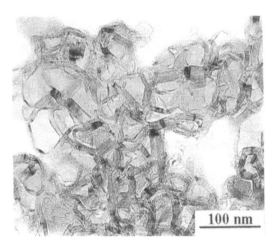

FIGURE 13. TEM imaging of polyhedral pores formed after 2000°C (contrasted bright-field); from
Villey (1979).

P_1 increases. P_1 is the probability to find a pair of layers in the graphite order. It is one only for
natural graphite because natural graphite is not formed only by thermal process but by stress-
graphitization (Bonijoly *et al.*, 1982). Inside a natural graphite crystal, mosaic domains
often called crystallites can be recognized, inside which the AB ordering is established
everywhere. In all other carbons, graphitization is only a statistically homogeneous ordering.
Domains of coherence (inside which the amplitude of the scattered wavelets combine with
each other) are characterized by their diameter and thickness. Inside them, pairs of layers
are AB stacked, other pairs are still turbostratic and are randomly distributed. The optimum
graphitization is reached for P_1 max. about 0.9 (Iwashita and Inagaki, 1993).

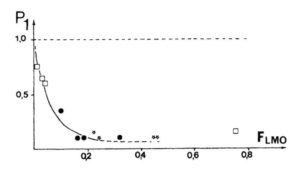

FIGURE 14. Ability to graphitize (P_1) versus the factor F_{LMO}; from Joseph and Oberlin (1983) (F_{LMO} increasing corresponds to final LMO and grain sizes decreasing). Compare to Figure 9.

As cross-linking increases, lamellae are replaced by porous carbons and the grain sizes (flat faces of the pores) decreases. P_1 maximum and $\overline{d}_{002}$ minimum reached at high temperature, respectively decreases and increases. All intermediates are observed between graphitizing and non-graphitizing carbons (Monthioux et al., 1982; Oberlin, 1989). Since grain size is inherited from LMO diameter and since LMO final size is a function of the factor $F_{LMO} = ((O+S_r)/H)_{at.}$. P_1 is a function of F_{LMO} (Figure 14, Joseph and Oberlin, 1983).

The remarkable changeability of carbons is now explained. It is due to that of their early chemical composition, since the latter introduces increasing cross-linking. Correspondingly, the type of liquid crystals obtained is therefore changed from A, to B, to C and to domains with digitized contours.

It is remarkable to note that cross-linking is not necessarily a precursor intrinsic characteristic. It can be obtained experimentally by increasing oxidation, if it is applied before solidification (Oberlin, 1989; Joseph and Oberlin, 1983; Oberlin et al., 1998), so as to go back further from graphitizing down to non graphitizing carbons, including all intermediates. Reciprocally, hydrogenation (under pressure with or without catalysts) necessarily applied before solidification, may improve materials such as coals or oils residues, from non graphitizing to graphitizing products (van Krevelen, 1993; Vogt et al., 1986; Lambert et al., 1995). Depending on the intensity of the oxidation or hydrogenation treatments, all intermediates can be reached.

9.4 ANISOTROPIC PITCHES

The liquid crystals formed during primary carbonization were characterized by their insolubility in quinoline (α resins, see Table 1). As long as they are in their pitch matrix, they are mixed with β and γ resins. Considered as colloidal systems, they are expected to be able to reversibly disperse in the matrix. This is obtained by a strong stirring (gas-sparge) of Brooks and Taylor mesophase spheres in residual matrix (Chwastiak, 1980). The new pitch obtained is 100% optically anisotropic (it resembles a coarse mosaic). It is 100% β resins quinoline soluble and toluene insoluble. Figure 15 is orthogonal 002 dark-field of a thin

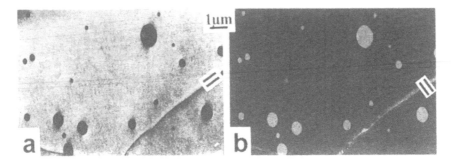

FIGURE 15.   TEM imaging of gas-sparge anisotropic pitch. (a) and (b) are orthogonal 002 dark-field; from Lafdi *et al.* (1991).

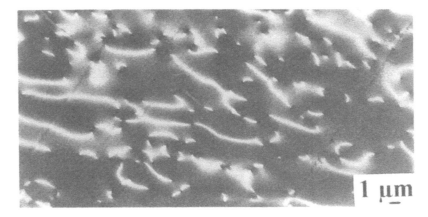

FIGURE 16.   TEM imaging of β resin anisotropic pitch obtained by fractionation (002 dark-field); from Lafdi and Oberlin (1994b).

section of a gas-sparge pitch (Lafdi *et al.*, 1991; Lafdi and Oberlin, 1994a). Anisotropic main component contains dispersed droplets themselves weakly anisotropic. They are partially soluble in toluene. This association common to all gas-sparge pitches is due to heterogeneity of mesophase dispersion.

Fractionation of a soaked pitch to get β resins is an other way to prepare 100% anisotropic pitches (Diefendorf and Riggs, 1980). Figure 16 shows that, this kind of pitch is devoid of droplets but very rich in disclinations (sinuous bright contours in 002 dark-field image, Lafdi and Oberlin, 1994b).

These new kind of pitches have common characters. They are viscoelastic so that they acquire easily flow induced orientations (they are spinnable). They have swelling ability and are thermotropic. Two specific features contribute to distinguish them from liquid crystals (anisotropic mesophase spheres A, B, C and bodies with digitized contours). They are entirely quinoline soluble and they have a specific long range statistical orientation (SO) different from local molecular orientation. Instead of a reduced misorientation of BSU relative to the preferred orientation plane (LMO), a majority of BSU are oriented and distributed at random everywhere in the bulk. However, a relatively large number of them are parallel to a preferred direction. It is improper to call them mesophasic or mesogenic pitches, which unfortunately often implies that they derive from demixtion, coalescence and solidification of Brooks and Taylor mesophases A. Their total anisotropy is due to stresses and is not obtained progressively. It would be more right to call them anisotropic pitches or β resin pitches.

9.5 APPLICATIONS

Two categories of carbons were above described. Those issued from liquid crystals which are α resins insoluble in quinoline, and those issued from other colloidal systems (anisotropic pitches), which are β resins soluble in quinoline. This strong variability is the origin of all applications of carbons.

In the first category, the initial liquid crystals (A, B, C or bodies with digitized contours) are entirely dependent on the degree of cross-linking get by the precursors at LMO occurrence. Correspondingly, LMO final size governs the ability to graphitize. Graphitizing carbons are mainly used as industrial cokes (1100–1200°C) for making electrodes artefacts (steel or aluminium anodes) or synthetic more or less graphitized powders. The basic properties needed are thus electrical and thermal conductivities (lamellar textures). The non-graphitizing carbons (such as glassy carbons or cellulose-based products) are more efficiently used as resistors, thermal or electrical insulators in powder, felt or fibers.

Among the intermediates, mostly issued from natural products, coals of increasing rank (van Krevelen; 1993) exhibit a great variety of graphitizability (Bensaid, 1983; Rouzaud, 1984) directly connected to their wide applications such as cathodes for aluminium processing, blast furnace cokes, ... Kerogens, i.e. parent rocks of oil (Durand, 1980; Oberlin *et al.*, 1973, 1980; Bonnamy, 1987; Oberlin, 1989), are also of great importance, since their origin and degree of coalification determine the amount and the composition of the various oils that they produce. Kerogens which get large LMO when cokefied, have high oil potential. Kerogens getting the smallest LMO are those having the lowest oil potential. Poorly coalified kerogens which have not yet reached LMO stage, provide heavy oils. Highly coalified kerogens showing natural LMO give light oils. Kerogens are thus finger-prints of the production of the various oils.

The specific properties of *anisotropic pitches* allow them to be used as precursors for high performances carbon fibers (Oberlin *et al.*, 1998). After spinning and moderate cross-linking by air (to avoid softening), carbonized anisotropic pitches provide graphitizing fibers. However highest graphitization degree P_1 belongs to gas-sparge process, whereas a lower value of P_1 corresponds to fractionated pitch (the persistence of a high concentration of disclinations prevents grain formation). As a consequence, the Young's modulus, as well as the electrical and thermal conductivities, are maximum for the gas-sparge pitches but the tensile strength is a little better for the fractionated pitches (Oberlin *et al.*, 1998).

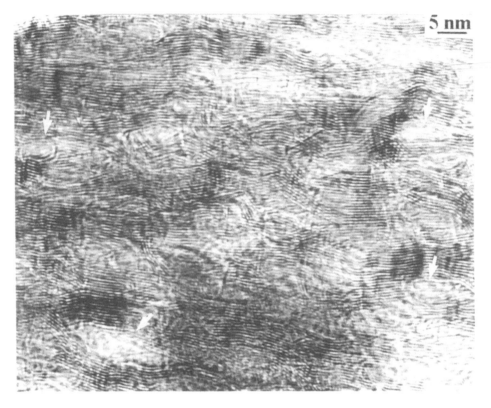

FIGURE 17. TEM imaging of flattened pores in anthracite (002 lattice fringes), from Oberlin and
Terrière (1975).

9.6 STRESS GRAPHITIZATION*

A very peculiar mechanism must be mentioned here. It is the stress graphitization (Inagaki
and Meyer, 1997). It provides a sudden transformation into graphite. Shear stresses introduce
strains into porous carbons by flattening the pores (Figure 17) (Oberlin and Terrière, 1975;
Bustin *et al.*, 1995). At the plasticity occurrence (2000°C) the more fragil pore walls
suddenly break, providing simultaneously lamellae and perfect layers i.e. materials almost
or even entirely graphitized.

Stress graphitization is the origin of natural graphite, from coals strained by tectonic
stresses (Bonijoly *et al.*, 1982). In the same manner, thermally treated anthracites (Oberlin
and Terrière, 1975) and non graphitizing carbons heat-treated under pressure (Kumiya *et al.*,
1968; de Fonton *et al.*, 1980) suddenly transform into graphite. Thin polyimide films also
could turn to graphite in a similar way (Inagaki *et al.*, 1993; Bourgerette *et al.*, 1995).

*This paragraph is written with M. Inagaki (see chapters 8 and 12).

FIGURE 18. Optical micrograph of the cross-section of a composite (carbon fiber-filler and resin-matrix), heat-treated at 2800°C and etched, from Yasuda *et al.* (1980).

Since the maximum degree of graphitization is very high, these films are particularly good host graphite for intercalation compounds.

In addition, during heat-treatment of carbon fibers and thermosetting resins to produce carbon-carbon composites, the stresses developed at the interface between filler-fiber and matrix-resin lead to the orientation of BSU around fibers developing optical anisotropy (Inagaki and Kamiya, 1973; Hishiyama *et al.*, 1974; Yasuda *et al.*, 1980). At high temperature a graphite-like phase is formed (Figure 18).

9.7 CONCLUSION

The versatility of carbons is now demonstrated to be due to the three dimensional arrangements of BSU (Local Molecular Orientation), stabilized at the end of the primary carbonization. LMO are under the dependence of the elemental compostion (F_{LMO}) at the LMO occurrence, properties are thus widely different. If BSU are almost parallel in extended domains as lamellae, the material is ductile, compact, thermal and good electrical conductor in the lamellae plane (graphitizing carbons). If BSU are almost at random, the product is fragile, bad electrical conductor and non graphitizing. Reciprocally, if a given arrangement is predetermined, the properties could be fixed at will. This flexibility itself is due to the fact that as long as carbonaceous matters are not yet brittle solids, they are systems able to provide a great number of liquid crystals or other colloidal phases.

Acknowledgements

The authors thank Thomas Cacciaguerra for his help to illustrate this chapter.

References

Auguié, D., Oberlin, M., Oberlin, A. and Hyvernat, P. (1981) *Carbon*, **19**, 227.

Bacon, G.E. (1950) *Acta Cryst.*, **3**, 137.

Bensaid, F. (1983) Doctorat d'Etat Thesis, Université d'Orléans.

Billard, J. (1980) in *Chem. Phys. Series 11*, Springer Verlag (Berlin) *Discotic phases, a Review*, pp. 383–395.

Bonijoly, M., Oberlin, M. and Oberlin, A. (1982) *Intern. J. Coal Geol.* **1**, 283.

Bonnamy, S. (1987) Doctorat d'Etat Thesis, Université d'Orléans.

Bonnamy, S. (1999) Part I and Part II, *Carbon*, **37**, 1691; *Carbon*, **37**, 1707.

Boulmier, J.L. (1976) Doctorat d'Etat Thesis, Université d'Orléans.

Bourgerette, C., Oberlin, A. and Inagaki, M. (1995) *J. Mater. Res.*, **10**, 1024.

Bourrat, X. (1982) Doctorat 3 ème cycle, Université d'Orléans.

Bourrat, X., Oberlin, A. and Escalier, J.C. (1987) *Fuel*, **66**, 542.

Brooks, J.D. and Taylor, G.H. (1968) in *Chemistry and Physics of Carbon*, vol. **4**, P.L. Walker Jr. ed., p. 339, Dekker NY.

Bustin, R.M., Rouzaud, J.N. and Ross, J.V. (1995) *Carbon*, **33**, 679.

Chermin, H.A.G. and van Krevelen, D.W. (1957) *Fuel*, **36**, 85.

Chwastiak, S. (1980) US Patent 4.209.500.

Deurbergue, A. and Oberlin, A. (1991) *Carbon*, **29**, 621.

Diefendorf, R.J. and Riggs, D.M. (1980) US Patent 4.208.267.

Donnet, J.B. and Qin, R. (1993) *Carbon*, **31**, 7.

Dryden, I.G.C. (1952) *Fuel*, **31**, 176.

Endo, M., Takeuchi, K., Takahashi, K., Oshida, K., Dresselhaus, M.S. and Dresselhaus, G. (1995) *22nd Biennial Conf. on Carbon*, San Diego, pp. 340–341.

Durand, B., in *Kerogen*, B. Durand ed., 549p. (1980) Technip Paris.

Espinat, D. and Ravey, J.C. (1993) *Proceed. SPE Intern. Symp. on Oil-field Chemistry*, New Orleans, p. 365, Soc. Petr. Eng. Inc.

Fishbach, D.B. (1971) in *Chemistry and Physics of Carbon*, vol. **7**, P.L. Walker Jr. ed., p. 1, Dekker NY.

Fitzer, E., Mueller, K. and Schaefer, W. (1971) in *Chemistry and Physics of Carbon*, vol. **7**, P.L. Walker Jr. ed., p. 238, Dekker NY.

Fitzer, E. and Weisenburger, S. (1976) *Carbon*, **14**, 195.

de Fonton, S., Oberlin, A. and Inagaki, M. (1980) *J. Mater. Sci.*, **15**, 909.

Franklin, R.E., *Acta Cryst.*, (1950) **3**, 107; (1951) **4**, 250, and (1957) **10**, 359.

Friedel, G. (1922) *Annales de Physique*, **18**, 273–474.

Garza-Gomez, A. (1982) Thèse Docteur Ingénieur, Université d'Orléans.

Goma, J., Oberlin, M. and Oberlin, A. (1980) *Thin Solid Films*, **65**, 221.

Graham, T. (1861) *Phil. Trans. Roy. Soc. (London)*, **151**, 183.

Hishiyama, Y., Inagaki, M., Kimura, S. and Yamada, S. (1974) *Carbon*, **12**, 249.

Ihnatowicz, M., Chiche, P., Deduit, J., Pregermain, S. and Tournant, R. (1966) *Carbon*, **4**, 41.

Inagaki, M. and Kamiya, K. (1973) *Carbon*, **11**, 429.

Inagaki M. and Meyer R.A. (1999) in *Chemistry and Physics of Carbon*, P.A. Thrower and L.R. Radovic, ed., Dekker NY, vol. 26, pp. 149–244.

Inagaki, M., Oberlin, A. and Noda, T. (1975) *Tanso*, **81**, 68.

Inagaki, M., Takeichi, T. and Hishiyama, Y. (1993) in *New fonctionality materials*: Preparation of highly crystallized graphite films from polyimides (T. Tsuruta, M. Doyama and M. Seno eds.), vol. C, Elsevier.

Iwashita, N. and Inagaki, M. (1993) *Carbon*, **31**, 1107.

Joseph, D. and Oberlin, A. (1983) *Carbon*, **21**, 559 and 565.

van Krevelen, D.W. (1993) *Coal Typology, Chemistry, Physics Constitution*, 3rd ed., 979pp., Elsevier Amsterdam – see also *Coal* (1961) 1st ed.

Kumiya, K., Mizutani, M., Noda, T. and Inagaki, M. (1968) *Bull. Chem. Soc. Japan*, **41**, 2169.

Lafdi, K., Bonnamy, S. and Oberlin, A. (1991) *Carbon*, **29**, 831.

Lafdi, K. and Oberlin, A. (1994a) *Carbon*, **32**, 11.

Lafdi, K. and Oberlin, A. (1994b) *Carbon*, **32**, 61.

Lambert, F., Conard, J., Pépin-Donat, B. and Bonnamy, S. (1995) *Proceeding 22nd Biennial Conf. on Carbon*, July 16–21, San Diego (ACS ed.), pp. 248–249.

Maire, J. and Mering, J. (1970) in *Chemistry and Physics of Carbon*, vol. **6**, P.L. Walker Jr. ed., pp. 125–190, Dekker NY.

Marchand, A., Delhaes, P., Pacault, A., Zanchetta, J. and Uebersfeld, J. (1965) in *les Carbones* (ed. A. Pacault, vol. 1, pp. 405-521, Masson Paris, 2nd ed. 1997.

Monthioux, M., Oberlin, M., Oberlin, A. and Bourrat, X. (1982) *Carbon*, **20**, 167.

Mrozowski, S. (1979) *Carbon*, **17**, 227.

Mrozowski, S. and Gutsze, A. (1977) *Carbon*, **15**, 335.

Nguyen, H.T., Destrade, C. and Gasparoux, H. (1979) *Physics Letters*, **72A** 3, 251.

Oberlin, A. (1984) *Carbon*, **22**, 521.

Oberlin, A. (1989) in *Chemistry and Physics of Carbon*, vol. **22**, P.A. Thrower. ed., p. 1, Dekker NY.

Oberlin, A., Bonnamy, S. and Lafdi, K. (1998) in *Carbon Fibers* (ed. J.B. Donnet), Dekker, pp. 85–159.

Oberlin, A., Bonnamy, S. and Rouxhet, P. (1999) in *Chemistry and Physics of Carbon*, L.R. Radovic and P.A. Thrower, ed., Dekker NY, vol. 26, pp. 1–148.

Oberlin, A., Boulmier, J.L. and Durand, B. (1973) *Adv. Org. Geochem. Proceed.*, p. 15, Technip Paris ed.

Oberlin, A., Boulmier, J.L. and Durand, B. (1974) *Geochem. Cosmochem. Acta*, **38**, 647.

Oberlin, A., Boulmier, J.L. and Villey, M. (1980) in *Kerogen*, B. Durand ed., pp. 191–241, Technip Paris.

Oberlin, A. and Terrière, G. (1973) *J. Microscopie*, **18**, 247.

Oberlin, A. and Terrière, G. (1975) *Carbon*, **13**, 367.

van Olphen, H. (1963) in *Introduction to clay colloid chemistry*, J. Wiley and Sons, pp. 1–15.

Pacault, A. (1971) in *Chemistry and Physics of Carbon*, vol. **7**, P.L. Walker Jr. ed., p. 107, Dekker NY.

Pfeiffer, J.P. and Saal, R.N.J. (1940) *J. Phys; Chem.*, **44**, 139.

Riggs, D.M. and Diefendorf, R.J. (1983) *Proceeding 16th Biennial Conf. on Carbon*, July 18–22 San Diego (ACS ed.), pp. 24–25.

Rouxhet, P.G., Villey, M. and Oberlin, A. (1979) *Geochem. Cosmochem. Acta*, **43**, 1705.

Rouzaud, J.N. (1984) Doctorat d'Etat Thesis, Université d'Orléans.

Rouzaud, J.N. and. Oberlin, A. (1989) *Carbon*, **27**, 517.

Saadaoui, A., Roux, J.C. and Flandrois, S. (1993) *Carbon*, **31**, 481.

Souma, I., Shioyama, H., Tatsumi, K. and Sawada, Y. (1995) *Proceeding 4th Japan Intern. SAMPE Symp.*, Sept. 25–28.

Villey, M. (1979) Doctorat d'Etat thesis, Université d'Orléans.

Vogt, D., Rouzaud, J.N. and Oberlin, A. (1986) *Fuel Processing Technology*, **12**, 63.
Vorpagel, E.R. and Lavin, J.G. (1992) *Carbon*, **30**, 1033.
Warren, B.E. (1941) *Phys. Rev.*, **59**, 693.
White, J.C. and Zimmer, J. (1978) *Carbon*, **16**, 469.
Yasuda, E., Kimura, S. and Shibusa, V. (1980) *Trans. Jap. Soc. Compo. Mater.*, **6**, 40.

10 Preparation and Properties of Mesophase Pitches

ISAO MOCHIDA, YOZO KORAI, YONG-GANG WANG and SEONG-HWA HONG

Kyushu University, Kasuga, Fukuoka 816, Japan

10.1 INTRODUCTION

Numbers of organic materials including polymers, fossils and biomass can exhibit a liquid state when they are heated beyond a certain temperature; such a temperature is defined as the melting point for pure compounds. The mixtures of organic species can also lead to such a liquid state according to the melting and the dissolution of the compounds. Softening point, instead of melting point, which depends on the method of measurement, is also defined with more complex mixtures.

Some organic species undergo chemical condensation reactions while they are liquid or solid before their vaporization, raising their melting and boiling points. The condensation reactions compete with vaporization, keeping the condensed material in liquid or solid phase. A part of polymeric substances is liberated in forms of fragment molecules through the cracking and/or pyrolytic reactions at high temperatures. The repetition of the condensation and pyrolytic reactions leads finally to carbonaceous polymers: such products are called cokes or carbons. When the process progresses in the liquid phase until the final solidification, it is defined as the liquid phase carbonization, being compared to the gas and solid phase carbonization (Mochida, 1990).

The products are often unwanted in the conversion of hydrocarbons, however, cokes of desired structure and properties are designed to be produced as the functional material in some cases. Although many of low molecular weight organic species vaporize before the carbonization takes place to give coke, specially under atmospheric pressure, some particular organic compounds as well as coal, petroleum residue, and thermoplastic polymers give coke through their liquid phase at reasonable yields of 20 ~ 70 wt%.

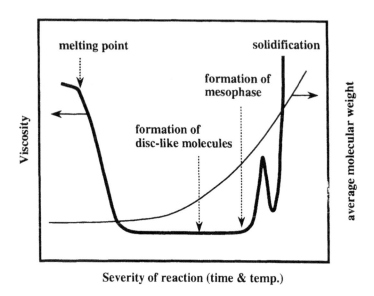

FIGURE 1. Changes of molecular weight and viscosity during the liquid phase carbonization.

Condensation and dehydrogenation leading to the polyaromatic species are the primary reactions of the carbonization process. These reactions increase first the melt viscosity of the pitch-like product, finally leading to non-fusibility, and non-solubility characteristics. The coke consists of polyaromatic planes, which stack to variable extent reflecting the graphitizing ability. The stacking of aromatic planes is observed anisotropic under optical microscope with polarized light.

Finally the aim of this chapter will be to describe this liquid phase carbonization called mesophase pitch in relation with the general topic of liquid crystals.

10.2 CHEMISTRY OF MESOPHASE PITCHES

10.2.1 *Phenomenological Aspects of Liquid Phase Carbonization*

Changes of molecular weight and viscosity during the liquid phase carbonization of pitches are illustrated in Figure 1: viscosity of pitch decreases with rising temperature, forming an isotropic liquid phase. Constituent molecules in the isotropic liquid phase undergo condensation reactions where their average molecular weight increases with further increase of the carbonization temperature. The large aromatic nuclei produced through the repeated condensation reaction in the liquid phase carbonization interact with each other through $\pi - \pi$ orbitals, forming molecular assembly of layer stacking which segregates from the matrix of smaller molecules to produce anisotropic spheres as discovered first by Brooks and Taylor (1968). The anisotropic spheres grow in diameter, up to 50 μm, absorbing molecules of large size, coalesce towards a bulk mesophase, then form isochromatic regions in spite of an increasing viscosity. The isochromatic region spreads to govern the whole system and finally solidifies into an anisotropic coke. Temporary decrease in viscosity of the anisotropic pitch can be observed, being related to the slip of stacked planar structure by the external shear force at the viscosity measurement; this is related to the existence of disclination defects which play a crucial role in rheology (White and Buchler, 1986). This series of steps leading to optical anisotropy are followed in the quenched carbonization product as shown in Figure 2, although some effects due to the quenching may be included. Hot-stage microscopy shows a similar progress during primary carbonization (Lewis, 1978; White and Buchler, 1986). Thus, it is phenomelogically concluded that anisotropic graphitizable coke is produced through several successive steps of anisotropic (mesophase) spheres.

Another scheme of anisotropic coke is observed with semi-anthracite. A loose stacking of aromatic planes achieved at the coal bedding is transformed directly into the correct stacking in the anisotropic coke (Mochida et al., 1981). Such a scheme is also observed in the carbonization of some polymers under stress where the aromatic planes are forced to be aligned (Inagaki et al., 1976).

10.2.2 Chemical Scheme and Intermediates of Liquid Phase Carbonization

The chemical schemes of carbonization have been investigated. The molecular structure of the carbonization intermediates has been revealed to strongly influence the development of optical anisotropy in the resultant coke (Mochida et al., 1975a, 1976 or 1977). The carbonization intermediates have been identified in the pyrolysis of acenaphthylene, which provides a rare example of atmospheric carbonization of pure organic chemical (Fitzer et al., 1971; Mochida and Marsh, 1979). Its carbonization scheme is illustrated in Figure 3: the intermediates labelled II, III and VI are proposed formulae based on chemical analyses.

Comparison of the carbonization reactivities of acenaphthylene (I) and decacyclene (VI) indicates the interaction among the carbonization intermediates. Decacyclene is found always as an intermediate in the pyrolysis of acenaphthylene. However, decacyclene alone is stable even at 500 C, whereas acenaphthylene develops optical anisotropy at much lower pyrolysis temperature (Mochida et al., 1977). These results suggest a role of reactive intermediates which are not included in the single pyrolysis of decacyclene. A minor component may affect the final properties of coke at the intermediate steps of the carbonization.

Molecular structure and constitution of the anisotropic mesophase spheres have been analysed (Mochida et al., 1978a). Chemical analyses were carried out after the sphere is solubilized by non-destructive hydrogenation or reductive alkylation. A structure model is illustrated in Figure 4 where aromatic sheets of 0.6 to 1.5 nm in diameter are linked directly or through methylene bonds to form polyaromatic polynuclear molecules with molecular weight ranging from 400 to 4000. Such a broad distribution of molecular weight is important since the large molecules may be responsible to accommodate smaller molecules in the ordered stacking through $\pi - \pi$ interactions and to be insoluble in the matrix. The small molecules are expected to behave as a solvent to assure the fusibility of the whole sphere if heated, exhibiting a liquid crystal behavior.

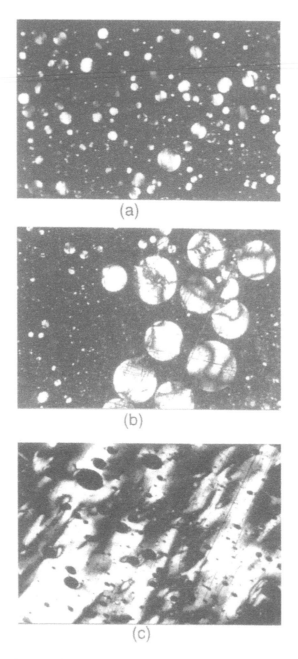

FIGURE 2. Development of optical texture in liquid phase carbonization (a) small anisotropic sphere, (b) coalescence and growth of sphere, (c) anisotropic domain.

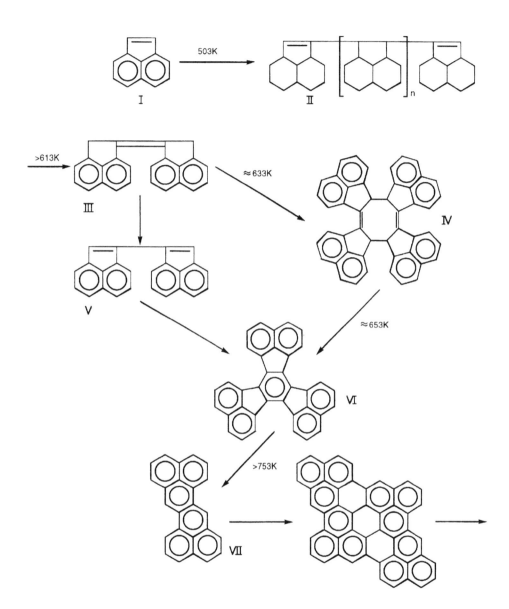

FIGURE 3. Carbonization scheme of acenaphthylene.

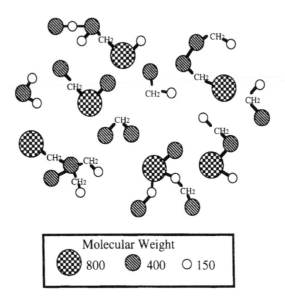

FIGURE 4. A model chemical structure of quinoline insoluble mesophase molecules.

10.2.3 Catalytic Carbonization in the Liquid Phase

Figure 5 describes reaction schemes for naphthalene carbonization catalyzed by metallic potassium or by aluminum chloride (Mochida, 1990). Such catalysts produce isotropic and anisotropic carbons, respectively. The intermediate structures are similar except for more naphthenic structure induced in the aluminum chloride-catalyzed carbonization. The role of naphthenic structures leading to optical anisotropy has been recognized in many examples, and their introduction can improve the anisotropic development (Mochida *et al.*, 1985). Higher solubility, lower softening point, and higher solubility of the intermediate molecules may be obtained by the aid of partial naphthenic structure.

Aluminum chloride also polymerizes the nitrogen containing aromatic compounds. The reaction schemes of quinoline and isoquinoline catalyzed by $AlCl_3$ are illustrated in Figure 6. Flow domain texture prevailed in the coke from quinoline while isoquinoline formed very fine mosaic texture (An *et al.*, 1995). Quinoline derived pitch retained nitrogen, hydrogen, and ring structure, leading to many naphthenic groups. The formation of condensed rings appears rather small, principally oligomers of aryl-aryl bond being major components. More longer reaction time and higher reaction temperature increased the yield of the pitch, the insoluble fraction of larger molecular weight, and more number of quinoline units by maintaining the above structural characteristics. In contrast, isoquinoline lost nitrogen and hydrogen considerably during the oligomerization, hence significant degradation of its aromatic character is suggested. Both nitrogen as well as carbon in quinoline can be cationic to form oligomers at high yield, while the latter intermediate loses nitrogen as NH_3 through the C-N bond fission induced by the reactive naphthenic hydrogens, reducing the

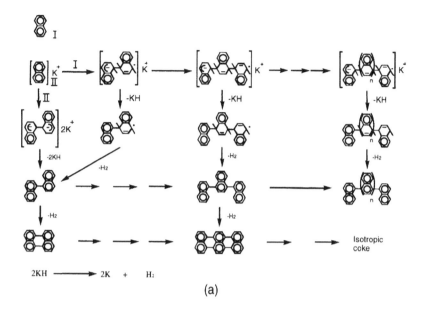

(a)

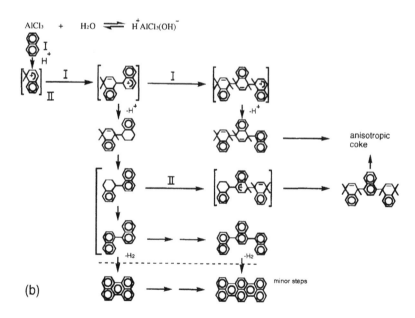

(b)

FIGURE 5. Reaction schemes of naphthalene carbonization catalyzed by metallic potassium (a) and aluminum chloride (b).

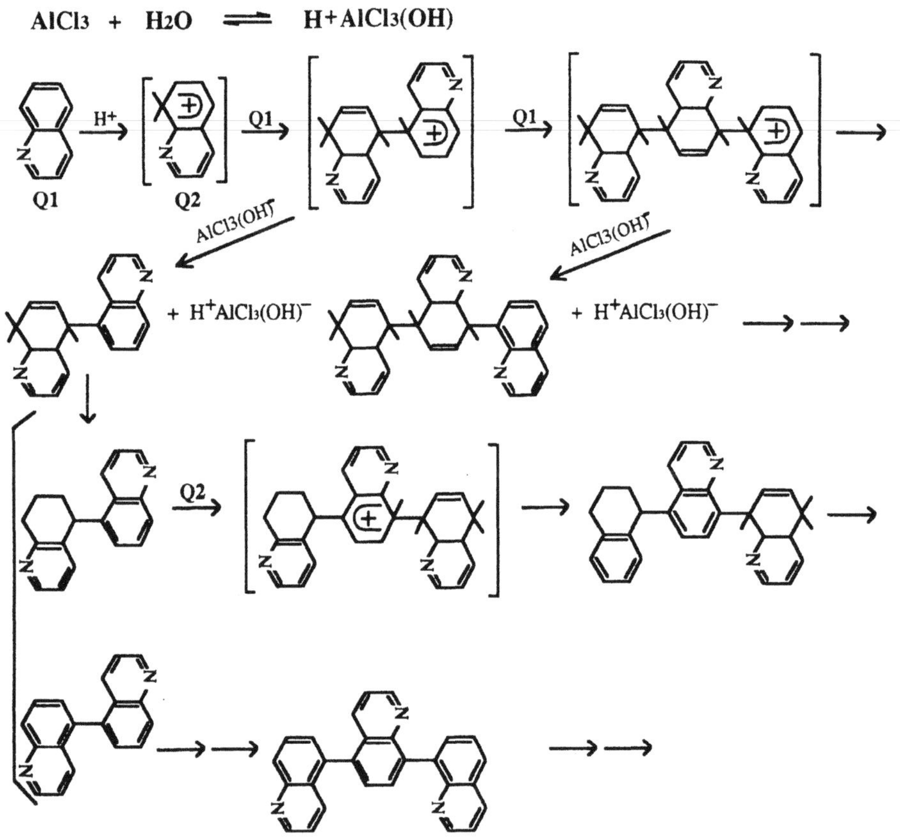

FIGURE 6. Reaction scheme of quinoline catalyzed by aluminum chloride.

pitch yield. The latter reaction removes both nitrogen and naphthenic hydrogen, leading to early solidification due to the presence of highly aromatic structures.

10.2.4 Design of Liquid Phase Carbonization

Based on the carbonization mechanism, anisotropic development from a given starting material can be designed through several approaches. Such methods have been practiced in the coking of coal and needle coke production.

The first method is the design of carbonization conditions. Pressure and temperature programs in carbonization can control the structure of resultant coke through controlling the carbonization reaction and phase.

As mentioned in the carbonization of acenaphthylene and decacyclene, carbonization reactions are often governed by a minor component. Marsh and Diez (1994) defined such situation as the dominant partner effect. A suitable additive of small amount can effectively govern the carbonization reaction to produce a desired optical texture. The aromaticity of additive to solubilize the partner and its hydrogen-donating ability to maintain the viscosity lower are recognized as being important for its modifying ability to develop optical anisotropy.

The extent of modification is also governed by the chemical reactivity of the principal carbonizing substances. It was found that oxygen-containing functional groups are believed influential on such susceptibility. Such groups should be removed by the hydrogen donating additive to develop the optical anisotropy (Korai and Mochida, 1983).

The molecular size distribution of aromatic and paraffinic components in the feed stock is another factor to define the carbonization phase and segregation of respective components. Thus, the balance of modifying ability and reactivity is important to design the liquid phase carbonization.

Another approach of modifying principal carbonizing substances includes the separation of undesirable components and chemical modification of the intermediate as well as starting materials. Anti-solvent technique was applied to remove quinoline insoluble particles from coal tar pitch for the needle production in the delayed coking.

Various kinds of chemical modifications, thermal and catalytic treatments and their combination are applicable to modify the carbonization properties of carbonizing substances. These treatments include dealkylation, cracking, condensation, hydrogenation and hydrocracking (Mochida *et al.*, 1978b, 1980). Diels-Alder reactions attract interests for changing the structure of carbonizing substances including mesogen molecules. Less soluble fractions in the mesophase pitch were effectively solubilized by such reactions. Sterically hindered structure of the carbonizing intermediate for the stacking is most interesting to know how the properties of the liquid crystal mesophase and its derived carbon are modified.

Introduction of functional groups at the terminal positions of molecules is also possible. Hydroxyl, carbonyl, carboxyl and amino groups on the aromatic rings can modify extensively the interactive forces among the component molecules (Otani *et al.*, 1970).

10.3 STRUCTURE AND PROPERTIES OF MESOPHASE PITCHES

The progress in the preparation of a mesophase pitch for producing a carbon fiber is shown in Figure 7. How to prepare or to get the most suitable raw material for the precursor of mesophase pitch is key for the preparation of spinnable mesophase pitch.

Otani *et al.* (1970) first found that the mesophase pitch could be produced by the heat-treatment of tetrabenzophenazine (TBP) at temperature above $500°C$ and this mesophase pitch could be spun into the fibers of rather larger diameter (20 μm) at $410 \sim 440°C$. It is noteworthy to point out that the suppression of subliming TBP is a key for the preparation of spinnable mesophase pitch. The requirement of high spinning temperature and the high cost of TBP as starting material made it not a feasible precursor for commercial fiber manufacture.

Singer and co-workers (Barr *et al.*, 1976), at Union Carbide Co. (UCC) found the process to produce a mesophase pitch with a softening point below $350°C$ by heating petroleum pitch under rapid nitrogen flow that removes the low molecular weight non-mesogen species that

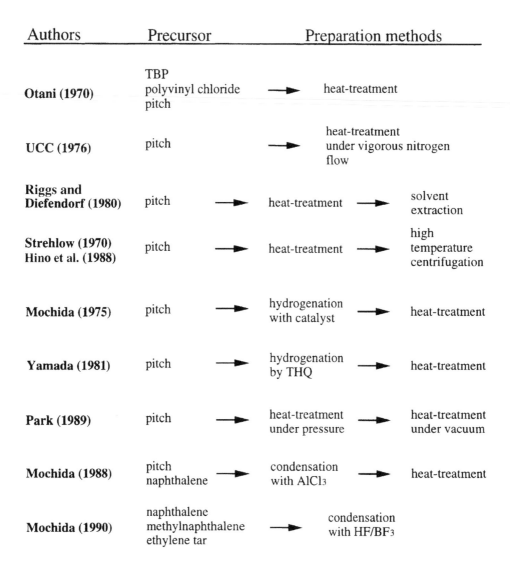

Authors	Precursor	Preparation methods
Otani (1970)	TBP polyvinyl chloride pitch	⟶ heat-treatment
UCC (1976)	pitch	⟶ heat-treatment under vigorous nitrogen flow
Riggs and Diefendorf (1980)	pitch	⟶ heat-treatment ⟶ solvent extraction
Strehlow (1970) Hino et al. (1988)	pitch	⟶ heat-treatment ⟶ high temperature centrifugation
Mochida (1975)	pitch	⟶ hydrogenation with catalyst ⟶ heat-treatment
Yamada (1981)	pitch	⟶ hydrogenation by THQ ⟶ heat-treatment
Park (1989)	pitch	⟶ heat-treatment under pressure ⟶ heat-treatment under vacuum
Mochida (1988)	pitch naphthalene	⟶ condensation with $AlCl_3$ ⟶ heat-treatment
Mochida (1990)	naphthalene methylnaphthalene ethylene tar	⟶ condensation with HF/BF_3

FIGURE 7. Progression of preparation methods of mesophase pitches.

tend to form isotropic phase. The produced mesophase pitch showed better spinnability and higher solubility. However the long heat-treatment time promotes the condensation reaction even at relatively low temperature of 380 ~ 430°C; consequently, the resultant mesophase pitch obliged to be spun around 350°C. The high temperature that was required for spinning promoted the polymerization thus increasing the viscosity of pitch resulting in some unstable spinning. The process indicates that the condensation of aromatic constituents should be carefully controlled. How to remove non-mesogen without polymerizing larger molecules in the mesogen is another key for production process. This is because the starting pitch consists of a large variety of species of which reactivities are very different. Condensation of low reactive species may induces the excess condensation of reactive species into infusible substances.

Riggs and Diefendorf (1980) at Exxon, introduced the solvent extraction technique to concentrate the suitable fraction of mesophase pitch. Molecular weight distribution and the properties of mesophase pitch can be controlled by the selective extraction with different solvents.

Since the mesophase is more dense than the isotropic phase, mesophase can precipitate slowly to the bottom of a vessel from the isotropic matrix phase at the fused stage. Strehlow (1970) first reported that the mesophase could be separated from the isotropic phase during the heat-treatment of coal tar pitch, using high temperature centrifugation. Hino *et al.* (1988) at Tonen, prepared the mesophase pitch as a precursor for carbon fiber by the high temperature centrifugation technique.

Mochida found that naphthenic and short alkyl chains groups are essential for the mesophase pitch to keep its low softening point for a stable spinning and a reasonable stabilization reactivity. Yamada *et al.* (1981) and Mochida *et al.* (1975b) hydrogenated the pitch to improve the mesophase pitch by introducing naphthenic group into the precursor pitch by the aid of hydrogen donor solvent (tetrahydroquinoline) or catalyst, respectively, before the preparation of the mesophase pitch. Some of the introduced naphthenic groups still remained in the mesophase pitch. The pending problem is the high cost of aromatic hydrogenation.

Park and Mochida (1989) proposed a two stage production process of mesophase pitch where the mesogen molecules were produced in the first stage under pressure at higher temperature, and then they were efficiently concentrated under vacuum in the second stage.

Recently, the present authors developed the catalytic condensation reactions of aromatic compounds using HF/BF_3 which is recovered easily by distillation for a repeated use (Mochida and Korai, 1986; Mochida *et al.*, 1989; Mochida, 1990; Korai *et al.*, 1991). Catalytic condensation proceeds in a similar scheme to that catalyzed by $AlCl_3$. Typical structures of mesogen molecules derived from aromatic compounds are illustrated in Figure 8. All mesogen molecules carry a lot of naphthenic structure and inherit chemical structure of the starting compound. Such structural analyses of the mesophase pitch allow a correlation of structure with bulk properties, which suggests the control of properties by the structural design of mesogen molecules. The structural change of isotropic to mesophase pitches is also clarified based on analyses of a series of isotropic as well as liquid crystal pitches. The mesophase pitch derived from the single species of an aromatic hydrocarbon is expected to be controlled in its properties which reflect the structure of monomer, the polymeric unit and the manners of polymerization, although it carries components of variable molecular weights which distribute in a range of 300 to 2000 as revealed by gel permeation chromatography (GPC) technique for example. A variety of mesophase pitches are now available by changing aromatic monomers, co-oligomer and mesophase pitch blending.

The molecular assembly of mesogen molecules found in the liquid crystal phase is analyzed by fused ^{13}C-NMR and X-ray diffraction (Korai and Mochida, 1992). Layered stacking of aromatic planes and their thermal behavior with temperature were indicated by chemical shift of ^{13}C-NMR and peak width of X-ray diffraction. Figure 9 illustrates fused

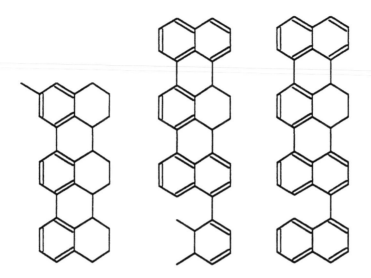

FIGURE 8. Typical structures of intermediates derived from naphthalene with HF/BF$_3$.

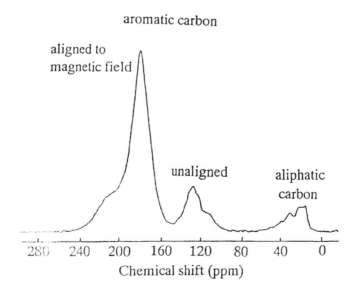

FIGURE 9. Fused state ^{13}C-NMR spectra of naphthalene derived mesophase pitch.

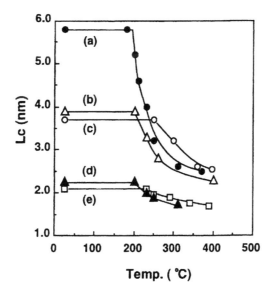

FIGURE 10. Stacking thickness of molecules in the mesophase pitches derived from (a) methyl-naphthalene, (b) petroleum pitch, (c) coal tar pitch, (d) naphthalene and (e) anthracene.

^{13}C-NMR of naphthalene mesophase pitch, which shows a resonance of 180 ppm. The orientation of aromatic planes along the magnetic field is suggested (Nishizawa and Sakata, 1991). Dissolution of mesophase pitch in the solvent shifted the aromatic resonance to 120 ppm, which is usually observed with aromatic carbon of free rotation (Nishizawa and Sakata, 1991). The orientation appears possible only when the aromatic planes are stacked. An isolated plane of large mesogen molecules in the isotropic matrix fails to be oriented to the magnetic field. It is noted that the mesogen molecule carries some alkyl or naphthenic groups.

Mesophase pitches prepared from aromatic hydrocarbons exhibit the definite 002 diffraction in X-ray diffraction (XRD), which allows the calculation of a stacking height for aromatic planes. The height of stacking varied according to the temperature of measurement and constituent molecular species as shown in Figure 10 (Korai and Mochida, 1992). At high temperature the stacking is lost as expected due the molecular mobility. It must be noted that the mesophase pitch exhibits the molecular ordering in its molten state, a liquid crystal organization being definitely indicated.

The second point of interest is the strong influence of molecular structure on the height of stacking. Methylnaphthalene derived mesophase pitch exhibited much higher stacking of 6 nm (around 18 aromatic planes) than other mesophase pitches. Degree of such stacking should influence various properties of the pitch such as the viscosity and the molecular ordering through the spinning in the resultant carbon fiber. Detail studies on the design of constituent molecules to provide excellent properties of the carbon fibers appear very promising in the future.

The role of molecular structure and its distribution have been discussed for the spinning, stabilization and fiber properties in the carbon fiber manufacture to provide bases for the design of mesogens in the mesophase pitch. It should be noted that the stacking will be changed according to experimental conditions such as temperature, pressure and shear. As a liquid crystal, the stacking of planar molecules to form the cluster ($\sim$10 nm) and domains ($\sim$100 nm) as the sub-units of molecular assemblies and their deformation according to the temperature and shear are expected to influence the rheological properties (Yoon *et al.*, 1993). Details of nanoscopic structure of mesophase pitch and carbon fiber are described below.

10.4 MESOPHASE FIBERS

10.4.1 *Mesoscopic Structure of Mesophase Pitch and Its Derived Carbon Fibers*

As described above, the mesophase pitch consists of oligomeric aromatic compounds, which are stacked at their aromatic planes to form the cluster unit, as revealed by XRD (Fortin *et al.*, 1994) and NMR (Korai and Mochida, 1992). The detailed structure of mesogenic molecules are also analyzed by TEM (Mochida *et al.*, 1996). Optical microscopy clarifies the region of aromatic oligomers in the same alignment (isochromatic unit), of which size spun to several hundred micrometers. The shape and size of isochromatic units are changed by stirring, spinning and temperature as for a thermotropic liquid crystal. The change of the chemical composition by adding or extracting the particular components as well as the temperature modifies the isochromatic unit and content of anisotropy, suggesting also its lyotropic liquid crystal nature but needs further developments to be clarified.

The resulting pitch based carbon fiber has been reported to carry macro, meso and microscopic structures. The optical microscope is believed to reveal the isochromatic units in the transverse section of the carbon fiber, which are very much reduced in their size because of the spinning. Recently high resolution microscope revealed a series of mesoscopic unit at nanometer scales in the fiber. Thus, radial, random and onion-skin like arrangements have been reported (Yoon *et al.*, 1992). The origin and relation to the property of such mesoscopic units are now under investigation.

Figure 11 illustrates a series of high resolution scanning electron microscope (HR-SEM) photographs of as-received, THF, and pyridine extracted naphthalene derived mesophase pitch. The mesophase pitch showed the textureless feature of its as-received surface. However the solvent extraction closed up the rod-like microdomain in the insoluble fraction, of which size and shape were rather uniform according to the extraction extent. There are many pores among the insoluble units, however no pore is found inside the units. Such features are similar to channeling structure of blended polymers alloy, where non-miscible components form their own self-assemble units.

Figure 12 illustrates a series of HR-SEM photographs of longitudinal surfaces on the as-spun, THF, and pyridine extracted fiber observed from the tilted view (10°). The as-spun fiber showed corrugated texture, but neither fibril nor pleats of ordered alignment. The solvent extraction was found to induce the fibril (100 nm widths) and pleat structure (30 nm widths) in the as-spun fiber.

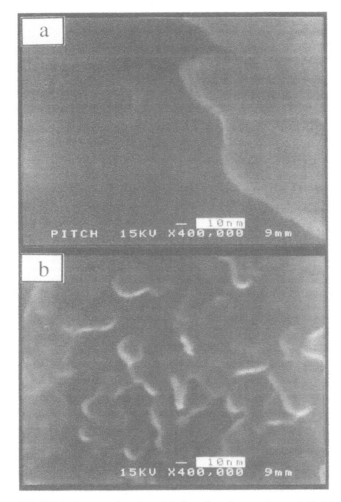

FIGURE 11. HR-SEM photographs of naphthalene-based mesophase pitch (a) and its insoluble fraction of tetrahydrofuran (b).

Figure 13 illustrates HR-SEM photographs of the transverse cross-sectional surface of as-spun and extracted fiber. The as-spun fiber revealed spherical or elliptical grains of 50–100 nm size. Extracted fibers showed microdomains of 50 nm of which size and shape were similar to that in insoluble fractions of mesophase pitch.

Figure 14 illustrates HR-SEM photographs of the longitudinal surfaces of the stabilized, carbonized and graphitized fibers. Although the stabilized fiber shows the wave-like surface as observed on the as-spun fiber, the carbonized and graphitized fibers close up the fibril and pleat units. They have basically same shapes as-observed to those in the extracted spun fiber. The sizes appear to be reduced by the graphitized fibers. The transverse cross-sectional surface of the carbonized fiber shows domains of 30 nm length and of 100 nm thickness.

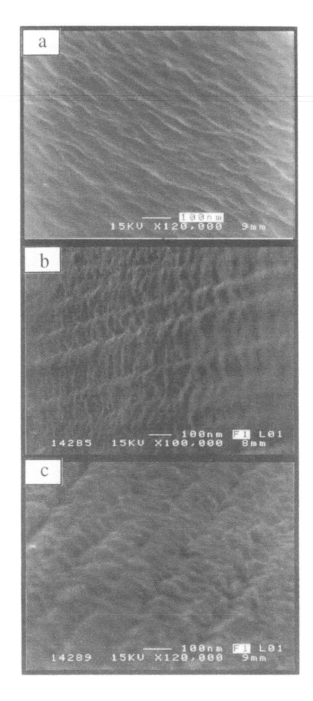

FIGURE 12. HR-SEM photographs from longitudinal surfaces on the as-spun (a), THF (b) and pyridine (c) extracted fibers observed from tilted view (10°).

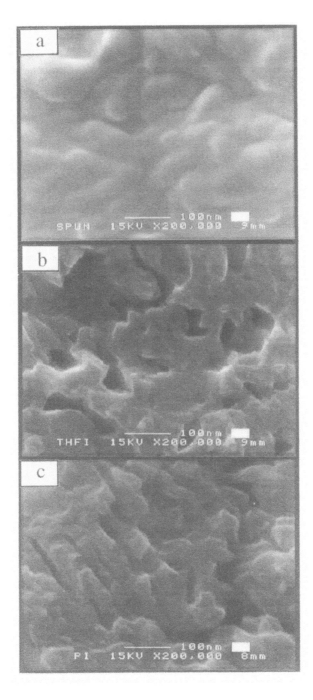

FIGURE 13. HR-SEM photographs of transverse cross-section of the as-spun (a), THF (b) and pyridine (c) extracted fibers.

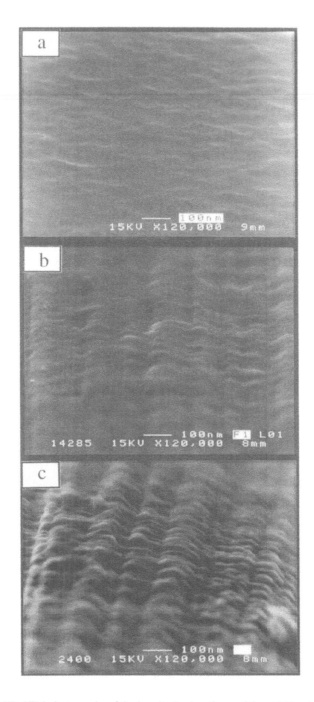

FIGURE 14. HR-SEM photographs of the longitudinal surfaces of the stabilized (a), carbonized (b) and graphitized fibers (c) observed from the tilted view (10°).

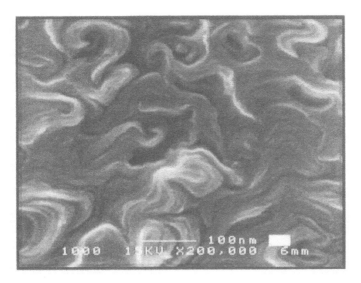

FIGURE 15. HR-SEM photograph of transverse cross-section of the carbonized fiber.

Figure 15 illustrates the transverse cross-section of the carbonized fiber. This particular fiber exhibited alignment of domains. The domains have shapes of sheet-like, bent and looped as observed at the graphitized fiber.

10.4.2 Formation Mechanism of Microstructures

Figure 16 (A)–(C) shows schematic pictures of such microstructures and their formation mechanism from aromatic planar molecules. Such microstructures should inherit the assembly units of aromatic molecules (Mochida *et al.*, 1996) and the microdomain in the mesophase pitch and their rearrangement through their deformation during the spinning. Hence, the structure observed in the fiber reflects the original self-assembly of mesogen molecules.

The liquid crystal mesophase pitch reserves micro-domain units in its melt as well as in solid state. Such a unit is composed of planar stacking units of planar molecules by alignment in the same direction. Such micro-domains are closely packed in the solid- or melt-state. Since the shear stress deforms the dimension and shape of micro-domains of the mesophase pitch, melt spinning through the capillary aligns the micro-domains parallel to the fiber axis, forming fibrils. The deformed micro-domains of the mesophase pitch are found to correspond to the pleat units in the resultant fiber.

The graphitic units in the fibril is also induced by the alignment of molecular stackings in a micro-domain. Some folded units may form topological defect which are disclinations (Bourrat *et al.*, 1990). The small value of the planar stacking of mesogen molecules, La (110), in the mesophase pitch grows into 30–60 nm of the hexagonal plane in the resultant graphitized fiber after graphitization at 2500°C, its growth being limited by the length of

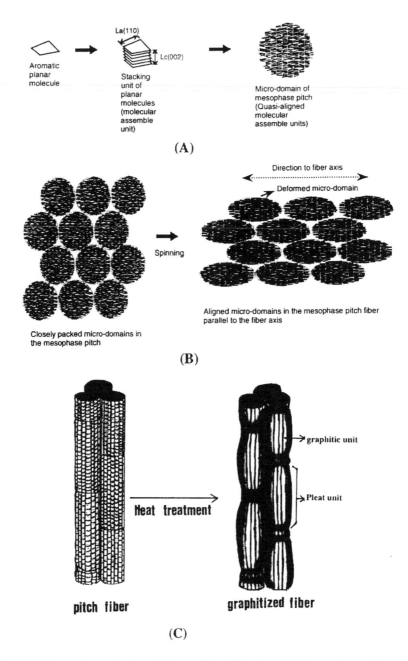

FIGURE 16. Formation mechanism of the graphitized fiber from liquid crystal mesophase pitch. (A) Formation of the molecular stacking and microdomain unit from aromatic planar molecules in liquid crystal mesophase pitch. (B) Deformed alignment of microdomains in the liquid crystal mesophase pitch under shear. (C) Formation of the graphitic and pleat unit in the graphitized fiber.

pleat unit. The thickness of the graphitic unit appeared to be about 2-4 nm, reflecting the height of molecular stackings, Lc (002), in the precursor mesophase pitch.

10.5 MESOCARBON MICROBEADS

10.5.1 Preparation of Meso-carbon Microbeads

Mesocarbon microbeads have been prepared by the following procedures.

(1) Coal tar or petroleum pitch is heated to produce mesophase spheres in the isotropic matrix. Before the extensive growth and coalescence of the spheres, the carbonization is quenched and spheres are extracted from the isotropic matrix (Yamada et al., 1974).

(2) Mesophase pitch or its soluble component is dispersed into spheres as the emulsion in the proper media. The dispersed spheres are extracted (Kodama et al., 1988, Yoon et al., 1992).

(3) Pitch or polyvinyl chloride is carbonized with polyethylene under very high pressure to prepare carbonized spheres (Inagaki et al., 1981, 1983).

Such procedure provided ca. 10% of single spheres. Kawatetsu and Osaka Co. commercialize the mesocarbon microbeads from the coal tar pitch basically according to the first procedure. The low yield is ascribed that the only heaviest fractions produce spheres in the early stage of the carbonization before the coalescence takes place when the rest is carbonized.

Procedure 1 is modified by heat-treatment under vacuum to increase the yield of spheres (Park et al., 1984).

Both isotropic and anisotropic pitches have been commercially available from naphthalene precursor. The isotropic pitch consists of naphthalene oligomers with rather narrow molecular weight distribution. Homogeneous nucleation and uniform growth of spheres increased their yield and uniformity in their diameter. The spheres are separated from the isotropic matrix by extraction with THF or pyridine. The yields of isolated sphere was as high as 34% as THF insoluble and 20% as pyridine insoluble (Korai et al., 1996a).

The synthetic naphthalene mesophase pitch of lower softening point is more easily dispersed in a variety of solvents. Indeed, the synthetic isotropic pitch is an excellent matrix; spheres are floating in the matrix, the specific gravity and melt viscosity of both pitches being compatible. Rapid agitation tends to reduce the mean size of the spheres, while increasing their number.

The influences of infusible substances on the development of the microstructural anisotropy have been reported extensively in the carbonization of coal tar pitch which carries primary quinoline-insoluble component as the non-fusible particles. Such non-fusible particles initiate the radical reactions to develop the onset of spheres earlier and prohibit their growth and coalescence by absorbing on their surface a matrix of increasing viscosity. Thus quinoleine insoluble (QI) in the sphere is revealed to enhance the non-fusibility and reduce the graphitizability in the successive carbonization and graphitization.

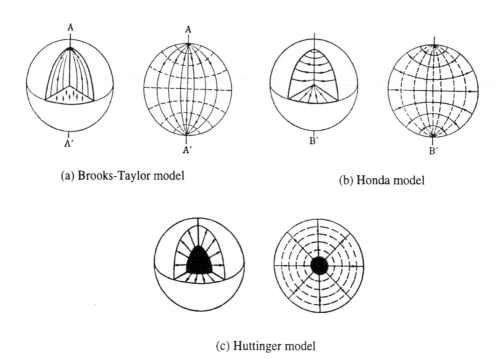

(a) Brooks-Taylor model

(b) Honda model

(c) Huttinger model

FIGURE 17. Structure of meso-carbon microbeads.

   The present authors have reported that 1% addition of carbon black is favorable to obtain single spheres at the high yield from the synthetic isotropic pitches (Korai *et al.*, 1996b). The introduction of carbon black can decrease the size of spheres dispersed in the isotropic matrix, and it is effective to disperse the mesophase pitch in the isotropic phase.

10.5.2 Structure of Meso-Carbon Microbeads

The structure of meso-carbon microbeads has been assessed by an optical, scanning and transmission electron microscopes. The optical microscope suggested the alignment of mesogen molecules as well as hexagonal plane according to the optical anisotropy in the sphere. Brooks and Taylor described the alignment as illustrated in Figure 17 (Brooks and Taylor, 1965) where other typical alignments are also reported (Hishiyama *et al.*, 1982).

   Figure 18 exhibits the microstructure of mesocarbon microbeads. Lower magnification of SEM illustrates the spherical and elliptical shapes of the mesocarbon microbeads. However, high resolution SEM clarifies that they consist of a number of microdomain units, of which their size are about 50 nm long and 100 nm in diameter.

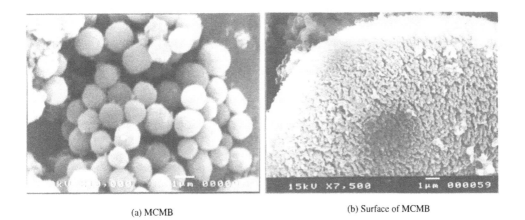

(a) MCMB

(b) Surface of MCMB

(c) With carbon black

FIGURE 18. Morphology of meso-carbon microbeads prepared from AR-isotropic pitch by pyridine extraction in the absence and presence of carbon black.

10.6 APPLICATIONS OF MESOPHASE PITCHES

Mesophase pitch now available in industrial scales at reasonable price is under development for a variety of high functional materials and devices. Typical examples of its applications are summarized below.

10.6.1 Pitch Based Carbon Fiber

Continuous and short carbon fibers have been manufactured in industrial scales. The continuous carbon fiber from the mesophase pitch has been recognized to show both very high Young's modulus and thermal conductivity. Application of the fibers utilizing such characteristic properties is now expanding, in particular better tensile and especially compressive strengths are looked for. High purity, variety of molecular species, controlled properties of melt fluidity and stabilization reactivity as well as molecular stacking of the synthetic mesophase pitch are expected to solve the problems. Spinning technologies to control the micro-structure in the transverse and axial surfaces are being studied, utilizing the liquid crystal nature of the mesophase pitch on molecular basis. Blending of two mesophase pitches brings in the concept of pitch alloys of variable microtexture. One extreme is to spin the mesophase spheres-containing isotropic pitch.

Application of mesophase pitch based carbon fiber in construction technology is rapidly growing to simplify its processes and to improve drastically the structural stability because of high strength, stiffness, and their balanced properties.

Very high thermal conductivity superior to those of metals such as silver and copper is another characteristic of mesophase pitch based carbon fibers. Effective devices of thermo-management such as heat-exchange, heat-release, and heating wall can be designed, using the fiber as the filler of the composite. Novel synthetic mesophase pitch is a superior precursor of carbon fibers for such applications.

Short carbon fibers by melt blow spinning from mesophase pitch has been commercialized since the mesophase pitch of low softening point is now available by synthetic procedure. High graphitizability and "PAN-AM" type transverse alignment are appreciated as an anodic carbon of high performance for the lithium battery.

10.6.2   *Meso-Carbon Microbeads*

Graphitizable carbon has been generally recognized as a poor precursor for activated carbon of large surface area. However meso-carbon microbeads (MCMB) have been reported to give very large surface area by steam or catalytic activation.

Meso-carbon microbeads are utilized in application of column packing beads, high density graphites, and active carbons of very large surface area. They have been prepared through the heat treatment of coal tar and solvent extraction of anisotropic spheres in the isotropic matrix. Such procedures suffer high cost which prevents the expansion of its applications. Recently meso-carbon microbeads are highly appreciated as the anodic carbon of high performance for the lithium battery which allows a little higher price.

MCMB offer high resistance to very acidic and very basic reagents, low expansion and shrinkage coefficients and the ability to withstand an operating temperature of 250°C (Lebedev *et al.*, 1986). They can maintain their spherical shape for densely-packing when they are used for the column packing in liquid chromatography, showing excellent performance for separating aromatic and polar compounds.

MCMB exhibited self-cohesiveness and a controlled fusibility when they are heat-treated under inert atmosphere. They can shrink homogeneously and partly fuse at surface, producing the densely-packing. Hence, MCMB can be used for the unique precursor for high density-high strength carbonaceous artifacts with a bulk density of above 2.0 g/cm^3 after heat-treatment at high temperature from coal tar pitches and asphalt (Hishiyama *et al.*, 1982).

For MCMB consisted of the Brooks' and Taylor organization (see Figure 7), the complicated micropores were developed when they are activated by active reagent because the reactant can insert the interlayer of spheres. The developed micropores provided high surface active carbons: Osaka Gas Co. have got 4000 m^2/g of specific surface.

Recently, MCMB have been applied for the anode materials for lithium batteries, showing high charge/discharge capacity. Heat-treated at 700°C they showed the highest discharge capacity of 750 mAh/g at the current density of 0.1 mA/cm^2 (Mabuchi, 1994), which is more than twice the theoretical capacity based on a stoichiometric GIC, LiC_6. When MCMB are heat-treated at 2800°C the discharge capacity is about 250 mAh/g under a current of 2.0 mA/cm^2. Even though the mechanism of MCMB as anode materials are not so clear, their laminar microstructure may be related to the specific capacity. Their morphology is one of the important factors to determine its specific capacity.

10.6.3 Binder for Carbon-Carbon Composites

The preparation involves a series of steps: impregnation of the binding pitch into a two-dimensional carbon fiber preform, oxidative stabilization, and successive carbonization. The size of the anisotropic domains in the coke generated from the mesophase pitch can be controlled by the addition of phenolic resin and carbon black particles if required. The resulting composites are expected to have higher mechanical strength.

10.6.4 Binder for Firebricks

We also have developed technology using the mesophase pitch as a binder for the MgO high-temperature brick used by the steel industry to line furnaces. Testing of this technology on an industrial scale is now in progress.

10.6.5 Carbon Foams

Heating liquid AR mesophase pitch, first pressurized with an inert gas, to 600°C while releasing the pressure produces a carbon foam product. It has been proposed to use nitrogen gas, followed by a stabilization and a carbonization to produce the carbon foam (Kanno *et al.*, 1994). Density and pore size of the foam can be controlled by the initial pressurization, release pressure, and heating rate; it can be further graphitized if required. Mixing of chopped carbon fiber as a filler improves the strength of such materials. In addition to the mechanical strength, such foams are expected to have other desirable properties such as high insulating capacity, high chemical resistivity, resistance to air oxidation, good acoustic quality, and light weight.

10.6.6 High-Temperature Lubricant

AR mesophase pitch is considered as a stable liquid between 300 and 400°C. As such it may be a useful high-temperature liquid lubricant or a component of a graphite-based solid lubricant (Tomari *et al.*, 1993).

10.7 CONCLUSION

The chemical preparation, structures, properties, and applications of the mesophase pitches are reviewed in this chapter. The advantages of synthetic mesophase pitches derived from aromatic hydrocarbons are emphasized. This is an intermediate state in the liquid phase carbonization leading to graphitizable carbons. Hence, its chemistry has been focused in the production of needle and blast furnace coke. Its molecular components of polyaromatic, polynuclear polymers in spider wedge forms are recognized for a wide variety of molecular weight. The deformation, coalescence, and rearrangement in the successive steps of the primary carbonization are keys to lead to graphitizable carbons with preferential orientation in micrometer dimension. Isolated spheres and spinnable mesophase pitches are now appreciated as unique precursors for high performance carbon materials such as high density — high strength artifact and carbon fibers. The oligomers of aromatic hydrocarbons have been proved to form some kind of thermotropic/lyotropic liquid crystals. Their molecular structures, which inherits that of the starting aromatic spheres, influences the molecular stacking organization. The change of stacking according to temperature and shear is basically reversible, although the rapid cooling can quench some structural disorder. Indeed, it has been shown that a variety of mesophase pitches derived from selected aromatic hydrocarbons exhibit unique characteristics with the possibility of a molecular design function for a selected application.

References

An, K.H., Korai, Y. and Mochida, I. (1995) *Carbon*, **33**, 1069.

Barr, J., Chwastiak, S., Didchenko, R., Lewis, I. and Singer, L. (1976) *Appl. Polym. Sym.*, **29**, 161.

Bourrat, X., Roche, E.J. and Lavin, J.G. (1990) *Carbon*, **28**, 236.

Brooks, J.D. and Taylor, G.H. (1965) *Carbon*, **3**, 185.

Brooks, J.D. and Taylor, G.H. (1968) *Chem. and Phys. of Carbon*, Marcel Dekker Inc. **4**, 243.

Fitzer, E., Mueller, K. and Schaefer, W. (1971) *Chem. and Phys. of Carbon*, Marcel Dekker Inc. **7**, 237.

Fortin, F., Yoon, S.-H., Korai, Y. and Mochida, I. (1994) *Carbon*, **32**, 979.

Hino, T., Naito, T., Tsushima, H. and Nomura, T. (1988) Japan Patent 63-120112.

Hishiyama, Y., Yoshida, A. and Inagaki, M. (1982) *Carbon*, **20**, 79.

Inagaki, M., Ishikawa, M. and Naka, S. (1976) *High Temp. – High Press.*, **8**, 279.

Inagaki, M., Kuroda, K. and Sakai, M. (1981) *High Temp. – High Press.*, **13**, 207.

Inagaki, M., Kuroda, K. and Sakai, M. (1983) *Carbon*, **21**, 231.

Kanno, K., Yoon, K.E., Fernandez, J.J., Mochida, I., Fortin, F. and Korai, Y. (1994) *Carbon*, **32**, 801.

Kodama, M., Fujiura, T., Esumi, K., Meguro, K. and Honda, H. (1988) *Carbon*, **26**, 595.

Korai, Y. and Mochida, I. (1983) *Fuel*, **62**, 893.

Korai, Y., Nakamura, M. and Mochida, I. (1991) *Carbon*, **29**, 561.

Korai, Y. and Mochida, I. (1992) *Carbon*, **30**, 1019.

Korai, Y. and Mochida, I. (1994) *Carbon*, **32**, 1182.

Korai, Y., Ishida, S., Yoon, S.H., Wang, Y.G., Mochida, I., Nakagawa, Y., Matsumura, Y., Sakai, Y. and Komatu, M. (1996a) *Carbon*, **34**, 1569.

Korai, Y., Wang, Y.G., Yoon, S.-H., Ishida, S., Mochida, I., Nakagawa, Y. and Matsumura, Y. (1996b) *Carbon*, **34**, 1156.

Lebedev, Y.A., Krekhov, A.P., Chuvyrov, A.N. and Gilmanova, N.Kn. (1986) *Carbon*, **24**, 719.

Lewis, I.C. (1978) *Carbon*, **16**, 503.

Mabuchi, A. (1994) *Tanso*, **165**, 298.

Marsh, H. and Diez, M.A. (1994) in *Liquid crystalline and mesomorphic polymers*, Eds., V.P. Shibaev and L. Lam, vol. 7, pp. 231–257, Springer-Verlag, New-York.

Mochida, I., Nakamura, E., Maeda, K. and Takeshita, K. (1995a) *Carbon*, **13**, 489.

Mochida, I., Kubo, K., Fukuda, N., Takeshita, T. and Takahishi, R. (1995b), *Carbon*, **13**, 135.

Mochida, I., Nakamura, E., Maeda, K. and Takeshita, K. (1976) *Carbon*, **14**, 123.

Mochida, I., Maeda, K. and Takeshita, K. (1977) *Carbon*, **15**, 17.

Mochida, I., Maeda, K. and Takeshita, K. (1978a) *Carbon*, **16**, 459.

Mochida, I., Ando, T., Maeda, K. and Takeshita, K. (1978b) *Carbon*, **16**, 453.

Mochida, I. and Marsh, H. (1979) *Fuel*, **58**, 626.

Mochida, I., Ando T., Maeda, K. and Takeshita, K. (1980) *Carbon*, **18**, 319.

Mochida, I., Korai Y., Fujitsu H., Takeshita K., Komatsubara Y. and Koba Y. (1981) *Fuel*, **60**, 1083.

Mochida, I., Sone, Y. and Korai, Y. (1985) *Carbon*, **23**, 175.

Mochida, I., Korai, Y. (1986) *Petroleum derived carbons*, ACS Symposium Series **303**, 31.

Mochida, I. (1990) *Chem. and Eng. of Carbon*, Asakura-shoten. Vol. 87.

Mochida, I., Shimizu, K., Korai, Y., Sakai, Y. and Fujiyama, S. (1996) *Chem. Letters* 1893.

Mochida, I., Yoon, S.-H., Takano, N., Fortin, F., Korai, Y. and Yokogawa, K. (1996) *Carbon*, **34**, 941.

Nishizawa, T. and Sakata, M. (1991) *Fuel*, **70**, 124.

Otani, S., Okubo, Y. and Koitabashi, T. (1970) *Bull. Chem. Soc. Jpn.*, **43**, 3291.

Park, Y.D., Korai, Y. and Mochida, I. (1984) *High Temp. – High Press.*, **16**, 689.

Park, Y.D. and Mochida, I. (1989) *Carbon*, **27**, 925.

Riggs, D. and Diefendorf, R.D. (1980) U.S. Pat. 4208267.

Strehlow, R.A. (1970) U.S. Oak Ridge Nat. Lab. Report Nber ORNL-4622, Molten Salt Reactor Program, Semi-Annual Progress Report, Period Ending August 31, 135–141.

Tomari, Y., Mochida, I. and Lino, M. (1993) *Tribologist*, **38**, 1097.

White, J.L. and Buechler, M. (1986) Petroleum Derived Carbons, *ACS Symposium Series*, **303**, 62.

Yamada, Y., Imamura, T., Kakiyama, H., Honda, H., Oi, S. and Fukuda, K. (1974) *Carbon*, **12**, 307.

Yamada, Y., Matsumoto, S., Fukuda, K. and Honda, H. (1981) *Tanso*, **107**, 144.

Yoon, S.-H., Park, Y.D., Mochida, I. (1992) *Carbon*, **30**, 781.

Yoon, S.-H., Korai, Y. and Mochida, I. (1993) *Carbon*, **31**, 849.

11. Amorphous and Non-Crystalline Carbons

J. ROBERTSON

Engineering Department, Cambridge University, Cambridge CB2 1PZ, UK

11.1 INTRODUCTION: TYPES OF NON-CRYSTALLINE CARBONS

There are now three well established crystalline phases of carbon; diamond, graphite and crystalline C_{60} molecules. These phases are distinguished by their bonding; sp^3 in diamond, sp^2 in graphite, and a combination of intra-molecular sp^2 and inter-molecular van der Waals bonding in C_{60}. There is also a large variety of non-crystalline carbons due to the different forms of bonding and different degrees of disorder, as summarised in Table 1.

Graphite which is sp^2 bonded has a hexagonal layer structure and is an anisotropic semi-metal. Diamond which is sp^3 bonded has the cubic structure and is a wide band gap semiconductor with very high hardness. A similar distinction is found in their amorphous carbons, between those which are sp^2 and sp^3 bonded.

When atoms form a solid, the discrete energy levels of the electrons become a series of broad bands of states separated by band gaps. If the solid is a crystal, the periodicity and k-space cause the density of electron states of these bands to have sharp features called van Hove singularities (Elliott, 1990). If a crystal becomes disordered, it will still retain its short-range order, as this is determined by its chemical bonds, but it will loose to a varying degree its long-range order. The overall disposition of bands and any band gaps depends on the short range order, so that the loss of long range order does not remove band gaps, but leads to a smoothing out of the sharp features of the density of states (DOS) and a tailing of band-edge states into the gaps. Thus, the general classification into metals and semiconductors depends on the bond type and not on the presence of long-range order.

The sp^2 bonded amorphous (a-) and non-crystalline carbons have a zero or small band gap and are not mechanically hard. The vast majority of non-crystalline carbons known until

TABLE 1. Characteristic properties of different forms of carbon.

	% Carbon sp^3	% hydrogen	Density, gm/cm^3	Optical gap, eV	Hardness, GPa
Graphite	0	0	2.267	−0.04	
Diamond	100	0	3.515	5.5	100
C$_{60}$	0	0	1.1	1.8	0
Polyethylene	100	67	1	6	low
glassy C	ε	0	1.3–1.55	0.01	2–3
evap. a-C	0	0	1.9–2.0	0.4–0.7	2–5
sputtered a-C	<5	0	2.0–2.4	0.7	5-10
ta-C	80	0	2.9	2.4	80
a-C:H	25–70	25–65	0.9–2.2	1–4	2–20
ta-C:H	60	20–25	2.6	2.6	60

1970 fall into this category. The various types in this class can be distinguished by how they were made and their medium range order. They can be further subdivided according as to whether they will eventually graphitize under heating into graphite (graphitizing carbons), or whether they become locked into a more topologically layered disordered state known as glassy carbon (non-graphitizing carbon).

Those sp^2 bonded carbons with little medium range order can be called amorphous. The best known form of amorphous (a-) sp^2 carbon is that formed by evaporation, for example from a carbon arc or by electron beam evaporation. This material is fully sp^2 bonded and has a small band gap of about 0.5 eV. It may contain some graphitic sheets, because the condensing atoms from evaporation tend to grow back into graphitic sheets. Sputtered a-C is also essentially fully sp^2 bonded, containing less than 5% sp^3 sites (Pan et al., 1991), but now it has much more homogeneous disorder (Li and Lannin, 1990), because the sputtering process involves atom by atom depositions. This a-C also has a band gap of 0.5 to 0.7 eV.

Amorphous carbons containing some sp^3 bonding have only been recognised as such since 1971. These solids have been called diamond-like carbon (DLC) because the presence of some sp^3 bonding confers the diamond-like properties of a band gap, chemical inertness, and mechanical hardness (Robertson, 1986, 1991). These materials generally only exist as a thin film formed by deposition from the vapour, rather than as bulk solids. It is also useful to include in the diamond-like carbons the closely related hydrogenated amorphous carbons (a-C:H).

The composition on these various forms of a-C and a-C:H can best be displayed on a ternary phase diagram as shown in Figure 1. This shows how C-H compounds cover most of the available phase space. The hydrogen content can be very large, but it is still sensible to class these materials as carbons because of their common behaviour. However, if the H content is too large, it is not possible to form a continuous network of C-C bonds needed to form a solid.

It is well known that diamond is slightly less stable that graphite, but only by an excess of cohesion energy of 0.03 eV at 0°K, so the barrier to diamond formation is mainly kinetic

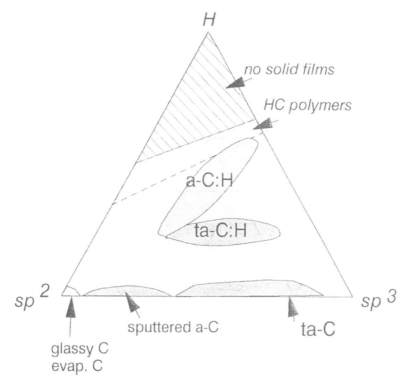

FIGURE 1. Ternary phase diagram of structural types in the C-H system.

not thermodynamic. The sp^3 bonding is also less stable in the disorder phases. The energy difference is now much greater, up to 0.5 eV. This is mainly disorder energy and arises from the distortions of the strong, highly direction covalent carbon bonds in the disordered bond networks.

It is so far not possible to produce a pure sp^3 bonded a-C which would correspond to 'amorphous diamond' and be analogous to amorphous silicon. The highest sp^3 content attained so far is about 85%. This material is often called tetrahedral amorphous carbon (ta-C) (McKenzie, 1996). Ta-C is grown by a number of special deposition techniques involving energetic ion or plasma beams, such as the mass selected ion beam (MSIB) (Lifshitz, 1996), the filtered cathodic vacuum arc (FCVA) and pulsed laser deposition (PLD) (Voevodin and Donley, 1996). The energetic beam deposition is needed to stabilise the metastable sp^3 bonding. To illustrate this, Figure 2 plots the sp^3 fraction, density, and band gap of ta-C deposited by the FCVA method as a function of the ion energy. It is seen that the sp^3 bonding reaches a maximum at some optimum ion energy of about 100 eV, and that all the properties such as density, band gap, elastic modulus and mechanical hardness correlate directly with the sp^3 fraction.

The most common form of DLC is a-C:H. This is typically grown by plasma enhanced chemical vapour deposition (PECVD) methods from a hydrocarbon plasma. The properties

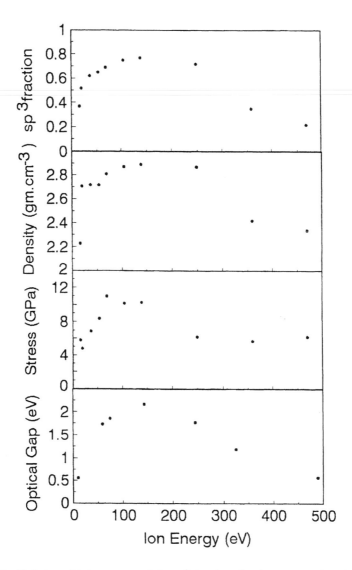

FIGURE 2. Variation with ion energy of the sp³ fraction, density, stress and optical gap of ta-C deposited at room temperature by the FCVA method. Data from Fallon *et al.* (1993)

of these films also depend on the ion energy in the plasma, or in practice the electrode bias voltage which controls the ion energy (Koidl *et al.*, 1989). Figure 3 shows the sp^3 fraction, density, band gap, and hydrogen content plotted against the deposition bias voltage. It is seen that the hydrogen content declines strongly as the bias voltage increases, and that the sp^3 content declines as well. This combination changes the relative amounts of sp^3 C-C bonds, sp^2 C-C bonds and C-H bonds, so the properties vary in a more complicated manner.

At low bias voltages, the high hydrogen and high sp^3 content creates a rather polymeric a-C:H which is mechanically soft and has a wide optical gap of 2–4 eV. At moderate bias voltages, the moderate H and sp^3 contents leads to the most diamond-like material with the high density and hardness, but with actually rather moderate band gaps of 1.1–2 eV. At even higher bias voltages, the low H and sp^3 content leads to a more 'graphitic' type carbon and a narrow gap (0.7–1.1 eV). Similar a-C:H can be formed by the sputtering of graphite in an argon /hydrogen plasma. A notable feature is that sp^3 bonding is much denser than sp^2 bonding, while hydrogen lowers the density. Thus, in Figure 2 the most diamond-like material corresponds to a maximum in the density and mechanical hardness. However, the absolute values of density and hardness are still quite low compared those of diamond itself (Table 1). Harder, higher sp^3 content a-C:H films can be prepared by low-pressure PECVD methods such as the plasma beam source, and these films are termed ta-C:H, by analogy to ta-C (Weiler *et al.*, 1996).

11.2 AMORPHOUS SOLIDS – SOME BACKGROUND

A notable feature of the disordered carbons is that they are all to some degree semiconductors. The presence of long-range disorder has specific effects on the DOS of semiconductors, it causes a smoothing of the density of states and band states to tail into the band gap. Broadly speaking, there are now two types of states, the extended band-like states in the main bands, and the states at the band edges which are localized and do not conduct electricity at $0°$ K (Elliott, 1990). The two types of states are separated by a defined energy called a mobility edge. The localized states usually extend right across the band gap to form a continuum. Now there is no band *gap* as such in the density of states, just a mobility gap between the valence band and conduction band mobility edges.

The band gaps quoted for amorphous semiconductors would normally be mobility gaps. However, gaps are determined more simply by optical means. The optical absorption spectrum of an amorphous semiconductor does not have a sharp threshold corresponding to the band edge of a crystal, but consists of a broader absorption edge. Two definitions of optical gap are commonly used, the E_{04} gap at which the optical absorption coefficient reaches 10^4 cm^{-1}, and the Tauc gap E_g found from fitting the energy dependence of the absorption coefficient for α above 10^4 cm^{-1} to

$$(\alpha E)^{1/2} = B(E - E_g) \tag{1}$$

This formula assumes the bands still retain a parabolic density of states.

The structure of a covalently bonded amorphous semiconductor can be represented, as in a glass, by the continuous random network model, in which all atoms have their chemically preferred valence or coordination number given by the 8-N rule, and the disorder is in the bond lengths, bond angles and topological disorder of the ring statistics. Real amorphous semiconductors differ from this ideal, because a small fraction of atoms also have the wrong coordination number. This is usually an under-coordination, as in broken or 'dangling' bonds.

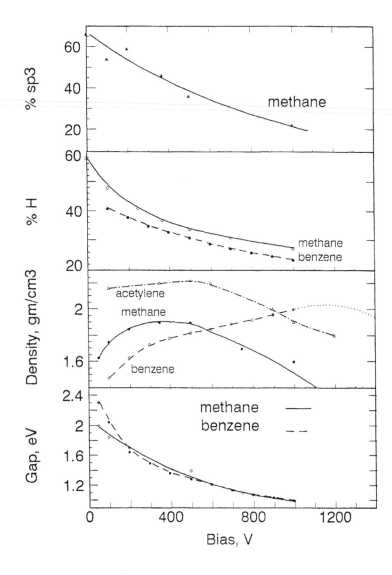

FIGURE 3. Variation of the sp^3 fraction, hydrogen content, density and optical gap with bias voltage of a-C:H deposited by PECVD at room temperature, from different source gases. Data from Tamor *et al.* (1991), Tamor and Vassell (1994) and Koidl *et al.* (1989).

In chemical bonding terms, amorphous semiconductors are covalently bonded and the valence band is formed of the filled bonding states and the conduction band is formed of the empty antibonding states. In carbon, we must distinguish between the sp^3, sp^2 and sp^1 sites. At sp^3 sites, the four valence electrons of carbon each form tetrahedrally-directed σ bonds to four adjacent atoms. At sp^2 sites, three of the four valence electrons form three σ bonds

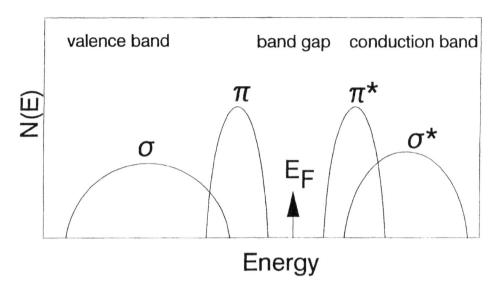

FIGURE 4. Schematic diagram of the density of states N(E) of carbon systems, with the valence and conduction bands of σ and π states.

to three adjacent atoms within a plane, and the fourth valence electron forms a weaker σ bond, with its orbital lying normal to the σ bonding plane. The strong σ bonds of sp^3 and sp^2 sites form the occupied bonding (σ) states in the valence band and empty antibonding (σ^*) states in the conduction, separated by a wide band gap, as shown in Figure 4. The π bonds are weaker and these give rise to occupied π and unoccupied π^* states which lie largely within the $\sigma - \sigma^*$ gap (see Dresselhaus *et al.*, Chapter 2).

11.3 EXPERIMENTAL DETERMINATIONS OF ATOMIC STRUCTURE

The main parameters which define the atomic structure of an amorphous carbon are the fraction of sp^3 or sp^2 bonding and the fraction of hydrogen. It is also of interest to know if the sp^2 sites are arranged in a particular way, for example as aromatic or olefinic groups. The hydrogen content can be determined by nuclear techniques such as nuclear reaction analysis or energy recoil detection analysis (ERDA) or by combustion analysis. The most quantitative method for the sp^3 content is nuclear magnetic resonance (NMR). However, NMR it not a generally usable method as it requires large or ^{13}C enriched samples, or special pulsed NMR techniques.

The most applicable technique is electron energy loss spectroscopy (EELS)(Fallon *et al.*, 1993). EELS measures the spectrum of energy loss of an electron passing through the sample and can be performed on a dedicated instrument or as an attachment to a transmission electron microscope. The EELS spectrum is the excitation spectrum from the core levels to

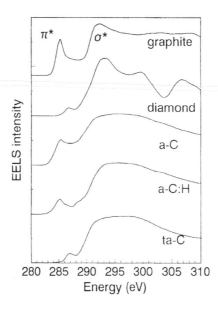

FIGURE 5. Comparison of the EELS spectrum of graphite, diamond, sputtered a-C, a-C:H and ta-C, showing π^* peak at 285 eV and σ^* step at 290 eV. Data from Fink (1983) and Fallon (1993).

the empty conduction band states and therefore, as the core levels are sharp, it consists of a series of conduction band DOSs. The EELS spectrum of carbon, with only a single 1s core level, consists of a peak at 285 eV due to excitations to empty π^* states of sp^2 sites and a step at 290 eV due to excitations to empty σ^* states of sp^2 and sp^3 sites. The sp^2 fraction is given by the ratio of the 285 eV peak to the 290 eV step, compared to its ratio in 100% sp^2 bonded graphite. The spectra in Figure 5 show that the expected trend in the size of the π^* peak between graphite, diamond, a-C:H and ta-C. Other techniques such as Auger and Electron Spectroscopy for Chemical Analysis (ESCA) can be used, but they suffer because the difference between sp^2 and sp^3 signatures is smaller.

The structure of carbons can be determined in more detail by diffraction using either electrons, x-rays or neutrons. The diffraction pattern is then processed and Fourier transformed to give the radial distribution function (RDF), which is the probability of finding another atom at a given distance from a chosen atom. The position of the first peak in the RDF gives the average bond length, the area under the first peak gives the number of nearest neighbours, and the ratio of second peak to first peak distances gives the bond angle. The bond length of diamond is 1.54 Å, of graphite is 1.42 Å and in ethylene is 1.33 Å, so the first peak gives considerable information on the bonding type. In the case of ta-C, the first peak consists of a single broad peak at 1.54 Å (Gilkes *et al.*, 1995), showing that there is considerable bond length disorder, with the sp^2 bonds contributing to the broadening. One very high resolution study of a-C:H was able to resolve a split first peak due to the separate contributions of sp^3 and sp^2 bonds, and it also found that the sp^2 bonds tended to have the shorter bond length of olefinic chains rather than aromatic groups (Walters and Newport, 1996).

Further characterisation of atomic structure is possible from the vibrational spectra, by infra-red (IR) and Raman spectroscopy. A vibrational mode is IR active if it has a dipole moment. C-H bonds have a small dipole moment and the IR spectra are an important means to determine the configuration of C-H bonding in a-C:H. In elemental systems, IR activity requires the absence of a centre of symmetry. For elemental crystals, this requires unit cells with 3 or more atoms, so Bernal graphite with 4 atoms per cell has two weakly IR active modes. In a random network, the loss of periodicity means that all modes are in principle IR active and the IR spectrum is related to the vibrational density of states (VDOS).

Raman is the inelastic scattering of light by phonons, by the induced polarisation of that vibration. In a random network, the loss of periodicity means that all modes are Raman active. Thus, the Raman and IR spectra of a-Si or a-Ge both give the VDOS, but each weighted by a different matrix element.

Raman is a widely used non-destructive technique to characterise bonding in carbon phases because of the differences between their spectra, as summarised in Figure 6. The Raman of diamond consists of a single peak at $1332 \ cm^{-1}$ corresponding to the zone centre phonon. The spectrum of graphite consists of a single peak at $1581 \ cm^{-1}$ also corresponding to the zone centre phonon. The spectrum of microcrystalline or disordered graphite consists of two peaks, a $1590 \ cm^{-1}$ peak denoted the G peak, and a second peak around $1350 \ cm^{-1}$ denoted D (for disorder) which is a disorder-activated zone-boundary mode. The intensity ratio of the D and G peaks, I(D)/I(G) is found to vary inversely with the in-plane coherence length of the microcrystals, L_a (Tuinstra and Koenig, 1970). The Raman spectra of glassy carbon is similar to that of microcrystalline graphite with a short L_a, as seen in Figure 6. The spectrum of sputtered a-C (essentially sp^2 bonded) is seen to be an extreme version of microcrystalline graphite, with strong and very broadened G and D peaks. The I(D)/I(G) ratio is actually over one, but the D peak is lower because it is broader than the G peak. Generally, the spectra of such carbons can be reasonably well represented by two gaussian G and D peaks, rather than a spectrum derived from the overall VDOS.

The Raman spectra of the two diamond-like carbons in Figure 6 are still dominated by the G peak of the sp^2 sites, because the sp^2 sites in DLC have a roughly 50 times higher Raman matrix element than the sp^3 sites (Schroder et al., 1990). The spectra also show a peak or shoulder in the D peak position. Indeed, the spectra of a-C:H can also be fitted by two gaussian G and D peaks. The positions and widths of the G and D peaks is a useful empirical guide to the bonding in a-C:H (Tamor and Vassell, 1994). The G peak is found to fall with increasing sp^3 content. This is related to the lowering of the Raman mode from graphite to diamond, but this occurs at a slower rate because the scattering is still coming only from the sp^2 sites. Figure 7(a) shows that the G peak position in a-C:H decreases with increasing optical band gap, which is equivalent to it decreasing with increasing sp^3 content, as the gap increases with sp^3 content (see Figure 13).

Figure 7(b) shows that the G peak width passes through a maximum at a gap of 1.3 eV. The G width is proportional to the internal stress and in fact is a good guide to the degree of diamond-like bonding-note that the G width reaches a maximum where the density is a maximum in Figure 2.

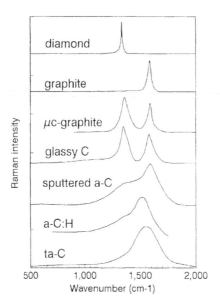

FIGURE 6. Comparison of the Raman spectra of diamond, graphite, microcrystalline graphite, glassy carbon, sputtered a-C, a-C:H and ta-C. Data from Robertson (1986) and Prawer (1996).

It was noted that the intensity ratio I(D)/I(G) varies inversely with the in-plane correlation length L_a. This dependence cannot continue indefinitely, and indeed I(D)/I(G) is found to reach a maximum for $L_a \sim 1$ nm, and then decrease (Schwan *et al.*, 1996), as in Figure 8.

The Raman spectra of ta-C is also dominated by the G peak, due to its residual sp^2 sites, but there is no clear D feature. It is found that this spectra is best fitted by a single skewed lorentzian rather than by gaussians (Prater *et al.*, 1996). The lorentzian is essentially symmetric at the maximum sp^3 content and the skewness increases with sp^2 content.

To understand why the Raman spectra of disordered carbons is not so simply related to the VDOS, it is necessary to consider Raman as a resonant process, in which the scattering occurs between real electron states in the valence and conduction band. Photon energies in the visible spectrum lie within the forbidden gap for diamond or a sp^3 network and so do not to first order excite Raman. Graphite is a semimetal, but the only allowed vertical transitions are for $\pi - \pi^*$ transitions along the K-H zone boundary. Thus, photons only couple to phonons at the zone boundary. Such modes give rise to the D peak. The absence of available transitions at sp^3 sites is the cause of their very low Raman matrix element. More generally, the Raman spectrum of disorder carbons tend to reflect resonantly excited modes rather than the VDOS.

It is possible to minimise the resonant effects in Raman by using UV excitation. UV photons excite above the $\sigma - \sigma^*$ band edge and over most of the zone, so now sp^2 and sp^3 sites have a similar Raman matrix element. Thus UV Raman allows the first direct probe of sp^3 sites. The UV Raman spectra of ta-C shows a G peak at 1600 cm^{-1} and a peak at 1200 cm^{-1} due to its sp^3 sites (Gilkes *et al.*, 1997).

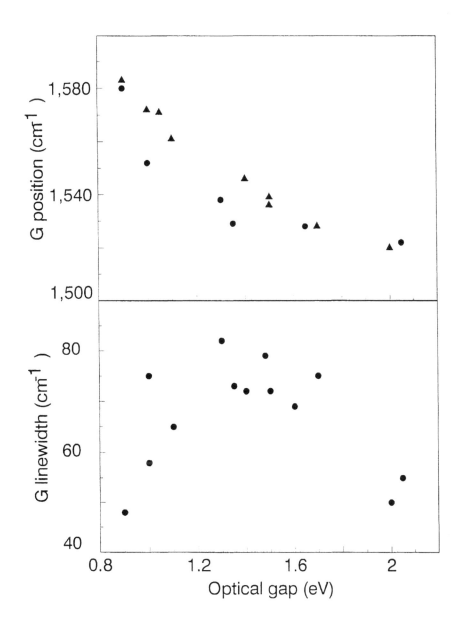

FIGURE 7. Variation of the position and width of the Raman G peak with optical gap for a-C:H films. The optical gap is correlated to sp^3 fraction.

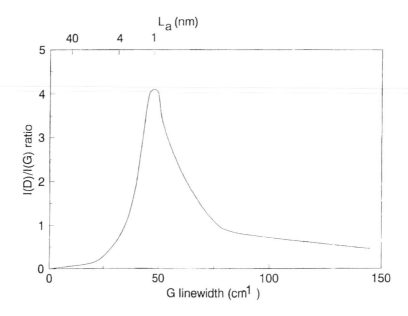

FIGURE 8.   Correlation of Raman G linewidth with I(D)/I(G) ratio, showing the inverse dependence for larger grains on the right, and the turnover for small grains.

11.4 ELECTRONIC STRUCTURE MODELS

The basic electronic structure of amorphous carbons was summarised in Figure 4. The strong σ bonds of sp^3 and sp^2 sites form the occupied bonding (σ) states in the valence band and empty antibonding (σ^*) states in the conduction, separated by a wide band gap. The π bonds of sp^2 and any sp^1 sites give rise to occupied π and unoccupied π^* states which lie largely within the $\sigma - \sigma^*$ gap.

The bonding in amorphous carbons has been analyzed in more detail by model calculations and by large scale simulations. The π states of sp^2 sites lie closest to the Fermi level E_f and, by perturbation theory, these states tend to control the electronic and structural properties of non-crystalline carbons (Robertson and O'Reilly, 1987). The simplest way to proceed further is to decouple the π and σ states. This is possible if the sp^2 sites are relatively planar, as occurs in glassy C and evaporated a-C. The σ states form 2-centre pair bonds which give the structural skeleton of the network. The σ bonds control the short range order (SRO) of the network, that is bond lengths and bond angles. In contrast, the π states can form both pair bonds and multi-centre bonds. The nature of the π bonding means that, for a given proportion of sp^2 and sp^3 sites, the sp^2 sites will tend to form clusters within an sp^3 bonded matrix.

This effect is seen if the π states are analyzed by the Huckel approximation, as summarised in Figure 9. This retains only the nearest neighbour π transfer integral which we define as γ for one-electron energies and γ' for total energies. γ' can be estimated from the free

group	name	E_{tot}/γ'
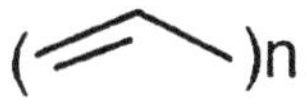 $(\diagup\diagdown)n$	polyacetylene	1
$(\,\flat\,-)n$	carbynes	2
	benzene	1.333
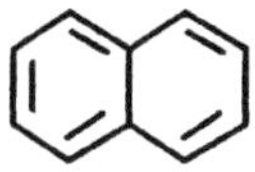	napthalene	1.368
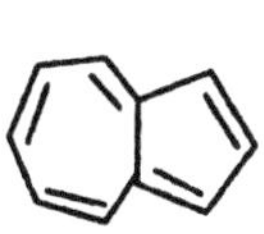	azulene	1.336
	quinoid	1.24
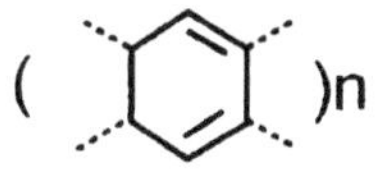 $(\quad)n$	polybenzoid	1.403
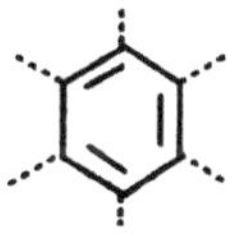	graphite	1.616

FIGURE 9. Dependence of the normalised π binding energy per site in the Huckel approximation of various configurations of sp^2 sites in a-C.

energies of ethane and ethene as $\gamma' = 1.4$ eV. An isolated π orbital is quite unstable, as it has no π bond and no π binding energy. The π orbital gains a considerable energy, γ' per site, if it pairs up with another π state to form a double bond. The π binding energy is maximised if the π orbitals are aligned in one direction so the π cluster is planar. The sites will gain more energy ($\gamma'/3$ per site according to Huckel theory) if three pairs join to form a planar 6-fold or benzene ring. Further energy is gained if the 6-fold rings fuse into clusters of aromatic rings. The clusters are more stable if they are compact graphitic clusters rather than row-like polybenzoid clusters. Other configurations which mix rings and chains such as quinoid structures are less stable than simple aromatic rings.

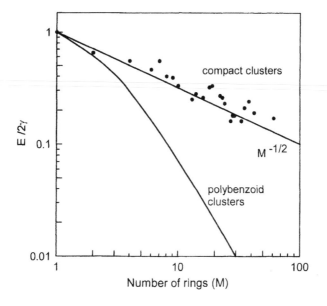

FIGURE 10. Variation of the band gap of compact ring configurations and chain (polyacetylene) configurations, in the Huckel approximation. The line shows the $M^{-1/2}$ law.

The π bonded clusters with an even number of sites have an optical gap. The gap has been calculated for aromatic cluster configurations in the Huckel approximation, as a function of the number of rings M, and is shown in Figure 10. It is found that the gap for planar compact clusters of aromatic rings follows the relation

$$E = 2\gamma/M^{1/2} \text{ eV} \qquad (2)$$

The band gap of other planar configurations of sp^2 sites is generally much smaller. The value of γ has been estimated from the band structure of graphite as 3.16 eV. The gap in these clusters can be considered to arise from the confinement of the π states; disorder has opened up a gap in these systems, rather than washed one out.

This Huckel model provides a useful description of non-crystalline carbons in the case where the disorder is low, so the sp^2 sites are likely to lie parallel. This picture is consistent with changes occurring during the carbonization of carbonaceous solids, in which further heating causes gradual graphitisation. The Huckel model turns out to be less valid for the more disordered DLCs, as will be seen shortly.

These ideas were refined into the cluster model of amorphous carbon (Robertson, 1986, 1992). This proposed that in a network containing a given fraction of sp^2 and sp^3 sites, the sp^2 sites would tend to form sp^2 bonded clusters within a sp^3 bonded, possibly hydrogenated, matrix. There would be a functional splitting in the role of sp^2 and sp^3 sites; the π states of the sp^2 clusters would control the electronic properties, while the C-C coordination of the sp^3 bonded matrix would control the mechanical properties.

11.5 ELECTRONIC STRUCTURE SIMULATIONS

The atomic structure depends primarily on the sp^3 fraction and hydrogen content. The atomic structure of H-free ta-C had assumed that the sp^3 sites form a four-fold coordinated continuous random network, analogous to a-Si, in which the bond length and bond angle vary about the crystalline value, and the atoms form rings of 5 atoms and upwards. A first assumption was that the sp^3 sites of ta-C would behave rather like the sp^3 sites in a-Si, in that topological disorder would give rise to 5- and 7-membered rings in addition to 6-fold rings, but that low order 3- and 4-membered rings would be inhibited because of their large bond angle distortion. In fact, Marks *et al.* (1996) found that ta-C may possess 3- and 4-fold rings, the analogue of cyclopropane and cyclobutane, and that this would account for the anomalously low area of the second neighbour peak in the radial distribution function of ta-C noted by Gilkes *et al.* (1995).

It turns out that the configuration of sp^2 sites of diamond-like carbons is not as simple as described by the Huckel model of section 11.4. A number of detailed simulations have been carried out in recent years which give a more accurate view of the electronic structure (Galli, 1989; Frauenheim *et al.*, 1994; Stephan *et al.*, 1994; Drabbold *et al.*, 1994). Their main conclusion is that the sp^2 sites form much smaller clusters that in the Huckel model, and that the sp^2 sites tend to form chain (olefinic) groups rather than ring (aromatic) groups. The tendency for sp^2 sites to pair up is still very strong, but further development of clusters is less strong.

The reasons for the failure of the cluster model for DLC is firstly that the Huckel theory overestimates the stabilisation of aromatic rings. The stabilisation energy of benzene is $\gamma'/3 = 0.47$ eV per site in the Huckel model, but only 0.22 eV if evaluated directly from the free energies of ethene and benzene. The second, perhaps more fundamental point, is that DLC is a relatively high disorder system (Robertson, 1995). The disorder potential in most amorphous solids is actually quite low, and the observed atomic configurations tend to be those of lowest free energy. Diamond-like carbon differs in that the intense ion bombardment used during deposition to stabilise sp^3 bonding creates strong disorder. This disorder opposes the energy gain from clustering, as shown in Figure 11.

The strong disorder in DLC is seen from two measures of disorder, the much wider optical absorption edge of a-C:H seen in Figure 12 and the greater Raman line width seen in Figure 7(b). Thus, the disorder opposes the clustering, so that only relatively small clusters are observed in practice.

The structure of diamond-like carbon is now less simple to describe. The sp^2 sites are more disordered. The π orbitals on adjacent atoms are less likely to lie parallel. The sp^2 sites tend to form small chain-like clusters, not aromatic rings. The distortions mean that it is less valid to treat the σ and π states as separable, as the $\sigma - \pi$ interaction can be considerable. The band gap is no longer given by the simple Huckel formula (2) valid for planar aromatic clusters. The gap is now controlled by the disorder of π states, rather than cluster size.

Figure 13 shows the density of states (DOS) of calculated for three periodic models of amorphous carbon of varying density (Robertson, 1997). The fraction of sp^3 sites in these models is found to increase in proportion to the density. This is expected, as diamond is 50% more dense than graphite. The total DOS is broken down into those from σ and π

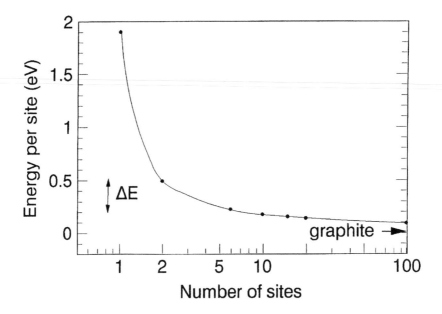

FIGURE 11. Energy gain per site for π states versus sp^2 cluster size.

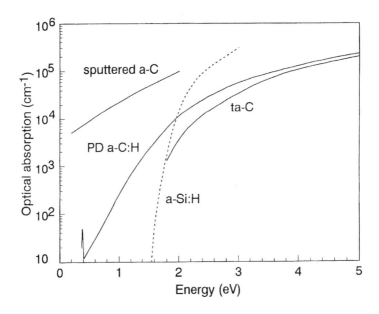

FIGURE 12. Comparison of the optical absorption edges of sputtered a-C, a-C:H, ta-C and a-Si:H, showing the much broader absorption edges of amorphous carbons.

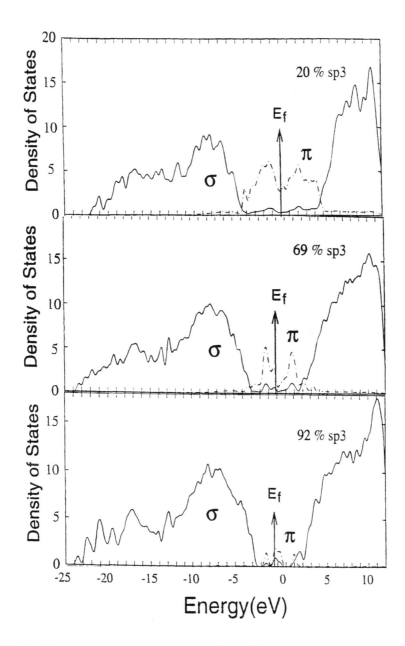

FIGURE 13. Calculated density of states of periodic models of a-C, with various fractions of sp³ sites. Solid lines show σ states and dashed lines π states.

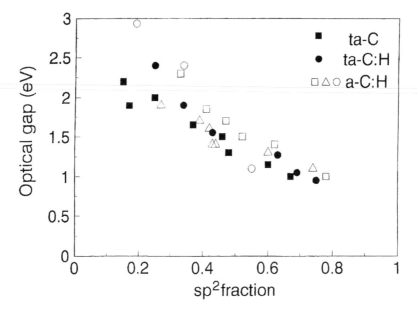

FIGURE 14. Experimental variation of optical gap versus carbon sp^2 fraction for a-C:H, ta-C and ta-C:H.

states. There is also a contribution from defect gap states, discussed later. The DOS is seen to follow the scheme shown in Figure 4, with π states around the Fermi level E_F.

11.6 EXPERIMENTAL MEASUREMENTS OF THE ELECTRONIC STRUCTURE

It is found that in practice the optical gap of all types of diamond-like carbon depends on the sp^2 fraction of carbon sites (Robertson, 1996). Figure 14 shows that the optical gap of ta-C, ta-C:H and plasma deposited a-C:H varies almost linearly with sp^2 fraction, from a very wide gap at low sp^2 content, to a minimum gap of about 0.7 eV, typical of sputtered a-C.

The valence band density of states (DOS) can be determined by photoemission (Wesner *et al.*, 1983; Schafer *et al.*, 1996). In photoemission, a monochromatic light source is used to excite electrons from the occupied valence states into the vacuum and their energy spectrum gives the valence DOS. Figure 15 compares the photoemission valence DOS of graphite, diamond, sp^2 bonded a-C and a-C:H, referred to the Fermi energy at 0 eV. The DOS of diamond consists of a peak at low binding energy due to its p states and double peak at higher binding energies due to s states. The semiconducting nature of diamond is evident from the region of zero DOS between the valence band top to 0 eV. The graphite spectrum is clearly that of a metal. It shows a similar peak structure to diamond, with an additional shoulder at low binding energy due to the π states. The spectrum of sp^2 bonded a-C is

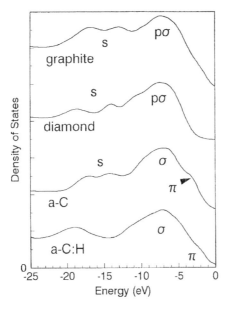

FIGURE 15. Comparison of the ultraviolet photoemission spectra (UPS) of graphite, diamond, a-C and a-C:H. Data from Wesner *et al.* (1983).

surprisingly similar to that of graphite, except that it shows a small semiconducting gap. The spectrum of a-C:H shows some differences. The shoulder of π states on the low energy slope of the valence band is weak. The twin peaks of the s band have merged into a single peak, because of the presence of odd membered rings. The conduction band DOS shown in Figure 5 has already been described in detail .

The wide band optical spectra also give information on the electronic states. Figure 16 compares the imaginary part of the dielectric function ϵ_2 of various phases. The ϵ_2 of diamond shows a single peak structure due to excitations of σ states to empty σ^* states, with a main peak at 13.6 eV. The ϵ_2 of graphite shows a two peak structure, the lower features below 8 eV are due to excitations of π states to π^* conduction states while the upper peak around 13 eV is mainly due to excitations from π states to π^* states, as in diamond. There can be no $\sigma - \pi^*$ cross excitations in graphite because these states lie in perpendicular directions. The deep valley in the ϵ_2 of graphite is a notable feature and was believed to separate the π and σ excitations. However, it is now recognised that the oscillator strength of π states is not exhausted below 8 eV and contributes weakly across the upper peak.

The ϵ_2 of a-C:H shows the same two peak structure (Fink *et al.*, 1983). The lower peak is due exclusively to π excitations and indicates the importance of these states. It is notable that ϵ_2 never reaches high values in a-C:H because its oscillator strength is spread over a wide energy range. The ϵ_2 of ta-C shows an almost single peaked structure with only very weak π excitations (Xiong *et al.*, 1993). The main peak is due to excitations of σ states, and it now occurs at lower energies than in diamond, because of the loss of the k-selection rule. This same lowering occurs in the spectrum of a-Si.

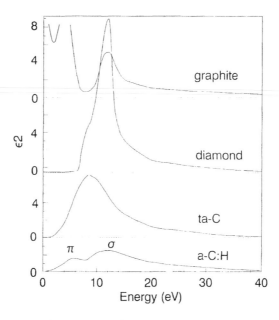

FIGURE 16. Comparison of the ϵ_2 spectra of graphite ($E \perp c$), diamond, ta-C and a-C:H. π excitations peak at 6–8 eV, σ excitations peak at 10–12 eV. Data from Fink (1983) and Xoing *et al.* (1993).

11.7 LOCALIZED STATES

The discussion so far has emphasised the relationship of the electronic structure to the atomic structure. Future developments of a-C require a deeper understanding of the transport and recombination of carriers. The presence of disorder creates localized states below the band edges which tail into the band gap. One distinguishes between band-like extended states and localized states which are separated by a mobility edge. One further distinguishes between tail states which lie next to the band edges and defect states lying deep in the gap which arise from defect configurations such as dangling bonds.

A practical definition of the localization of a state is the inverse participation ratio, P. P is defined for a state $\psi = \sum c_i \phi_i$ as (Elliott, 1990),

$$P = \sum c_i^4 \tag{3}$$

P is a measure of the localization of a state, so that $P = 1$ for a state localized on one site, and $1/N$ for a state delocalized over N sites.

In a typical σ bonded network like a-Si, the disorder localizes only states at the edges of each band. Figure 17 shows the partial DOS of a 128 atom a-C network containing 69% sp³ states, with contributions from σ, π and defect states. (Defect states are defined as π states on isolated 3-fold sites.) Figure 17 shows that P behaves very differently for each type of state. P has a low value for the σ states, and rises to about 0.1 for all the π states and

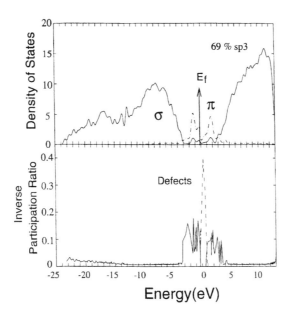

FIGURE 17. Calculated partial density of states and inverse participation ratio P, showing the increasing localization of σ, π and defect states.

to 0.4 for defect states. This shows that disorder produces a much wider range of localized states in the π states amorphous carbons than in a-Si. This is because π states are subject to dihedral disorder unlike σ states. The consequence is that the mobility gap tends to exceed the optical gap.

11.8 DEFECTS

Defects are certain atomic configurations which give rise to states deep in the band gap. Defect states are of critical importance because they are more strongly localized than tail states and they act as deep traps and non-radiative recombination centres. In a-Si:H and σ-bonded semiconductors, the main defects are 3-fold coordinated sites or dangling bonds. In a-C and a-C:H, dangling bonds would be isolated sp^2 sites. There is however, a wider range of possibilities because of the π bonding. Now any odd numbered cluster of sp^2 sites will produce a defect state near mid-gap (Robertson and O'Reilly, 1986). Simulations suggest that the defects in ta-C are isolated sp^2 sites which have not paired up as π bonds. Stephan et al (1994) found a gradual change in the character for two sp^2 sites, from dangling bonds, to π bond as the overlap of the dangling bonds increases.

The density of defects is generally much higher in ta-C, ta-C:H and a-C:H than in a-Si:H and it tends to decrease with increasing band gap, as seen in Figure 18 (Kleber *et al.*, 1991; Schutte *et al.*, 1993; Ristein *et al.*, 1995; Weiler *et al.*, 1996). Defects with an unpaired spin

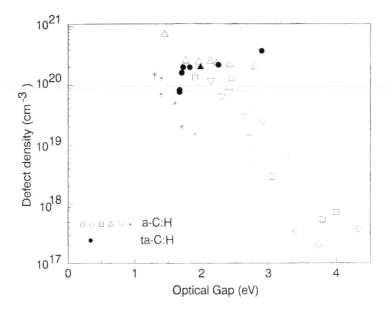

FIGURE 18. Paramagnetic defect density in various forms of a-C:H and ta-C:H, as a function of the optical band gap.

are generally identified structurally by electron spin resonance (ESR). ESR has been rather inconclusive in a-C:H on whether the centre is a single site or a cluster. This is because the main C isotope ^{12}C is a spin zero nucleus, so that there are no strong hyperfine lines to give structural data. C has a small spin-orbit interaction, so all intrinsic defects have a g-factor of 2.0028. (Other lines can occur particularly if the centre lies close to an oxygen impurity atom.) The width of the ESR line is attributed to exchange narrowing (Sadki *et al.*, 1993), as it does not correlate with the H content in a-C:H.

The more polymeric forms of a-C:H give strong room temperature photoluminescence (Schutte *et al.*, 1993, Rusli *et al.*, 1996). PL is believed to occur by the excitation of electrons from π to π^* states within a single cluster, which then recombine within the same cluster (Robertson, 1996), as shown in Figure 19. The strong localization of electron and hole are consistent with the short lifetime, weak temperature dependence, lack of electric field quenching and polarization memory of the PL in a-C:H. The thermal quenching of PL occurs because the carriers separate by thermally assisted hopping, to recombine non-radiatively at defects. The carriers can also tunnel apart and recombine non-radiatively at defects. The role of defects in PL remains contentious. It has been recently argued that defects are the dominant recombination centre, as in a-Si:H, as the PL efficiency is found to decrease at high defect densities as shown in Figure 20, but with a much higher critical density than in a-Si:H.

In narrower gap a-C such as sputtered a-C or evaporated a-C, the deep states control the electrical conductivity. They give rise to 'variable range hopping', in which the conductivity

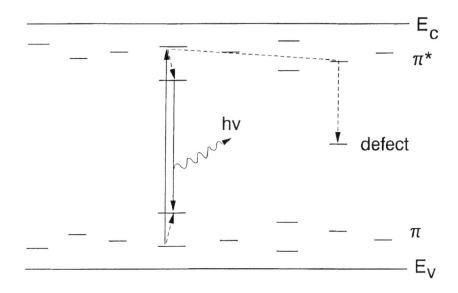

FIGURE 19. Proposed photoluminescence mechanism in a-C:H. Photons create electron-hole pairs, which either recombine to give PL, or tunnel or hop apart, and recombine non-radiatively at defects.

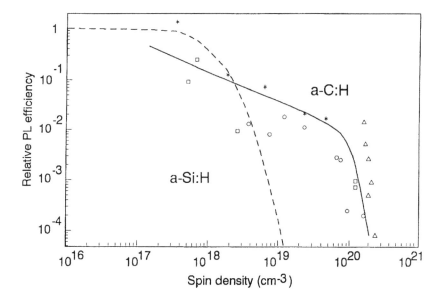

FIGURE 20. Correlation of PL efficiency with paramagnetic defect density, showing how the PL efficiency declines at high defect density.

shows a temperature dependence given by Mott's $T^{1/4}$ law (Hauser, 1977). In the wider gap a-C:H films, the Fermi level lies near mid-gap and conduction occurs by excitation to the valence or conduction band tail states, so that conduction occurs through the low mobility of the tail states. This gives rise to a 1/T Arrhenius type temperature dependence. So far, conduction by excitation to extended states has not been observed, presumably because of the wide mobility gap in these materials noted in Figure 17.

11.9 CONCLUSION

There are a wide variety of non-crystalline carbons with both sp^2 and sp^3 bonding. Those with sp^2 bonding are formed by evaporation or sputtering and are narrow band gap semiconductors with gaps of about 0.5 eV and electrical conductivity controlled by hopping between localized states. The amorphous carbons containing sp^3 bonds are called diamond-like carbons. There band gap can vary from 0.7 eV to almost 4 eV. Their sp^2 sites tend to form small clusters and their π states form the band edges and mid-gap states, and these control the electronic and optical properties. The connectivity of the sp^3 sites control the mechanical properties such as elastic modulus and hardness. Raman spectroscopy is a useful method to distinguish between the various forms of non-crystalline carbon.

References

Drabold, D.A., Fedders, P.A. and Strumm, P. (1994) *Phys. Rev. B*, **49**, 16415.

Dresselhaus, M.S. and Dresselhaus, Saito (1999) This volume chapter II.

Elliott, S.R. (1990) 'Physics of Amorphous Materials' (2nd, Longmans, London, 1990) p. 283.

Fallon, P.J., Veerasamy, V.S., Davis, C.A., Robertson, J., Amaratunga, G., Milne, W.I. and Koskinen, J. (1993) *Phys. Rev. B*, **48**, 4777.

Frauenheim, T., Blaudeck, P., Stephan, U. and Jungnickel, G., *Phys. Rev. B*, **48**, 4823 (1993); **50**, 6709 (1994); **50**, 7940.

Fink, J. *et al.* (1983) *Solid State Commun.*, **47**, 687.

Galli, G., Martin, R.M., Car, R. and Parinello, M. (1989) *Phys. Rev. Lett.*, **62**, 555.

Gilkes, K.W.R., Gaskell, P.H. and Robertson, J. (1995) *Phys. Rev. B*, **51**, 12303.

Gilkes, K.W.R., Sans, H.S., Batchelder, D.N., Robertson, J. and Milne, W.I. (1997) *App Phys Lett*, **70**, 1980.

Hauser, J. (1977) *J. Non-Cryst Solids*, **23**, 21.

Koidl, P., Wild, C., Dischler, B., Wagner, J. and Ramsteiner, M. (1989) *Mat. Sci. Forum*, **52**, 41.

Li, F. and Lannin, J.S. (1990) *Phys. Rev. Lett.*, **65**, 1905.

Kleber, R., Jung, K., Ehrhardt, H., Muhling, J., Breuer, K., Metz, H. and Engelke, F. (1991) *Thin Solid Films*, **205**, 274.

Lifshitz, Y. (1996) *Diamond Related Mats*, **5**, 388.

Marks, N., McKenzie, D.R., Pailthorpe, B.A., Bernasconi, M. and Parrinello, M. (1996) *Phys. Rev. Lett.*, **76**, 768.

McKenzie, D.R. (1996) *Rep. Prog. Phys.*, **59**, 1611.

Pan, H., Pruski, M., Gerstein, B.C., Li, F. and Lannin, J.S. (1991) *Phys. Rev. B*, **44**, 6741.

Prawer, S., Nugent, K.W., Lifshitz, Y., Lempert, G.D., Grossman, E., Kulik, J., Avigal, I. and Kalish, R. (1996) *Diamond Related Mats.*, **5**, 433.

Ristein, J., Schafer, J. and Ley, L. (1995) *Diamond Related Mats.*, **4**, 508.

Robertson, J. (1986) *Adv. Phys.*, **35**, 317.

Robertson, J. (1987) *and O'Reilly, E.P. Phys. Rev. B*, **35**, 2946.

Robertson, J. (1991) *Prog. Solid State Chem.*, **21**, 199.

Robertson, J. (1992) *Phys. Rev. Lett.*, **68**, 220.

Robertson, J. (1995) *Diamond Related Mats.*, **4**, 297.

Robertson, J. (1996) *Phys. Rev. B*, **53**, 16302.

Robertson, J. (1997) *Phil. Mag. B*, **76**, 335.

Rusli, Robertson, J. and Amaratunga, G. (1996) *J. Appl. Phys.*, **80**, 2998.

Sadki, A., Bounouh, Y., Theye, M.L. and von Bardeleben, J. (1996) *Diamond Related Mats.*, **5**, 439.

Schafer, J., Ristein, J., Graupner, R., Ley, L., Stephan, U., Frauenheim, T., Veerasamy, V.S., Amaratunga, G., Weiler, M. and Ehrhardt, H. (1996) *Phys. Rev. B*, **53**, 7762.

Schroder, R.E., Nemanich, R.J. and Glass, J.T. (1990) *Phys. Rev. B*, **41**, 3738.

Schutte, S., Will, S., Mell, H. and Fuhs, W. (1993) *Diamond Related Mats*, **2**, 1360.

Schwan, J., Ulrich, S., Batori, V., Ehrhardt, H. and Silva, S.R.P. (1996) *J. App. Phys.*, **80**, 440.

Stephan, U., Frauenheim, T., Blaudeck, P. and Jungnickel, G. (1994) *Phys. Rev. B*, **49**, 1489.

Tamor, M.A. and Vassell, W.C. (1994) *J. Appl. Phys.*, **76**, 3823.

Tamor, M.A. and Vassell, W.C., Carduner, K.C. (1991) *App. Phys. Lett.*, **58**, 592.

Tuinstra, F. and Koenig, J.L. (1970) *J. Chem. Phys.*, **53**, 1126.

Voevodin, A.A. and Donley, M.S. (1996) *Surface Coatings Technol.*, **82**, 199.

Walters, J.K. and Newport, R.J. (1995) *J. Phys. Cond. Mat.*, **7**, 1755.

Weiler, M., Sattel, S., Giessen, T., Jung, K., Ehrhardt, H., Veerasamy, V. and Robertson, J. (1996) *Phys. Rev. B*, **53**, 1594.

Wesner, D. *et al.* (1983) *Phys. Rev. B*, **28**, 2152.

Xiong, F., Wang, Y.Y. and Chang, R.P.H. (1993) *Phys. Rev. B*, **48**, 8016.

12. Physical Properties of Pregraphitic Carbons

MICHIO INAGAKI[1] and YOSHIHIRO HISHIYAMA[2]

[1]*Aichi Institute of Technology, Toyota 470-0392, Japan*
[2]*Faculty of Engineering, Musashi Institute of Technology, Setagaya-ku, Tokyo, 158 Japan*

12.1 INTRODUCTION

12.1.1 Pregraphitic Carbons

Carbon materials described in this chapter consist of carbon atoms bound by sp^2 hybrid orbital, which result in the formation of hexagonal layers of carbon atoms, known as graphite. Even though graphite structure which is fundamentally composed of hexagonal carbon layers stacked in parallel with a regularity of so-called ABABAB... is a thermodynamically stable phase, parallel stacking of these layers without any regularity (turbostratic structure) (Warren, 1941) is possible, particularly in the products just after carbonization at low temperatures of 1000–1500°C. These parallel-stacked carbon layers construct a structural unit in carbonized materials, and govern the further development in structure and also in properties, therefore, being called basic structural unit, BSU (Oberlin, 1989). By heat treatment to temperatures above carbonization temperature, these BSUs grow to form so-called crystallites in the field of crystallography. Further heat treatment up to 3200°C makes these crystallites grow in size, and also the stacking of carbon layers in each crystallite is improved to be in three-dimensional regularity, i.e. ABABAB... stacking, and this structural change is in a strict sense called graphitization. However, this structural development is known to depend strongly on the starting organic materials and their carbonization conditions, in other words, the size and local orientation of BSUs in carbonized materials, and also on the graphitization condition at high temperatures. The carbonaceous precursors have been classified schematically into two groups, graphitizing and non-graphitizing carbons, or soft and hard carbons, but it has to be pointed out that in many cases such rough classification is not appropriate, many carbon precursors showing an intermediate behavior between the two groups. Therefore, we have a wide range of structure of carbon materials, from completely turbostratic stacking to the three-dimensional graphite organization by selecting the precursor and the heat treatment conditions. Here, carbon materials of which structural change is on the way to reach perfect graphite are called pregraphitic carbons, especially those with a turbostractic structure.

In accordance with this strucural change, physical properties of pregraphitic carbons change strongly with their precursor and their graphitization process, particularly the maximum temperature of heat treatment (usually heat treatment temperature, HTT) (Kelly, 1981).

In this chapter, therefore, the physical properties of pregraphitic carbon materials are reviewed on representative precursors, such as pyrolytic carbons, carbon films, cokes and glass-like carbons, as a function of preparation conditions, by giving particular emphasis on galvanomagnetic and thermal properties. On other properties of graphitic materials, such as mechanical properties, the readers are asked to refer to other chapters in this book, for the physical definitions (Issi, 1998) and for the experimental techniques to corresponding papers.

12.1.2   *Dependence of Physical Properties on Preparation Conditions*

Most of physical properties, electrical resistivity ρ, Hall coefficient R_H and maximum transverse magnetoresistance $(\Delta\rho/\rho_0)_{max}$ (the absolute maximum value of the transverse magnetoresistance), absolute thermoelectric power S, specific heat C_p, thermal expansion coefficient α and thermal conductivity κ which are discussed in this chapter, for carbon materials are known to depend strongly on HTT, which are due to the structural change with HTT. These physical properties are tensors in principle. Carbon materials usually have a symmetry axis, such as the axis along the direction of pressing or molding for their production and that perpendicular to the deposition plane for pyrolytic carbons. Therefore, in these materials, the flow direction of the electrical and heat currents or the directions of the temperature gradient must clearly be indicated for the specimen to be measured on these properties. If we take note of the tensor character of transport properties in particular magnetoresistance, it is a powerful tool to characterize the structure and microtexture of pregraphitic carbons (Hishiyama, *et al.*, 1991) and, therefore, much work has been performed on various carbon materials, such as pyrolytic carbons, cokes, glass-like carbons and carbon fibers. Low temperature specific heat C_p depends also strongly on HTT, being also related to the presence of anomalous peaks at very low temperatures, which are also discussed in this chapter.

12.2   ELECTRIC AND GALVANOMAGNETIC PROPERTIES

12.2.1   *Cokes*

Figure 1 shows changes of ρ, R_H and $(\Delta\rho/\rho_0)_{max}$ at liquid nitrogen temperature against HTT for a petroleum coke which has a graphitizing nature (Inagaki and Hishiyama, 1994).

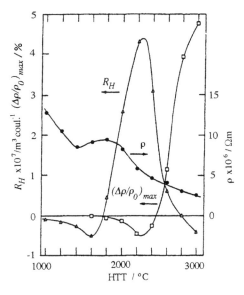

FIGURE 1. Dependence of electrical resistivity ρ, Hall coefficient R_H and maximum transverse magnetoresistance $(\Delta\rho/\rho_0)_{max}$ in a magnetic field of 1 T at 77 K on heat treatment temperature (HTT) for a coke (Inagaki and Hishiyama, 1994).

The measurements were carried out on small pieces of the cokes. The values of R_H and $(\Delta\rho/\rho_0)_{max}$ are those measured under an applied field of 1T. These changes measured on the coke are typical for carbon materials derived from different precursors.

Resistivity ρ exhibits a small hump around 1800°C and then decreases gradually with HTT increase. Hall coefficient which R_H is negative and field independent at low HTTs, becomes positive above 1800°C, reaches a maximum around 2200°C, and then decreases to be negative again and weakly field dependent above 2800°C. Maximum transverse magnetoresistance $(\Delta\rho/\rho_0)_{max}$ cannot be measured on the cokes heat-treated below 1600°C, but above this treatment temperature it shows a negative value, and through a minimum around 2200°C it starts to increase with HTT and becomes positive. The transition from negative to positive values in $(\Delta\rho/\rho_0)_{max}$ is studied by measuring its magnetic field dependence on the coke pieces heat-treated at different temperature; the results are shown in Figure 2 (Hishiyama et al., 1971).

These experimental results are explained qualitatively by a progressive change in energy band structure for carbon materials (Mrozowski, 1971); the model is schematically shown in Figure 3. However, the negative Hall coefficient observed at HTT below 1600°C cannot be explained by this model. At HTT around 1750°C, hydrogen atoms are completely eliminated to leave electron traps, most of which may locate at the edge of the hexagonal carbon layers. The Fermi level is sufficiently lowered from the top of the lower π electronic band, and the charge carriers are mainly holes. With increasing HTT, the growth of crystallites (layer size along a-axis and parallel stacking along c-axis) causes the number of traps to decrease

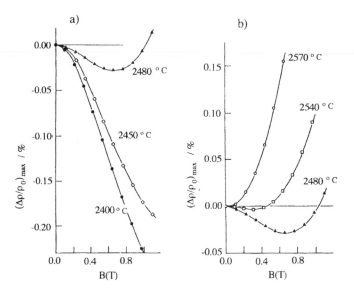

FIGURE 2. Dependence of maximum transverse magnetoresistance $(\Delta\rho/\rho_0)_{max}$ on magnetic field B for a coke (Hishiyama et al., 1971).

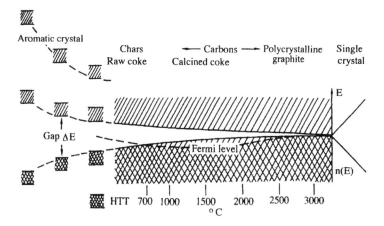

FIGURE 3. Schematic illustration of energy band structure as a function of heat treatment temperature HTT on a coke (Mrozowski, 1971)

and the Fermi level to rise, which leads to the decrease in concentration of holes and to the increase in R_H. Passing through the maximum in R_H and the associated minimum in $(\Delta\rho/\rho_0)_{max}$ at HTT around 2200°C, the upper and lower π bands start to be in contact, because of the improvement of stacking regularity, as for the graphitic ABAB... stacking. Since the positive magnetoresistance is theoretically due to the existence of two types of carriers, electrons and holes, the Fermi level must be located at the position where the electronic bands start to overlapp. The carriers are now electrons and holes of respectively the conduction and the valence bands. In the highly graphitized carbons, the concentration of electrons and holes are comparable, which gives the high value of $(\Delta\rho/\rho_0)_{max}$. A little higher mobility of electrons than those of holes will give a negative Hall coefficient R_H. Even though the values of these coefficients are approaching to those of graphite, they cannot reach the exact values which have been measured for the in-plane value of a single crystal.

Negative magnetoresistance is very characteristic of turbostratic carbons, and much experimental and theoretical work has been done (Yazawa, 1969; Hishiyama, 1970; Hishiyama et al., 1971; Bright and Singer, 1971; Delhaès et al., 1974; Bright, 1979; Bayot et al., 1989 & 1990), because the magnetoresistance for the materials with one type of carrier is classically expected to be zero or small positive.

For pregraphitic carbon blocks obtained from coke particles extruded with a binder pitch and heat-treated between 1700 and 3000°C after being baked at 1300°C, they were characterized by X-ray and maximum transverse magnetoresistance measurements that the layer order was turbostratic for the specimens heat-treated below 1900°C and then the graphitic structure was developped gradually with an HTT increase above 2100°C. Figure 4 shows experimental results of the absolute thermoelectric power S temperature dependences along the extrusion axis (Hishiyama and Ono, 1981). The experimental results on a kish graphite (KG) specimen, which exhibits single crystal behavior (Takezawa et al., 1971) is included in Figure 4 for comparison.

The turbostratic specimens show a positive thermoelectric power which increases almost linearly with temperature, indicating a majority hole-type conduction (Figure 4a). The slope of S versus temperature relation increases with the increase in HTT, in other words, the Fermi level rises with increasing HTT. The graphitic specimens show a negative peak in the range between 20 and 35 K, as observed in the KG specimen (Figure 4b). The negative peak is attributed to a phonon drag effect caused by the coupling between carriers and long wave-length in-plane phonons (Sugihara, 1970). It is important to note that the width of the peak for pregraphitic carbons is broader than that for the KG, in other words, the fall-off of the peak at the high temperature side is less rapid for pregraphitic carbons than KG specimen. This behavior can be explained by a Rayleigh scattering due to crystalline defects. As a general presentation of the thermoelectric power of the extruded coke, isotherms of S at 4.2, 77 and 280 K respectively are plotted as a function of HTT in Figure 5 (Hishiyama and Ono, 1981): the room temperature behavior is very similar to that of Hall coefficient shown in Figure 1.

12.2.2 Highly-Oriented Carbons

The dependence of electrical resistivity ρ, Hall coefficient R_H and maximum transverse magnetoresistance $(\Delta\rho/\rho_0)_{max}$ on HTT were studied on carbon films prepared from

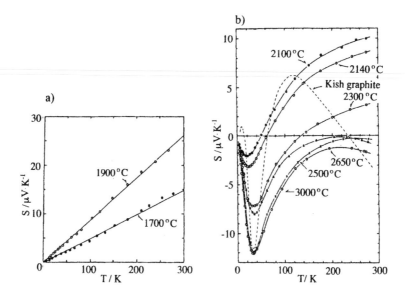

FIGURE 4. Dependence of absolute thermoelectric power S on temperature T for the extruded coke-based carbons with different heat treatment temperatures (Hishiyama and Ono, 1981).

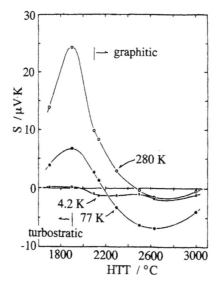

FIGURE 5. Changes of absolute thermoelectric power S at 4.2, 77 and 280 K with heat treatment temperature HTT for the extruded coke-based carbons (Hishiyama and Ono, 1981).

aromatic polyimide film Kapton, in which crystallites are highly oriented (Hishiyama *et al.*, 1997). In Figure 6, ρ and R_H at liquid nitrogen temperature are plotted as a function of HTT, ρ showing a bump and R_H a maximum around 2000°C; both are similar to those obtained for the coke as shown already in Figure 1. For the films heat-treated between 1600 and 2200°C, $(\Delta\rho/\rho_0)_{max}$ was measured as negative, but it suddenly changes to positive above 2300°C. The carbon films heat-treated above 2300°C showed highly anisotropic magnetoresistance, exhibiting a well-oriented texture. The abrupt changes in ρ and R_H, and also in structure were shown to be due to the departure of nitrogen atoms which were in substitution into the hexagonal carbon layers (Konno *et al.*, 1997).

Figure 7 shows the dependences of R_H and $(\Delta\rho/\rho_0)_{max}$ on magnetic field for the films heat-treated at different temperatures. R_H is negative at low HTTs and becomes positive above 1700°C, showing a maximum at about 2200°C, and then suddenly changes negative again. The films heat-treated above 2300°C show the magnetic field dependence of R_H similar to HOPG (Kaburagi and Hishiyama, 1995). At a constant magnetic field, the absolute value of $(\Delta\rho/\rho_0)_{max}$ increases with increasing HTT. The dependence of $(\Delta\rho/\rho_0)_{max}$ on magnetic field for the 2200°C-treated carbon film is different from others, showing the trend of saturation at higher fields. For HTT above 2300°C, the dependence of $(\Delta\rho/\rho_0)_{max}$ on magnetic field strength becomes pronounced.

The detailed studies on ρ, R_H and $(\Delta\rho/\rho_0)_{max}$ using commercially available pyrolytic carbon were performed, paying a particular attention to structural change with HTT (Kaburagi *et al.*, 1991). On Figure 8, the dependences of $(\Delta\rho/\rho_0)_{max}$ and R_H on magnetic field B measured at liquid nitrogen temperature are shown. $(\Delta\rho/\rho_0)_{max}$ increases its value with the decrease of d_{002} (the mean distance between graphitic planes) and changes its sign from negative to positive on the sample for d_{002} of about 0.3381 nm. R_H decreases with d_{002} and its sign changes also for the same d_{002} value. These changes in R_H and $(\Delta\rho/\rho_0)_{max}$ are very similar to those observed for cokes as shown in Figures 1 and 2, demonstrating the general trends for these galvanomagnetic properties.

12.2.3 Glass-like Carbons

In contrast to highly-oriented carbons discussed above, glass-like carbons have an isotropic and non-graphitizing nature.

The dependences of the electrical resistivity ρ measured at 20 and 300 K on HTT for a commercially available glass-like carbon are shown in Figure 9 (Yamagushi, 1963). With increasing HTT, ρ decreases at first, passes through a shallow minimum and then increases gradually.

The HTT dependences of R_H and $(\Delta\rho/\rho_0)_{max}$ measured for two glass-like carbons at different temperatures under an applied field of 0.65 T are shown in Figures 10a and 10b, respectively (Yamagushi, 1963; Baker and Bragg, 1983). Positive R_H and negative $(\Delta\rho/\rho_0)_{max}$ observed for the 3200°C-treated specimen are the characteristics of non-graphitizing carbon. R_H for glass-like carbons is independent of magnetic field and insensitive to temperature in the temperature range from 2.8 K to 300 K, but it is a strong function of HTT and the cross-over from negative to positive is observed on two samples at the same HTT of 1800°C.

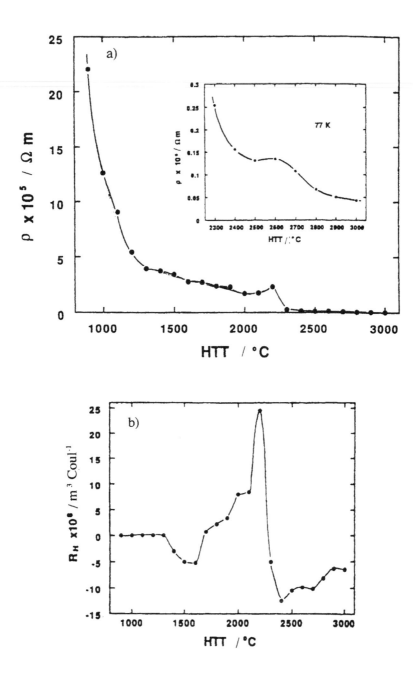

FIGURE 6. Electrical resistivity ρ and Hall coefficient R_H at liquid nitrogen temperature for Kapton-derived carbon films plotted as a function of HTT (Hishiyama *et al.*, 1997).

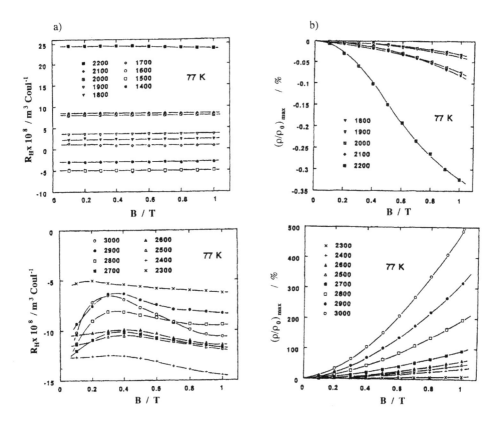

FIGURE 7. Dependences of Hall coefficient R_H and maximum transverse magnetoresistance $(\Delta\rho/\rho_0)_{max}$ at liquid nitrogen temperature on magnetic field B for Kapton-derived carbon films heat-treated at different temperatures (Hishiyama *et al.*, 1997).

Room temperature thermoelectric power S for a glass-like carbon was measured as a function of HTT (Yamagushi, 1963). The measurements of S were extended in the temperature range from 1.5 to 280 K (Kaburagi *et al.*, 1986; Baker and Bragg, 1983; Baker, 1983). The representative results are shown in Figure 11. Their temperature dependence can empirically be written as:

$$S = aT + bT^{1/2} + S_B, \tag{1}$$

where the first term is a strong-scattering metallic component, the second term is attributed to variable-range hopping and the third term is a "peaked" component. The constant a involves a factor $d \ln N(E_F)/dE$ for the extended states, where $N(E_F)$ is the density of states at the Fermi level and E the energy of the carriers. $dN(E_F)/dE$ gives the sign of the carriers, which change their sign from negative to positive above HTT of about 1500°C. The constant b provides information on the density of states at Fermi level for the localized states and the nature of S_B remains to be explored.

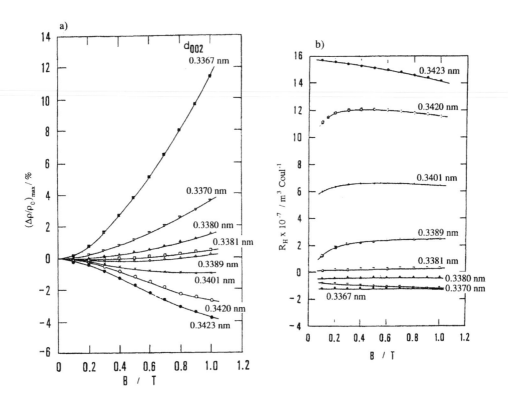

FIGURE 8. Dependences of the maximum transverse magneto-resistance $(\Delta\rho/\rho_0)_{max}$ and Hall coefficient R_H on magnetic field B measured at liquid nitrogen temperature with a parameter d_{002} (Kaburagi *et al.*, 1991).

12.3 THERMAL PROPERTIES

12.3.1 Specific Heat

The low temperature dependence of specific heat for metallic compounds is known to be fitted to the following relationships:

$$C = \gamma T + \alpha T^3 \tag{2}$$

This is the case for natural graphite in the very low temperature range between 0.4 and 1.5 K (Van der Hoeven and Keesom, 1963). The first term is believed to be associated with the electronic contribution, γ being a constant related to the density of states at the Fermi level $N(E_F)$. The second term gives the contribution of the lattice vibration, α being a constant related to the Debye temperature θ_D, through the ideal gas constant $R(\alpha = 234R/\theta_D^3)$ in the classical Debye model. For natural graphite these coefficients γ and α were obtained equal to 13.8 μJ mole^{-1} K^{-2} and 27.7 μJ mole^{-1} K^{-4}, respectively, by the extrapolation of

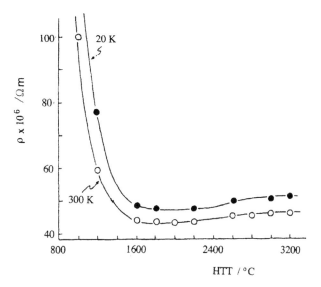

FIGURE 9. Changes of electrical resistivity ρ at 20 and 300 K on a glass-like carbon with heat treatment temperature HTT (Yamaguchi, 1963).

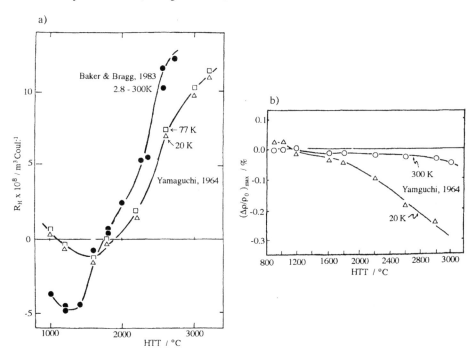

FIGURE 10. Dependence of Hall coefficient R_H and maximum transverse magnetoresistance $(\Delta\rho/\rho_0)_{max}$ in a magnetic field of 0.65 T at different temperatures of a glass-like carbon on heat treatment temperature HTT (Yamaguchi, 1963; Baker and Bragg, 1983).

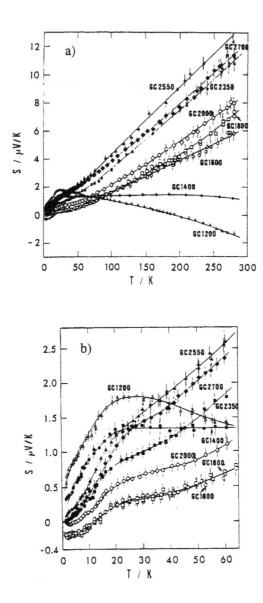

FIGURE 11. Absolute thermoelectric power S for glass-like carbons plotted as a function of temperature T (a), and low temperature details (b) (Kaburagi *et al.*, 1986).

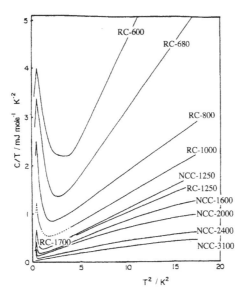

FIGURE 12. Specific heat curves obtained for two carbons (RC and NCC) heat-treated at various temperatures (Mrozowski and Vagh, 1976).

the C/T vs. T^2 line to zero Kelvin. The γ value agrees very well with the value predicted by the electronic band calculation for the conduction carriers. The value of α yields the Debye temperature θ_D of 413 K for 3D phonons.

Specific heat measurements on pregraphitic carbons were carried out on baked carbon rods (Delhaès and Hishiyama, 1970; Mrozowski and Vagh, 1976). The temperature dependences of specific heat on the specimens with different HTTs are shown in Figure 12. The result obtained for the sample graphitized at 3100°C agrees well with the curve obtained by Keesom and Pearlman for a similar polycrystalline graphite (Keesom and Pearlman, 1955). However, with the decrease in HTT, the contributions of the linear term as well as that of the non-linear term are particular. The value of the coefficient β is in all cases too large to be explained from the density of conduction carriers but rather due to structural defects and localized phonons. It was found that the temperature dependence of the non-linear term deviates from T^3 law in the higher temperature part of the range investigated because of the 2D vibrations of individual graphite-like planes. But the greatest deviation from the theoretical equation occurs at very low temperature as a peak around 0.6 K. This peak is supposed to be due to the structural defects. The height of the peak decreases continuously and its position shifts gradually to lower temperature with increasing HTT. The measurement was further extended to the 1250°C-treated sample (NCC-1250), Figure 13 showing the result in the temperature range of 0.07–1.00 K. Two peaks are observed at around 0.32 and 0.6 K (Vagh and Mrozowski, 1978); the former can be observed even for the sample heat-treated up to 3000°C, while the latter disappears by further heat treatment. These two peaks are ascribed to the localized spin centers related to the lattice defects.

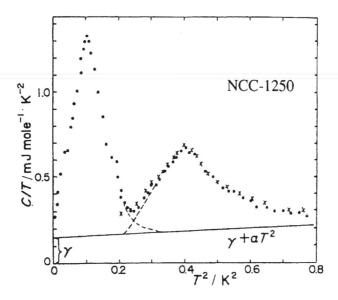

FIGURE 13. Specific heat data for 1250°C-treated sample (Mrozowski and Vagh, 1978). The line $\gamma + \alpha T$ and the data indicated by crosses are taken from the reference (Mrozowski and Vagh, 1976).

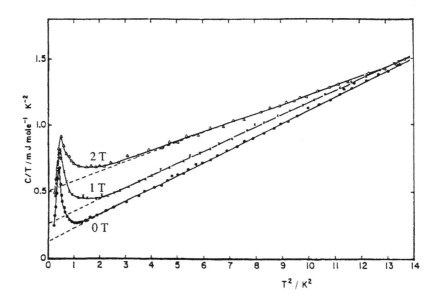

FIGURE 14. Specific heat data for 1250°C-treated sample in magnetic fieldsof 0, 1 and 2 T (Mrozowski and Vagh, 1975).

The measurement of the magnetic field dependence of the specific heat C_p in the low temperature range 0.4-4 K was also carried out for a 1250°C-treated sample (Vagh and Mrozowski, 1975). The results are shown on Figure 14 where C_p increases when an external magnetic field is applied. An artificial Schottky anomaly is created associated with the Zeeman effect on paramagnetic localized centers (P. Delhaès and F. Carmona, 1981). Beside the above result suggests that the peaks in the lower temperature range are related to some antiferromagnetic ordering associated with these localized spin centers (Mrozowski, 1979).

12.3.2 Thermal Expansion Coefficient

A single crystal of graphite is known to have a strong anisotropy in thermal expansion coefficient α. According to direct measurements of α on highly oriented pyrolytic graphite (HOPG) specimens which are similar to single crystal of graphite, $\alpha_{//}$ along the basal plane is extremely small and negative (i.e., shrinkage) of the order of 10^{-6} K^{-1} below 600 K, showing a minimum at about 210 K (Bailey and Yates, 1970; Harrison, 1977). After sign reversal it increases gradually together with the temperature. Room temperature value of $\alpha_\perp$ perpendicular to the basal planes is quite large, 26.3×10^{-6} K^{-1}, and positive in the whole temperature range, increasing gradually with the temperature.

On the bulk samples which are fabricated from coke particles by using binder pitches and thus the aggregates of anisotropic crystallites, however, the observed thermal expansion coefficient is positive above room temperature, the value being not as high as that perpendicular to the basal plane in HOPG, as shown in Figure 15 on the bulk specimens (Miyazaki, 1987). Not only the value of α but also its anisotropy depends strongly on the preparation conditions, particularly the forming process; the carbon formed by rubber pressing is almost isotropic, but those by either extrusion or compression process show a pronounced anisotropy, which is mainly due to preferred orientation of crystallites along the pressing direction.

These temperature dependences for the bulk specimens are explained by the relaxation of anisotropic large expansion perpendicular to the basal planes in each crystallites, which are constructed from graphite-like layers, by microcracks and micropores in the bulk. It has been pointed out that α_{obs} observed on the carbon blocks is approximated by the following simple equation:

$$\alpha_{obs} = A\alpha_c + (1 - A)\alpha_a, \tag{3}$$

where α_a and α_c are thermal expansion coefficients along and perpendicular to the basal planes of single crystal of graphite, respectively, and A is an experimental parameter which depends on the development of structure and preferred orientation of crystallites in the block (Mrozowski, 1956). In Figure 15, an anisotropy in α's parallel and perpendicular to the extrusion direction is observed, which is mainly due to the preferred orientation of crystallites formed during extrusion process in the production of the blocks.

12.3.3 Thermal Conductivity

Thermal conductivities of pyrolytic carbons along the direction parallel and perpendicular to the deposition surface $\kappa_{//}$ and $\kappa_\perp$ were extensively studied at temperatures ranging

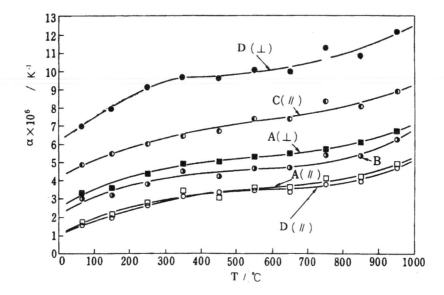

FIGURE 15. Dependences of thermal expansion coefficient α on coke-based carbons (Miyazaki, 1987) (by the courtesy of Prof. K. Kobayashi). A: Extruded and heat-treated up to about 2800°C, B: Rubber-pressed and heat-treated up to about 2800°C, C: Molded and heat-treated up to about 2800°C, D: Hot-pressed at 2200°C under 40 MPa with 5 wt% B_4C.

from 1.7 to 300 K (Klein and Holland, 1964). Some of these results are shown in Figure 16. The thermal anisotropy ratio $\kappa_{//}/\kappa_{\perp}$ is close to 3 in the liquid helium temperature region, but rises very rapidly with temperature up to about 125 at 300 K, reflecting highly oriented texture in the specimens. In the direction perpendicular to the basal plane, the Wiedemann-Franz ratio $\kappa_{\perp}\rho/T$ is at least three orders of magnitude larger than the Lorenz number, $L = 2.45 \times 10^{-8} V^2.K^{-2}$. It is reasonably concluded that inside the temperature range from 1.7 to 300 K the heat transport perpendicular to the basal plane proceeds solely through the lattice. However, the statement does not apply to the transport along basal plane; the Wiedemann-Franz ratio approaches to the Lorenz number with decreasing temperature below 4 K along the basal plane. The electronic contribution to $\kappa_{//}$ is estimated to be as much as 40% of the total thermal conduction at 2 K. At temperatures above 10 K, the Wiedemann-Franz ratio implies that phonons predominate, $\kappa_{\perp}\rho/T = 210L$ at 300 K, and the thermal conduction in graphite is interpreted mainly due to phonons.

The thermal conductivity κ of natural graphite parallel to the basal plane shows a maximum around 80 K, the value being 2.8×10^3 W.m^{-1}.K^{-1} (Smith and Rasor, 1956). At temperatures above the maximum, κ decreases rapidly with increasing temperature. The anisotropy ratio of κ along the basal plane to that perpendicular to the basal plane estimated from HOPG specimens is not smaller than 200.

There are many experimental results on thermal conductivity κ of commercially available and laboratory prepared pregraphitic carbons (Tyler and Wilson, 1953; Smith and Rasor,

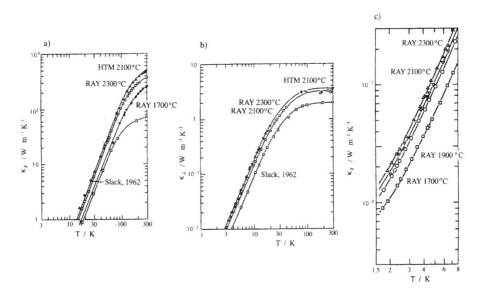

FIGURE 16. Thermal conductivities parallel and perpendicular to the basal plane $\kappa_{//}$ and $\kappa_{\perp}$ for the specimens deposited at different temperatures (Klein and Holland, 1964).

1956). The value of κ at a constant temperature increases with increasing HTT, reflecting a slight influence of the development of graphitic structure and their orientation. The mean values of κ for these carbons are much smaller than $\kappa_{\perp}$ measured perpendicular to the basal plane of natural graphite, even after heat treatment at $3100°C$.

Interpretation of the temperature dependence for pregraphitic carbons is simply described as follows (Tyler and Wilson, 1953). It is reasonably assumed that thermal conduction is due primarily to the energy transfer by phonons, except at low temperatures (Klein and Holland, 1964). Simple 3D Debye expression for lattice conductivity is written by the following equation:

$$\kappa = C_L v l / 3 \tag{4}$$

where C_L is the lattice specific heat, v the phonon velocity and l the phonon mean free path. For pure, single, non metallic crystals, l dominates in determining the temperature dependence of κ except at low temperatures. At temperatures above the Debye temperature $(T > \theta_D)$, l is inversely proportional to the temperature. Below the Debye temperature, C_L decreases with decreasing temperature, but κ increases due to the very rapid increase in l. At sufficiently low temperatures, l is limited by the scattering of the phonons at crystallite boundaries and κ reaches to the maximum value. For much lower temperatures, κ decreases with the temperature, being dominated by C_L term.

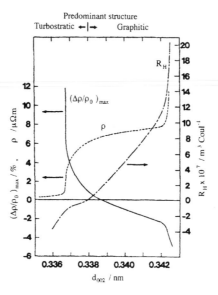

FIGURE 17. Electrical resistivity ρ, Hall coefficient R_H and maximum transverse magnetoresistance $(\Delta\rho/\rho_0)_{max}$ plotted against average interlayer spacing d_{002} on a pyrolytic carbon (Kaburagi *et al.*, 1991).

12.4 RELATIONSHIP BETWEEN PHYSICAL PROPERTIES AND STRUCTURAL PARAMETERS

As explained above and in other chapters, most of physical properties of pregraphitic carbons are more or less dependent on their structure and texture which are strongly governed by the precursors and their preparation conditions. Therefore, there must be certain relationships between physical properties and structure which are characterized by different parameters, degree of graphitization P_1, interlayer spacing d_{002}, crystallite sizes along a- and c-axis, L_a and L_c, which have to be independent from the kinds of carbon materials, as cokes, glass-like carbons, pyrolytic carbons, carbon fibers, carbon blacks, etc. for which the crucial factor is their spatial distribution. However, there have been only limited number of systematic studies on these relations.

In Figure 17, electrical resistivity ρ, Hall coefficient R_H and maximum transverse magnetoresistance $(\Delta\rho/\rho_0)_{max}$ are plotted against the average interlayer spacing d_{002} on a pyrolytic carbon (Kaburagi *et al.*, 1991). Since the pyrolytic carbon used was prepared by chemical vapor depositionn process at a rather high temperature 2200°C, R_H passed the maximum value already (see Figure 1); so, ρ and R_H decrease, and $(\Delta\rho/\rho_0)_{max}$ increases with d_{002} decrease. Below the d_{002}-values 0.338 nm, an anisotropic ratio ρ, determined from the anisotropy in magnetoresistance, starts to decrease, R_H and $(\Delta\rho/\rho_0)_{max}$ changes their sign, the former from positive to negative and the latter in the opposite way. This suggests that the graphitic structure becomes predominant in this pyrolytic carbon.

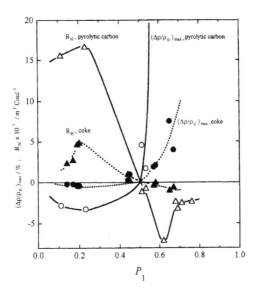

FIGURE 18. Dependences of Hall coefficient R_H and maximum transverse magnetoresistance $(\Delta\rho/\rho_0)_{max}$ on degree of graphitization P_1 for pyrolytic carbons and cokes (Iwashita *et al.*, 1997).

In order to understand the correspondence of these electric and galvanomagnetic properties on the development of graphitic structure in pregraphitic carbons, R_H and $(\Delta\rho/\rho_0)_{max}$ are plotted against the degree of graphitization P_1, which is the probability to find graphitic stacking order ABAB... of carbon layers in crystallites and determined from Fourier analysis of X-ray diffraction profile of *hk* lines, on two kinds of precursors, cokes and pyrolytic carbons in Figure 18 (Iwashita *et al.*, 1997). Though the absolute values of these parameters are different on cokes and pyrolytic carbons, because of the difference in preferred orientation scheme, the dependence of these parameters on P_1 is common; R_H decreases and $(\Delta\rho/\rho_0)_{max}$ increases with P_1. It has to be mentionned that R_H and $(\Delta\rho/\rho_0)_{max}$ change their sign at the P_1-value of 0.5, i.e., for the predominant structure changes from turbostratic stacking to graphitic one.

Indeed, the main conclusion of this presentation is to show that the transport electronic properties are much more sensitive than the thermal ones for these non-crystalline carbons. They are very good sensors to check the graphitization process in these materials.

References

Bailey, A. and Yates, B. (1970) *J. Appl. Phys.*, **41**, 5088.

Baker, D.F. (1983) Ph.D. Thesis, University of California Berkeley.

Baker, D.F. and Bragg, R.H. (1983) *J. Non-Crystal. Solids*, **58**, 57.

Bayot, V., Piraux, L., Michenaud, J.P. and Issi, J.P. (1989) *Phys. Rev. B*, **40**, 3514.

Bayot, V., Piraux, L., Michenaud, J.P., Issi, J.P., Lelaurain, M. and Moore, A. (1990) *Phys. Rev. B*, **41**, 11770.

Bright, A.A. (1979) *Phys. Rev. B*, **20**, 5142.

Bright, A.A. and Singer, L.S. (1971) *Carbon*, **17**, 59.

Delhaès, P. and Hishiyama, Y. (1970) *Carbon*, **8**, 31.

Delhaès, P., De Kepper, P. and Uhlrich, M. (1974) *Phil. Mag.*, **29**, 1301.

Delhaès, P. and Carmona, F. (1981) *Chemistry and Physics of Carbon*, **17**, 89, P.L. Walker and P. Thrower Editors, Marcel Dekker Inc., New York.

Harrison, J. (1977) *High Temp.-High Press.*, **9**, 211.

Hishiyama, Y. (1970) *Carbon*, **8**, 259.

Hishiyama, Y., Ono, A. and Hashimoto, M. (1971) *Jpn. J. Appl. Phys.*, **10**, 416.

Hishiyama, Y. and Ono, A. (1981) *Carbon*, **19**, 441.

Hishiyama, Y., Kaburagi, Y. and Inagaki, M. (1991) *Chemistry and Physics of Carbon*, 23, P.A. Thrower Ed., Marcel Dekker Inc., New York, 1.

Hishiyama, Y., Igarashi, K., Kanaoka, I., Fujii, H., Kaneda, T., Koidesawa, T., Shimazawa, Y. and Yoshida, A. (1997) *Carbon*, **35**, 657.

Inagaki, M. and Hishiyama, Y. (1994) *New Carbon Materials*, Gihoudo Shuppan, Tokyo.

Issi, J.P. (1998) Chapter 3 of this book *Graphites and its precursors*, Gordon and Breach Ed., Paris.

Iwashita, N., Hishiyama, Y. and Inagaki, M. (1997) *Carbon*, **35**, 1073.

Kaburagi, Y. and Hishiyama, Y. (1995) *Carbon*, **33**, 773.

Kaburagi, Y., Hishiyama, Y., Baker, D. and Bragg, R.H. (1986) *Phil. Mag. B*, **54**, 381.

Kaburagi, Y., Bragg, R.H. and Hishiyama, Y. (1991) *Phil. Mag. B*, **63**, 417.

Keesom, P.H. and Pearlman, N. (1995) *Phys. Rev.*, **99**, 1119.

Kelly, B.T. (1981) *Physics of Graphite*, Appl. Sci. Pub., London.

Klein, C.A. and Holland, M.G. (1964) *Phys. Rev.*, **136**, A576.

Konno, H., Nakahashi, T. and Inagaki, M. (1997) *Carbon*, **35**, 669.

Miyazaki, K. (1987) *Hitech Carbon Materials*, H. Honda and K. Kobayashi Ed., Kogyo Chosa Kai, Tokyo, 64.

Mrozowski, S. (1956) *Proc. First and Second Conferences on Carbon*, Univ. of Buffalo Press, Buffalo, New York, 31.

Mrozowski, S. (1971) *Carbon*, **9**, 97.

Mrozowski, S. and Vagh, A.S. (1976) *Carbon*, **14**, 211.

Mrozowski, S. (1979) *J. Low Temp. Phys.*, **35**, 231.

Oberlin, A. (1989) *Chemistry and Physics of Carbon*, **22**, 1, P.A. Thrower Ed., Marcel Dekker Inc., New York.

Smith, A.W. and Rasor, N.W. (1956) *Phys. Rev.*, **104**, 885.

Sugihara, K. (1970) *J. Phys. Soc. Jpn.*, **29**, 1465.

Takezawa, T., Tsuzuku, T., Ono, A. and Hishiyama, Y. (1971) *Phil. Mag.*, **19**, 623.

Tyler, W.W. and Wilson, Jr. A.C. (1953) *Phys. Rev.*, **89**, 870.

Vagh, A.S. and Mrozowski, S. (1975) *Carbon*, **13**, 301.

Van der Hoeven, B.J.C. and Keesom, P.H. (1963) *Phys. Rev.*, **130**, 1318.

Warren, B.E. (1941) *Phys. Rev.*, **59**, 693.

Yamaguchi, T. (1963) *Carbon*, **1**, 47.

Yamaguchi, T. (1963) *Carbon*, **1**, 535.

Yazawa, K. (1969) *J. Phys. Soc. Jpn.*, **26**, 1407.

Index

T - #0604 - 071024 - C0 - 254/190/15 - PB - 9780367397791 - Gloss Lamination